Sustainable Crop Production

Edited by
Mirza Hasanuzzaman, Marcelo Carvalho Minhoto Teixeira Filho, Masayuki Fujita and Thiago Assis Rodrigues Nogueira

Published in London, United Kingdom

IntechOpen

Supporting open minds since 2005

Sustainable Crop Production
http://dx.doi.org/10.5772/intechopen.83521
Edited by Mirza Hasanuzzaman, Marcelo Carvalho Minhoto Teixeira Filho, Masayuki Fujita and Thiago Assis Rodrigues Nogueira

Assistant to the Editor(s): Fernando Shintate Galindo

Contributors
Sebnem Seküre Ellialtioglu, Gülsün Elif Vural, Sinan Zengin, Esin Ari, K. Mehmet Tuğrul, Rajendran Kuppu, Mohan Easan, Romanus Osabohien, Toun Ogunbiyi, Adnan Mustafa, Tanveer Abbas, Muhammad Naveed, Qudsia Saeed, Azhar Hussain, Muhammad Nadeem Ashraf, Xu Minggang, Muhammad Kamran, Patrícia Vidigal, Maria M. Romeiras, Filipa Monteiro, Salih Kafkas, Jestinos Mzezewa, Leon Van Rensburg, Summy Yadav, Payal Modi, Akanksha Dave, Akdasbanu Vijapara, Disha Patel, Mohini Patel, Naoki Moritsuka, Graham Brodie, Dorin Gupta, Muhammad Jamal Khan, Ali Saleh Hassoon, Inas Abdulsattar Abduljabbar, Marcelo Carvalho Minhoto Teixeira Filho, Aeysha Khan, Kamran Azeem, Arshad Jalal, Charles Wortmann, Anthony Esilaba, Kayuki Kaizzi, Catherine Kibunja, Keziah Ndungu-Magiroi, Nouri Maman, Dubravka Savic, Zarko Ilin, Muhammad Aamir Iqbal, Mohammed Moro Buri, Nuhu Issaka, Salima Yousfi, José Fernando Marin Peira, Gregorio Rincón De La Horra, Pedro Vicente Mauri Ablanque, Fabiana Tonial, Francine Falcão De Macedo Nava, Ana Luisa Gayger, Talita Bernardon Mar, Tamanreet Kaur

First published in London, United Kingdom, 2020 by IntechOpen
IntechOpen is the global imprint of INTECHOPEN LIMITED, registered in England and Wales, registration number: 11086078, 7th floor, 10 Lower Thames Street, London, EC3R 6AF, United Kingdom
Printed in Croatia

British Library Cataloguing-in-Publication Data
A catalogue record for this book is available from the British Library

Additional hard and PDF copies can be obtained from orders@intechopen.com

Sustainable Crop Production
Edited by Mirza Hasanuzzaman, Marcelo Carvalho Minhoto Teixeira Filho, Masayuki Fujita and Thiago Assis Rodrigues Nogueira
p. cm.
Print ISBN 978-1-78985-317-9
Online ISBN 978-1-78985-318-6
eBook (PDF) ISBN 978-1-83880-899-0

For EU product safety concerns:
IN TECH d.o.o., Prolaz Marije Krucifikse Kozulić 3, 51000 Rijeka, Croatia, info@intechopen.com or visit our website at intechopen.com.

Meet the editors

Dr. Mirza Hasanuzzaman is a Professor of Agronomy at Sher-e-Bangla Agricultural University, Dhaka, Bangladesh. He received his PhD in Plant Stress Physiology and Antioxidant Metabolism from the United Graduate School of Agricultural Sciences, Ehime University, Japan, with a Japanese Government (Monbukagakusho: MEXT) Scholarship. Later, he completed his postdoctoral research at the Center of Molecular Biosciences (COMB), University of the Ryukyus, Okinawa, Japan, with a postdoctoral fellowship from the Japan Society for the Promotion of Science (JSPS). Subsequently, he joined the University of Tasmania as an adjunct senior researcher with an Endeavor Research Fellowship from the Australian government. Dr. Hasanuzzaman has conducted research in crop science with a focus on environmental stress physiology since 2004. He has published more than 100 articles in peer-reviewed journals and books. He has edited thirteen books and written thirty-five book chapters on important aspects of plant physiology, plant stress responses, and environmental problems in relation to plant species. Dr. Hasanuzzaman is a research supervisor of undergraduate and graduate students and has supervised twenty MS students to date. He is an editor and reviewer for more than fifty peer-reviewed international journals. He received the Publons Global Peer Review Award in 2017, 2018, and 2019. He is an active member of about forty professional societies and is acting Publication Secretary for the Bangladesh Society of Agronomy. He has been honored by different authorities due to his outstanding performance in different fields of research and education. He received the World Academy of Science (TWAS) Young Scientist Award in 2014. He has presented twenty-five papers and posters at national and international conferences in the United States, United Kingdom, Germany, Australia, Japan, Austria, Sweden, and Russia.

Marcelo Carvalho Minhoto Teixeira Filho has a master's Degree (2008) and a PhD (2011) in Agronomy from the Faculty of Engineering of Ilha Solteira (FEIS) at São Paulo State University (UNESP). He completed a short-term internship (2017) as Visiting Professor at Plant Nutrition Group, ETH Zürich, Switzerland. Since 2013, he has been Assistant Professor in the Department of Plant Protection, Rural Engineering and Soils (DEFERS) at UNESP where he teaches the discipline of Plant Nutrition. Since July 2018, he has served as editor-in-chief of the agronomic crop journal *Revista Cultura Agronômica*. He is associate member of the Brazilian Association of Scientific Editors (ABEC) and current president of the Permanent Research Commission (CPP) of FEIS-UNESP. His research interests include plant nutrition, fertilization, soil fertility, improved efficiency fertilizers, plant growth-promoting bacteria associated with fertilization reduction, *Azospirillum brasilense*, wheat, corn, soybean, sugarcane, and eucalyptus.

Dr. Masayuki Fujita is Professor in the Laboratory of Plant Stress Responses, Faculty of Agriculture, Kagawa University, Kagawa, Japan. He received his BS. in Chemistry from Shizuoka University, Shizuoka, and MAgr and PhD in Plant Biochemistry from Nagoya University, Nagoya, Japan. His research interests include physiological, biochemical, and molecular biological responses based on secondary metabolism in plants under various abiotic and biotic stresses; phytoalexin, cytochrome P450, glutathione S-transferase, and phytochelatin; and redox reactions and antioxidants. In the last decade, his work has focused on oxidative stress and antioxidant defense in plants under environmental stress. His group investigates the role of different exogenous protectants in enhancing antioxidant defense and methylglyoxal detoxification systems in plants. Dr. Fujita has supervised four MS students and thirteen PhD students. He has edited ten books and published about 150 papers and chapters in journals and books.

Thiago Nogueira earned his PhD in Agriculture and Environmental Chemistry from the Center for Nuclear Energy in Agriculture, University of São Paulo, in 2012. He was a postdoctoral associate at the University of Florida, USA, and a research scholar at Rothamsted Research, UK. He is Assistant Professor at São Paulo State University (UNESP) where he currently teaches "Soil Fertility" and "Soil Conservation and Management." He has published more than 150 peer-reviewed papers in reputed international journals and conferences, as well as two textbooks. He served as principal investigator on five government and industry research projects. His current research interests include nutrient management in agroecosystems, heavy metals in soils, waste management for sustainable agriculture, soil fertility, soil quality for food security, environmental chemistry, and fertilizers. He is an active member of a number of organizations including the Brazilian Society of Soil Science and the International Society of Trace Element Biogeochemistry.

Contents

Preface

Ensuring food security for the growing global population is a modern problem that must be addressed. Agronomists and plant biologists are responsible for improving crop production, which is the art and science of producing food by exploiting land and natural resources. However, various abiotic and biotic stresses as well as loss of soil productivity and natural biodiversity are hindering crop production. The success of crop production largely depends on how efficiently the crops are managed and cultivated. With the advancement of science and technology, many improved methods of crop production have been developed and are currently being practiced by farmers. However, due to the adverse effects of climate change, crop production warrants new methods and techniques to produce maximum output in a unit area. New approaches to soil management, crop husbandry, and water and nutrient management are being researched and adopted for crop production. Low-input agronomic practices are contributing to sustainable agriculture and food production. Considering these issues, researchers have been developing new approaches to make crop production more sustainable. Many agronomic strategies have also been developed to enhance stress tolerance in crops as well. This book provides a current and comprehensive overview of various crop production practices in the changing world.

Across twenty-one comprehensive chapters, this book details various soil and crop management issues, including modern techniques in enhancing crop production in the era of climate change. There are a few case studies and experimental evidence about these production systems in specific locations. The first section (Crop Production and Farming System) contains six chapters related to different issues and experimental evidence of managing farms and ways of crop production. As soil is at the heart of crop production systems, the second section (Soil Management) includes six chapters that present various soil management options, including nutrient management in different crops, different environments, and soil conditions. The final section (Sustainable and Advanced Technologies for Crop Production) includes eight chapters that discuss recent approaches for sustainable crop production, including biotechnology, nanotechnology, and precision agriculture.

We would like to give special thanks to the authors for their outstanding and timely work in producing such excellent chapters. We are very thankful to Nina Kalinic Babic, Author Service Manager at IntechOpen publishing, for her prompt responses during the acquisition. We believe that this book is useful for undergraduate and graduate students, teachers, and researchers, particularly in the fields of crop science, soil science, and agronomy.

Dr. Mirza Hasanuzzaman
Sher-e-Bangla Agricultural University,
Dhaka, Bangladesh

Dr. Masayuki Fujita
Kagawa University,
Japan

Dr. Marcelo Carvalho Minhoto Teixeira Filho and
Dr. Thiago Assis Rodrigues Nogueira
São Paulo State University,
Brazil

Section 1

Crop Production and Farming System

Chapter 1

Effect of Abiotic Stress on Crops

Summy Yadav, Payal Modi, Akanksha Dave, Akdasbanu Vijapura, Disha Patel and Mohini Patel

Abstract

Crop yield is mainly influenced by climatic factors, agronomic factors, pests and nutrient availability in the soil. Stress is any adverse environmental condition that hampers proper growth of plant. Abiotic stress creates adverse effect on multiple procedures of morphology, biochemistry and physiology that are directly connected with growth and yield of plant. Abiotic stress are quantitative trait hence genes linked to these traits can be identified and used to select desirable alleles responsible for tolerance in plant. Plants can initiate a number of molecular, cellular and physiological modifications to react to and adapt to abiotic stress. Crop productivity is significantly affected by drought, salinity and cold. Abiotic stress reduce water availability to plant roots by increasing water soluble salts in soil and plants suffer from increased osmotic pressure outside the root. Physiological changes include lowering of leaf osmotic potential, water potential and relative water content, creation of nutritional imbalance, enhancing relative stress injury or one or more combination of these factors. Morphological and biochemical changes include changes in root and shoot length, number of leaves, secondary metabolite (glycine betaine, proline, MDA, abscisic acid) accumulation in plant, source and sink ratio. Proposed chapter will concentrate on enhancing plant response to abiotic stress and contemporary breeding application to increasing stress tolerance.

Keywords: abiotic stress, drought, salinity, cold, heavy metals, morpho-physiological and biochemical changes

1. Introduction

Plants in their physical environment face several types of variation. Animals use techniques to prevent the impacts of this variation but plants fail because of the sessile nature of the growth habit. Plants therefore, rely on their internal processes to survive changes in the external environment. Plants are affected to function in an oscillating environment and normal external changes are countered by internal changes without any harm to growth or development. The possibility of abiotic or environmental stress is to cause physical harm to the plant due to serious or chronic adverse environmental circumstances. Any adverse influence of inanimate factors on living beings in a fixed setting is described as abiotic stress. To substantially impact the organism's demographic output or individual physiology, the non-living factor must alter the surrounding beyond its ordinary variation range. Due to the continuous climate change and environmental deterioration induced by human activity, physical surrounding stress has become a key threat to food security. Water deficit stress, salt stress, imbalances in nutrients (including mineral toxicity and deficiencies) and temperature extremes are significant environmental limitations on productivity of crops all over the world [1].

Plant growth and crop yield are majorly affected by cold, drought, salt, and heavy metals. Abiotic stress impacts plants to molecular levels from morphological levels and is visible at all phases of plant development where drought occurs [1]. There are three significant stages of plant: vegetative development, pre-anthesis and terminal phase that are impacted by the drought [2]. Plant physiological reactions to stress include wilting of the leaf, abscission of the leaf, decreased leaf region and decreased water loss through transpiration [3]. Under drought stress, crop development facilitates the issue of extreme water use in agriculture to a big extent. Turgor pressure is decreased, which is one of the most delicate physiological mechanisms that cause cell growth. Drought stress creates water flow disruption in higher plants from xylem to the neighboring elongating cells, thereby suppressing cell elongation. In addition, decreased leaf area, plant height, and development of crop result from drought pressure owing to cell elongation, impaired mitosis and expansion [1]. Abiotic stress resistance contains escape-avoidance and tolerance mechanisms. Detrimental impacts of stress can be decreased by osmotic modification, which helps with an active accumulation of solutes in the cytoplasm to maintain cell water balance [4].

Survival and geographical spreading of plants are also greatly affected by low temperatures. Significant loss of crop due to reduced plant growth and crop efficiency is usually caused by cold stress [5]. Cell and tissue dehydration and cellular water crystallization are caused by cold stress, thereby reducing plant growth and productivity. Reduced membrane conductivity, increased water viscosity, and hydro active stomata closure is inhibited resulting in water stress and increased leakage of electrolytes at low temperatures [6]. It also delays metabolism, dissipates energy, and causes free radicals to form as a result of oxidative stress [7].

For instance, up to 45% of the world's farming based land is encountered to frequent periods of time when there is scanty of rainfall in which 38% of the world's population resides and the world's mapped area is affected by salinity in more than 3106 km^2 area or about 6% of the entire area of land [8]. In addition, 19.5% of irrigated agricultural land is classified as saline. In addition, about 1% of world agricultural land is deteriorated by salinity (2 million ha) each year, resulting in decreased or no crop productivity [9]. Major abiotic stress affects the plants during their growth and development arises due to water limitation caused by inadequate rainfall, cold condition, and salinity. The worldwide land region impacted by drought, cold and salinity is 64, 57, and 6%, respectively, according to the FAO World Soil Resources Report, 2000. Comprehension of crop plant abiotic stress reactions has thus become component of plant studies to protect food security [10]. Abiotic stresses are interrelated and influence the relationships of plant water on the cellular and also the entire plant level, affecting certain and uncertain reactions resulting in a series of morphological, physiological, biochemical and molecular changes that affects the growth and development of plants adversely.

These abiotic stresses characterize the main cause of crop fiasco globally, introducing more than 50% of the average returns for significant crops [10]. Improving cultivation is therefore vital to fill the gap between population growth and food production, which is widening by initiating stress tolerance. Plants can introduce some molecular, cellular and physiological modifications to react to and acclimatize such stresses in order to deal with abiotic stress. Better knowledge of plant responsiveness to abiotic stress in both traditional and modern breeding application will assist to enhance stress tolerance. Studies with high stress tolerance on some wild plant species also make a significant contribution to our understanding of stress tolerance.

2. Types of abiotic stresses

Different abiotic stresses affect the plants due to global warming and fluctuations in the environmental conditions. Abiotic stresses like water scarcity, high salinity, extreme temperatures, and mineral deficiencies or metal toxicities significantly reduce the crop's productivity.

2.1 Drought stress

Drought is described as stress related to the water deficit. Drought is a climate word described as a period of moment with less rainfall. Drought stress in plants happens when environmental conditions result in a decrease in the quantity of water in the soil, leading in a constant loss of water through transpiration or evaporation. Water is a crucial element of plant survival and essentially needed for transportation of nutrients. Hence, deficiency of water leads to drought stress, which results in reduced vitality of plants [11]. Water stress may occur in plants due to high salt conditions. The soil water potential reduces because of elevated salt circumstances, because the osmotic potential of salt is smaller than water, which makes it hard for roots to absorb the soil's water. Also, owing to enhanced water loss via transpiration or evaporation, elevated temperatures can trigger drought stress. Not only greater temperatures, but reduced temperatures can also trigger stress from the water deficit. Lower freezing temperatures result in ice crystals being created in the extracellular spaces of plant cells, decreasing the water potential and leading to intracellular water efflux. Thus, in general, drought stress occurs due to various causes which further leads to the efflux of cellular water, leading to plasmolysis and thus causing cell death. Water deficit stress is damaging because it inhibits photosynthesis by affecting the thylakoid membranes. An increase in the toxic ions in all the cells of plants is the potential damage caused to the plants due to drought stress. Drought stress is therefore complicated abiotic stress that directly affects plant growth and advancement and leads to decrease in plant output.

2.2 Salinity stress

Salinity stress occurs due to an increase in salts contents in soil. Thus, an increase in salt content in soil is referred to as soil salinity or salinization [12]. It mainly occurs in arid as well as semi-arid environments where the plants have higher evaporation and transpiration rates compared to precipitation volume throughout the year. Salts in the soil may increase in the subsoil naturally which is referred to as primary soil salinity or it may be introduced due to anthropogenic conditions like environmental pollution which is called secondary soil salinity. Secondary soil salinity arises due to modification in soil content, an increase in fertilizers or due to the use of saline water in irrigation purposes [13]. Soil salinity is a global problem and a severe risk to the entire agricultural world as it reduces the output of plants. High salt concentration limits growth and development of crops in multiple ways. Two significant impacts in crops result from higher salt content: ionic toxicity and osmotic stress. The osmotic stress in plant cells is greater in ordinary circumstances than in soil. This increased osmotic pressure is used by plant cells to absorb water and the necessary minerals from the soil into the root cells. However, under circumstances of salt stress, the soils osmotic pressure solution surpasses the plant cells osmotic pressure owing to an increase in the salt content in the soil, thereby restricting plants 'ability to absorb water and minerals such as K^+ and Ca^{2+} while Na^+ and Cl^- ions move in the cells and have damaging effects on the cell membrane and metabolism in cytosol. Stress with salinity creates some adverse effects such as reduced cell growth, reduced membrane function, and reduced cytosolic metabolism and ROS production. High soil salinity adversely impacts plant production quality and quantity by inhibiting seed germination, damaging growth and development phases as a consequence of the combined impacts of higher osmotic potential and particular ion toxicity [14].

2.3 Extreme temperature stress (hot/cold)

Extreme temperatures are one of the prime causes of different abiotic stresses like drought. Increases or decreases in temperature, both undesirably affect the plant's growth, development, and yield. Cold stress occurs when plants are subjected to very low temperatures. Cold stress is a major abiotic stresses that reduce productivity of crops by affecting quality and life after harvest. Cold stress impacts all cellular function characteristics in crops. Plants are categorized into three kinds in reaction to cold temperatures: chilling delicate plants: plants that are highly damaged by temperatures above 0°C and below 15°C, chilling resistant plants: crops capable of tolerating low temperatures and wounded when ice formation begins in tissues and frost resistant plants: plants capable of tolerating exposure to very low temperatures. Cold stress causes injury to plants by changes in the membrane structure and decrease in the protoplasmic streaming, electrolyte leakage and plasmolysis which leads to cellular damage. The metabolism of the cells is damaged by an increase or decrease in respiration rate and depending on the intensity of the stress, synthesis of abnormal metabolites occurs due to abnormal anaerobic respiration. Due to cellular damage and altered metabolism, there is reduced plant growth, abnormal ripening of fruits, internal discoloration (vascular browning), and increased susceptibility to decay and also cause the death of the plant [15].

If crops are subjected to very elevated temperatures, heat stress happens. For adequate moment to cause permanent injury to functioning or development of plant. Heat stress is defined as elevated temperatures. High temperatures boost the rate of sexual development, which reduces the time needed to add photosynthesis to the production of fruit or seed. Also, high temperatures can cause drought stress due to increased water loss by transpiration or evaporation. High soil temperatures can decrease the emergence of plants. High temperature stress can influence seed germination, plant growth and development, and can trigger irreversible drought stress that can lead to death as well [16].

2.4 Metal stress

Heavy metal stress (HM) belongs to a group of non-biodegradables, determined inorganic chemicals having atomic mass more than 20 and a density exceeding 5 g cm^{-3} with toxic effects on cells and genes, which causes mutagenic impacts on crops by influencing and contaminating irrigation, soil, drinkable water, food chains and the surrounding environment [17, 18]. There are two categories of metals discovered in soils that are mentioned as vital micronutrients for standard plant growth (Fe, Mg, Mo, Zn, Mn, Cu, and Ni) and non-essential elements with unknown physiological and biological function (Ag, Cr, Cd, Co, As, Sb, Pb, Se, and Hg) [19]. Plant surfaces both underwater and above ground can take HMs. In the enzyme and protein structure, the vital elements play a main role. Plants need them in minute quantities for their metabolism, growth, and development; yet, the concentration of vital and non-essential metals is an only essential factor in the increasing crop cycle so that their excessive presence can cause a decline and inhibition of plant growth. HMs at poisonous concentrations hinder ordinary functioning in plants and act as an barrier to metabolic procedures in different ways, comprising the displacement or disturbance of protein structure construction blocks arising from the creation of blonds among HMs and sulfhydryl groups [20], interfering with functional groups of significant cellular molecules [21].

3. Major crops affected by abiotic stress

3.1 Chickpea

After dry beans and dry peas, chickpea is the third most significant food legume worldwide. It is grown on 12.4 million hectares, generating 11.3 million tons at an average output of 910 kg/ha, according to FAOSTAT information in 2012–2013. In chickpea manufacturing and productivity, climate change is a significant challenge (**Table 1**). Climate change's negative impacts seem to result from drought effects [15].

3.2 Wheat

Wheat is the major crop grown mainly in the Rabi season. Under various agro-ecological circumstances, it is commonly cultivated. Drought impacts vary from morphological to molecular. Many stages of plant development are influenced by drought. Drought has an impact on three major periods of plant development-vegetative, pre-anthesis and terminal stage. Physiological responses of plants to drought comprise leaf wilting, reduced leaf area, leaf abscission and thus reducing water loss through transpiration. Higher plants cell elongation is suppressed under serious water deficit by interrupted water flow from the xylem to the neighboring elongating cells. Cell elongation, impaired mitosis and expansion lead under drought to lower height of plant, leaf area, and crop development. There is a conservative water loss resulting in stomatal closure and disruption in cell structure as well as plant metabolism [22].

3.3 Maize

Maize (*Zea mays* L.) is one of the world's major staple food. It is used as cattle feed, meat supplement and also as biofuels. The crop is highly susceptible to elevated

Crops	Stress
Chickpea (*Cicer arietinum* L.)	Drought
	Cadmium (Cd) toxicity
	Copper (Cu) toxicity
Maize (*Zea mays* L.)	Drought
	Heat
	Arsenic (As) metal stress
Soybean (*Glycine max* L.)	Drought
Wheat (*Triticum aestivum* L.)	Drought
Rice (*Oryza sativa* L.)	Drought
	Heat
Black mustard (*Brassica juncea* L.)	Cadmium (Cd) toxicity

Table 1.
List of some of the crops that are affected by various abiotic stresses.

temperatures, resulting in significant crop yield losses [23]. Multiple abiotic stresses like salinity and drought can also affect the crop in semi-arid tropical regions [24]. Temperature increases above a threshold level have negative effects plant growth and development, that is, heat stress for an appropriate time span. Due to elevated temperature stress, the disruption in cellular homeostasis has the ability to cause retardation in plant growth, development and even death. High-temperature stress affects the phases of maize development differently. Drought stress resulted in maize yield loss ranging from 17 to 60% in Southern Africa. Sequential cycles of drought stress were subjected to maize inbred lines and their hybrid testcross progeny were at the seedling level. These cycles were done at diverse developmental stages of growth, that is, germination, survival and regeneration. The study revealed that the best parameter for secondary screening of maize under drought stress is seedling stage [23].

3.4 Soybean

A wealthy source of protein and edible oil is soybean (*Glycine max* (L.) Merr). Soybean is the major cultivated crop of the world (approximately 6% world's territory). Drought stress affects the plant's rate of germination and seedling vigor. Underwater scarcity, the length of hypocotyl, germination, and dry and new weight of root and stem are reduced while the length of the root is increased. Growth occurs through differentiation of cells and division of cells, which is negatively affected by water scarcity. Cell elongation decreases under drought conditions due to decrease in turgor pressure [25].

3.5 Rice

Rice (*Oryza sativa* L.) is a significant global staple food crop that provides food security and generates revenue, especially in developing countries. The anticipated global warming poses a severe danger to both rice manufacturing and the quality of the rice generated. In tropical and subtropical regions, temperature stress and drought are projected to increase to a greater degree, being the primary rice producing areas [26].

High temperature stress or drought conditions have adverse effects on plant development, including unalterable damage to plant growth and development, decreased photosynthesis [27], decreased amount of panicles in each plant and elongation of peduncle, limited pollen output, no pollen grain swelling, and reduced spikelet sterility. Low temperatures lead in inhibited growth of seedlings, decreased development of panicles, delay in heading, bad exertion in panicle, low fertility of spikelet and bad quality of grain. Water and temperature stresses, other than influencing development and grain yield, alter the chemical composition and quality of rice [28].

4. Effect of abiotic stress on crops

A complex set of biotic and abiotic pressures includes the natural environment of crops. Abiotic stresses are of greater importance because they include different environmental factors that cannot be prevented, that is, drought, salinity, cold, heat, metal, etc. The impacts of abiotic stress on crop manufacturing are hard to predict correctly. Plant reactions to abiotic stress are both dynamic and complicated and can be either elastic (reversible) or plastic (irreversible) [29].

4.1 Drought effect on crops

Drought stress affects the plants at all phenological developmental stages varying from morphological to molecular concentrations. In plants that determine yield, many physiological mechanisms are prone to drought [7]. Drought can trigger yield reductions in many plant species depending on their severity and period but the stress of drought after anthesis is detrimental to the output of grain regardless of its severity [30]. Prevailing drought stress limits the production of flowers and grain filling resulting in reduced quality and amount of grains. Micro and macronutrients like nitrogen (N), phosphorous (P) and potassium (K) are crucial for plant growth. Drought stress results in increased N significantly decreased P, in spite of this it has no definitive effects on K [31]. Overall, water deficit reduces nutrient accessibility in the root zone, absorption at root hair, translocation in xylem and phloem vessels leading to impaired metabolism of nutrients in cells and tissues [7]. The effectiveness of nutrient intake and utilization is also reduced due to less transpiration. Drought stress has significant on photosynthetic pigments like chlorophyll a, b, and carotenoid components and also impairs photosystem 1 and photosystem 2 [32]. It also reduces starch synthesis in plants by effecting Calvin cycle enzyme activity (Ribulose phosphatase). Plants can combat to drought stress by different mechanisms [33]. When the soil is scarce in water crops, their stomata tend to close, reducing CO_2 input into the leaves and spare more electrons for active oxygen species growth [34]. Environmental circumstances that improve the rate of transpiration also increase leaf sap pH, encourage abscisic acid deposition and at the same moment reduce stomatal behaviour [35]. Failure in Rubisco's activity restricts photosynthesis under very serious drought conditions [36]. Some studies revealed that under drought conditions the activity of the photosynthetic electron transport chain is finely adjusted to chloroplast CO_2 accessibility and photosystem II modifications [37]. The result of dehydration is a decrease in cell size. This improves the cellular content's viciousness. Protein-protein interaction increases outcomes in their aggregation and denaturation. An increased concentration of solutes may become toxic and may be detrimental to enzyme functioning, including photosynthetic equipment, leading in increased viscosity of cytoplasms [38].

4.2 Salinity effect on the crops

The magnitude of agricultural estate affected by high salinity is increasing worldwide as a result of both natural and agricultural occurrences such as irrigation schemes. In plant growth, salinity presents two primary concerns: osmotic stress and ionic stress. It also manifests oxidizing stress. The detrimental effects of salinity alter different physiological and metabolic processes of plants. Often, the answers to these modifications are accompanied by various symptoms such as decreased leaf area, increased leaf density and succulence, leaf abscission, root and shoot necrosis, and decreased internode lengths. Salinity stress inhibits growth and increases cell senescence during extended exposure. Inhibition of growth is the major injury resulting in other symptoms, while programmed cell death may also happen under serious salinity shock [39]. Abscisic acid synthesis is induced under salt stress that closes stomata during transportation to guard cells. Due to stomatal closure and inhibition of photosynthesis and oxidative stress, photosynthesis reduces. Osmotic stress can directly or indirectly inhibit the development of cells through abscisic acid metabolism and translocation. Potassium is not received by plant root surface due to excessive sodium ions near the root zone. Due to the comparable biochemical behavior of sodium and potassium ions, sodium has a powerful repressive impact on root potassium absorption. Deficiency of potassium predictably results in inhibition of growth as potassium maintains cell turgor, activity of enzyme and membrane potential as the most abundant cell cation. Once sodium enters the cytoplasm, the functions of many enzymes are inhibited. This inhibition also depends on the quantity of potassium present: the most deleterious is an elevated sodium/potassium ratio. Plant growth is decreased due to salinity related nutrient disturbances by altering accessibility, transport and partitioning of nutrients. High salt concentration can result in nutrient deficits or imbalances due to Na^+ and Cl^- competition with nutrients such as K^+, Ca^{2+} and NO_3^-. Under saline circumstances there are specific ion toxicity of Na^+ and Cl^- and ionic imbalances influencing biophysical components and/or metabolism of plant growth. Most of the crops combat salinity stress by deposition of low molecular weight organic solutes like linear polyols (sorbitol, glycerol or mannitol), amino acids (proline or glutamate) and betaine (betaine glycine or betaine alanine), cyclic polyols (inositol and other derivatives of mono- and dimethylated inositol) [40].

4.3 Cold effect on the crops

Cold stress is a significant abiotic stress which affects growth and development of crops, leading to loss of strength and lesions on the surface. These symptoms are triggered by changes in the physical and chemical organization of cell membranes, among other metabolic procedures [41]. It is estimated that rises in extreme temperature frequency, severity, and duration are a prevalent feature of our setting. Climate change controls greater changes in temperature, leading in frequent cold periods. Susceptible plants with cold temperatures have reduced growth and growth, restricted use of precious varieties, and reduced yields. Plants use separate strategies to cope with stressful conditions and integrate a variety of physiological, metabolic and molecular adaptations. These methods initially generate modifications to safeguard the plant, followed by cold acclimatization, which increases the survival of the plant under cold stress [9]. While a lot of these mechanisms are facilitated by transcription factors (TFs) that stimulate gene expression associated to stress, the transcription network is not restricted to the reaction of plants to cold [41, 42]. As a foremost element of plasma and endo-membranes, lipids play a significant organizational role in mitigating the effects of cold temperatures [43]. Cold stress decrease plants growth and development that affects the physical and chemical structure of the cell membrane, causes leakage of electrolytes, and reduces protoplasmic streaming and changes in the metabolism of cell [44, 45]. Additional cold reactions comprise changes in nucleic acid and protein synthesis, water and nutrient equilibrium, enzyme affinity and conformation and deficiency in photosynthesis, specifically down-regulation and photo-damage of Photosystem II (PSII) [16].

Changes in the structure of proteins and lipid membranes assist restore homeostasis of metabolites and are regarded a mechanism by which cells feel cold temperatures. For its metabolic and physical function, the liquid state of the plasma membrane is a structural and functional asset. When low temperatures are present, the plasma membrane transitions from a liquid state at elevated temperatures into a stiff gel stage [46]. Low temperature-mediated changes in the physical conformation of the membrane are mainly because of enhanced levels of unsaturated lipids, which increase the fluidity and stability of the membrane, enabling cells to adapt mechanically to cold [47].

4.4 Heat effect on the crops

The sequence of modifications in morphology, biochemistry, and physiology arising from high temperature stress also considerably disturbs growth and development of plant [48]. As a result of increasing atmospheric temperatures, heat shocks are currently primary limiting factors for crop productivity globally. This increasing temperature may result in changes in the phases of growth and distribution of agricultural plants [49]. High-temperature stress can cause serious protein damage, interrupt synthesis of protein, inactivate critical enzymes, and damage membranes. High temperature stress can have significant effects on the cell division process [50]. All of these harms can substantially restrict plant development and also promote oxidative damage. In addition, short exposure to elevated temperature in seed filling can lead to rapid filling, leading to low quality and lower yield. Under a restricted supply of water, the temperature rise is fatal. Overall, water loss due to heat stress is predominantly due to enhanced transpiration rate during the day, which eventually damages certain physiological procedures in crops. Heat stress also decreases the amount, weight and root growth and eventually decreases the accessibility of water and nutrients to the plant parts above ground [51, 52]. Light-dependent chemical reactions that happen in the stroma in the thylakoid and the carbon metabolism are the primary places of harm owing to elevated heat stress. Increased adjustment of PSII thermo-tolerance of PSII leaf temperature and density of photon flux [53]. The PSII is extremely temperature sensitive and significantly affects and even partly terminates its activity under elevated temperature stress [54]. Oxygen-evolving complex also experiences severe harm at elevated temperatures, which can lead to imbalanced electron flow to the PSII acceptor site [55]. At higher temperatures, the proteins D1 and D2 also suffer from denaturation [56]. High heat stress has a significant impact on the activity of significant enzymes like sucrose phosphate synthase, invertase, adenosine diphosphate-glucose pyrophosphorylase, and starch and sucrose synthesis [57]. The reduced CO_2 binding enzyme activation status, Rubisco, limits net photosynthesis in many species of plants. Although Rubisco's catalytic activity rises with greater temperatures, its low CO_2 affinity and O_2 binding ability limit the rise in net photosynthesis speed [58]. Despite all these negative photosynthesis impacts of elevated temperature, with elevated concentrations of CO_2 in the atmosphere, optimum photosynthesis temperature requirements are expected to rise [53].

4.5 Heavy metal

A prevalent and serious problem is heavy metal atmospheric pollution through human activities and/or natural processes. Often referred to as trace metals or heavy metals are potentially toxic elements. Trace metals are related to the trace quantities of components existing in soils. Heavy metals, a loosely specified group of components, constitute elements with an atomic mass exceeding 20 (excluding alkali metals) and specific gravity exceeding 5 [59]. Because of their differing solubility/bioavailability, heavy metals exist in different forms in soil. Many soil physicochemical characteristics change heavy metal geochemical conduct in soil, plant uptake, and effect on crop productivity. Excessive deposition of heavy metals in plant tissue is harmful to multiple biochemical, physiological, and morphological operations in plants either directly or indirectly and in turn affect crop productivity. Heavy metals decrease crop productivity by causing seed germination, accumulation, and re-mobilization of seed reserves during plant growth, germination and photosynthesis to deleterious effects on various plant physiological processes [60]. Heavy metal toxicity on the cell platform decreases the productivity of plants by forming reactive oxygen species, disrupting the redox equilibrium and causing oxidative stress. Metals mainly enter the plants through the root from the soil [61]. The cultivation of metals includes several processes, including desorption of metal from soil particles,, uptake of metals by roots, transportation of metals to plant roots and shoots [62]. Transport of heavy metal to aerial components of the plant is via the xylem and is most probably encouraged by transpiration [63]. The metals, after joining the central cylinder, move towards the aerial parts of the plant where evaporation of water occurs and metal stacks up through the water stream of the vascular system [64]. Only a slight percentage of heavy metals are translocated in most crops to the shooting tissues. In some cases, only if the plant is a hyper accumulator or chelate-assisted, there is sequestration of 95% or more of the metal in the plant's roots [65, 66].

5. Crops tolerance against abiotic stress

Plant resistance to abiotic stress includes escaping stress avoidance and tolerance. Escape: Before extreme stress begins, dry escape depends on efficient reproduction. The plants integrate brief life cycles with high growth rates and gas exchange, using the highest existing resources while soil moisture lasts. It also relies on escaping the unfavorable environmental conditions by shedding off the leaves, no germination, night-time closure of stomata, compact growth, that is, shortening of any plant part [67].

Avoidance: reversible physiological changes involve decreasing water loss (closing stomata, decreasing light absorption through rolling leaves and condensed canopy leaf region) and growing water absorption (increasing root investment, morphological changes occurring in crops to decrease transpiration, re-allocation of nutrients stored in older leaves and greater photosynthesis rates [68].

Tolerance: abiotic stress tolerance appears to be the consequence of cellular and molecular level coordination of physiological and biochemical changes. These changes may include osmotic adjustment, stiffer cell walls, or smaller cells [69]. Changes happening rapidly at the concentrations of mRNA and protein consequence in an intolerant state. Various morphological, physiological, biochemical and molecular modifications happen in crops in order to fight different abiotic stresses [70].

5.1 Morphological changes

Under stress, roots extend their length in the soil in order to seep water around themselves and absorb a sufficient amount of water to persist against stressed conditions. Due to an increase in length and more absorption of water through soil, roots biomass also increases in abiotic stress conditions. Shoot length is higher due to sufficient transpiration and translocation mechanism whereas in water deficit plants, shoot length observed was comparatively dwarf as plants need to overcome water and nutrient deficit conditions caused by drought. Shoot dry weight depends on the inner mass and tillers of wheat. Irrigated plants can accumulate water inside tissues due to a sufficient amount of landed water whereas water deficient cannot do so due to which inner mass decreases in case of water stress plants [23, 71].

5.2 Physiological changes

Plants facing abiotic stress, respond at the molecular, cellular and whole plant levels through a number of physiological modifications.

5.2.1 Relative water content (RWC)

A leaf's relative water content (RWC; or' relative turgidity') is measuring its real water content at complete turgidity relative to its peak water holding capacity.

RWC provides a measurement of the decline in leaf water content and may involve a degree of stress in water deficit and heat stress. RWC includes leaf water potential (another helpful estimation of plant water status) with the impact of osmotic adjustment, a powerful mechanism for preserving cell hydration as a measure of plant water status. The development of leaves relies on the water content and the rate of transpiration. With the absorption of water from roots and passing on to leaves, plants will have high rates of transpiration and water content in leaves will elevate effectively in irrigated plants unlikely in water deficit plants and water potential reduced in drought-stressed plants [67].

5.2.2 Relative stress injury (RSI)

It is the relative injury caused to plants under stressed conditions. Relative stress injury is actuated under stressed conditions providing a measurement of injury caused to plants. Plants under such stressed conditions try to become resistant towards the extraneous factors where plants activate some genes and provide tolerance towards the environment. Abiotic stress tolerance appears to be the result of cooperation at the cellular and molecular levels between physiological and biochemical alternation. These modifications may include more rigid cell walls, osmotic adjustment, or smaller cells. There are rapid changes in the mRNA and protein concentrations that result in tolerance towards stress [71].

5.2.3 Water use efficiency (WUE)

Efficiency in water use (WUE) is a crucial variable responsible for the productivity of plants under restricted supply of water. In agronomic terms, it is defined as the percentage of total dry matter (DM) generated (or harvested) to (or applied) used water. Physiologically speaking, nevertheless, WUE is well-defined as the proportion between the set carbon rate and the transpired water rate. The connection between water use and crop production rate is defined as water use efficiency. It is measured in terms of biomass generated by transpiration unit. Greater biomass generated by limited amount of water under stress circumstances is crucial for higher crop yield. Combined stress can also occur to the plant at same time, for example, water shortage can lead to drought and salinity stress simultaneously, uttermost significant factors limiting crop effectiveness and yielding worldwide. Drought resistance in plants can be improved by escaping or avoiding drought condition using WUE mechanism to maintain water level or growing drought tolerant plants [30, 71].

5.2.4 Osmotic adjustment

Osmotic adjustment (OA) is the net elevation of intercellular solutes in response to water stress that allows turgor to be conserved at a lower water potential. OA has been considered as the primary mechanisms in adaptation of plant towards drought as it promotes the tissue's metabolic activity and enables for regeneration but varies considerably between genotypes. The efficiency of plants in arid conditions has been linked with OA in many species such as sorghum, wheat and oilseed brassicas. High levels of ions can critically inhibit cytosolic enzymes of plant cells [72]. Throughout osmotic adjustment, ion accumulation appears to be limited to the vacuoles where ions are kept out of contact with cytosol or subcellular organelles [73]. Because of this ion compartmentation, other solutes such as sugar alcohol, amino acid proline must be assembled in the cytoplasm in order to preserve the cell's water potential balance [74].

5.3 Biochemical changes

Under stress, crops experience a number of cellular and molecular-level biochemical modifications.

5.3.1 Chlorophyll and carotenoid

Chlorophyll and carotenoid content depend on ATP, photosynthetic reactions, NAD. Chlorophyll cannot capture sunlight straight, so it gives sunlight to chlorophyll with the aid of carotenoid, which is an accessory pigment, and transfers it to photosystem I and photosystem II, which transforms light energy into chemical energy acquired in the form of ATP and NADPH. Now, with the help of end products of photosystems and fixed carbon dioxide, plants produce glucose. So, we can say that in wheat more the carotenoid present in the chloroplast, more will be the sunlight captured and thus more will be the chlorophyll [7].

5.3.2 Starch

Starch also evolves as a main molecule in enabling the response to abiotic stresses by plants like water deficit, salinity or extreme temperatures. When photosynthesis is limited under stress conditions plants have a tendency to use starch as energy source. Adverse effect of stress is reduced in plant by releasing some compatible solutes, osmoprotectants, derived sugars and other metabolites to encourage plant growth [75].

5.3.3 Amino acid

Under stress circumstances, amino acids such as proline and arginine play a significant role in controlling osmotic pressure. Proline acts as an osmoprotectant and its accumulation can lead to improved synthesis of cells and their reduced degradation. This behaviour of higher accumulation of proline is because of the expression of the gene encoding pyrroline-5-carboxylate synthase. Additional proof for proline's defensive function was discovered in transgenic crops, where proline overproduction improves tolerance to osmotic stress. In addition to proline, the reaction to osmotic stress also involves other amino acids. Arginine was found to operate as a compatible solution to enhance stress tolerance in leaves. The enzyme involved in arginine biosynthesis is enhanced under hyperosmotic circumstances. In addition, osmotic stress in sunflower and wheat causes enhanced expression of asparagine synthase genes. Glutamine synthase overexpression enhances tolerance of osmotic stress in rice. These findings indicate that changes in osmotic pressure-induced amino acid levels may be due to modified gene expression encoding the enzymes engaged in their metabolism [76].

5.4 Molecular changes

5.4.1 Late embryogenesis abundant proteins (LEA)

LEA proteins are the group of elevated molecular weight proteins that are abundantly present during early embryogenesis and collect in reaction to water stress during seed dehydration. There are different LEA protein groups. The proteins belonging to group 3 are considered to play a part in the sequestration of focused ions between these groups during cell dehydration. LEA proteins of group 1 are expected to have increased water-binding ability, whereas LEA proteins of group 5 are presumed to be appropriate ions during water loss [77].

5.4.2 Detoxifying genes

Also, there are certain detoxifying genes that help to combat abiotic stress. Plants can be protected from damage by increase tolerance towards stress by the accumulation of some attuned solutes and reactive oxygen species (ROS) are scavenged. This action helps to maintain protein structures and functions. The genes responsible for activation of these three enzymes: ascorbate peroxidase, glutathione peroxidase, and glutathione reductase have revealed to have some effect on various abiotic stresses [78].

5.4.3 Heat shock protein genes

An increase in the transcription of a set of genes by heat exposure or other abiotic stress in all species is a heavily maintained biological reaction. The reaction is promoted by the heat shock transcription factor (HSF) in the form of a monomeric, non-DNA binding type current in unstressed cells. It is caused by stress in the form of a trimeric shape that can bind heat shock gene promoters. Gene stimulation encoding thermal shock proteins (Hsps) is one of the most noticeable responses in organisms that are subjected to high molecular temperature [79].

5.5 In reaction to abiotic stress, various genetic mechanisms begin in the crops

5.5.1 ABA pathway

Many genes responsible for response to stress are triggered under abiotic stress conditions. Abscisic acid (ABA) is a main plant stress-signaling hormone and its accumulation automatically increases as the harsh conditions are faced by plant to fight the stress effect. Two pathways are triggered in plant under osmotic stress condition, that is, ABA-dependent and ABA-independent pathways. In ABA-dependent pathway, a mixture of transcription factors, ABRE binding protein/ABRE binding factors (AREB/ABFs) demonstrate critical functions. A cis-element, dehydration-response element/C-repeat (DRE/CRT) and DRE-/CRT-binding protein 2 (DREB2) transcription factors play a main part in the expression of ABA-independent genes in response to osmotic stress. Continuous increase in expression of AREB1/ABF2, AREB2/ABF4 and ABF3 is triggered by drought and salinity in vegetative tissues. Over-expression studies indicate that in conditions of drought stress, these three AREB/ABFs are useful signals from ABA regulators. As shown in the figure, AREB/ABF transcription factors result in gene expression of the genes involved in abiotic stress reaction and tolerance [80].

5.5.2 Cold stress pathway

CBF/DREB1 homologs have been acknowledged in various species. CBF/DREB1 may bind CRT/DRE cis-elements (A/GCCGAC) in the promoter area of COR genes to control the expression of COR genes belonging to the transcription factor family ERF/AP2 (ethylene-responsive factor/APETALA2). The CRT/DRE cis-acting components express the RD29A gene that is believed to be involved in abiotic stress reaction and tolerance [80].

5.5.3 SOS pathway

The SOS signaling path includes three significant enzymes, SOS1, SOS2, and SOS3. SOS1 protein codes for Na^+/H^+ anti-porter plasma membrane. This protein is essential for cell-level regulation of Na^+ efflux. Na^+ long-distance transportation from root to shoot is also facilitated. This protein's overexpression is related to plant salt tolerance [81]. Salt stress-stimulated signals from Ca^{2+} activate the SOS2 gene encoding serine/threonine kinase. This protein includes a correctly established catalytic N-terminal domain and a regulatory C-terminal domain. The SOS3 protein is the third protein involved in the SOS stress signaling pathway. It is a myristylated Ca^{+} binding protein and contains an N-terminus myristylation site. In salt tolerance, this site shows an important role. In the C-terminal regulatory domain of the SOS2 protein, FISL motif is present (also known as NAF domain) that is approximately 21 lengthy sequence of amino acids, and helps to interact with the Ca^{2+} binding SOS3 protein [82]. Kinase activation is the consequence of the SOS2–SOS3 protein interaction. The kinase caused then phosphorylated SOS1 protein thus enhancing its initially identified yeast transportation activity. SOS1 protein is defined by a long cytosolic C-terminal tail composed of a putative nucleotide binding motif and an auto inhibitory domain, which is roughly 700 amino acids long [83]. The target site for SOS2 phosphorylation is this auto inhibitory domain. In relation to salinity tolerance, it regulates trafficking of membrane vesicle, pH homeostasis and functions of vacuole. There is thus a significant increase in intracellular Ca^{2+} level with the increase in Na^{+} concentration, which in turn encourages its binding with SOS3 protein. Ca^{2+} controls homeostasis of intracellular Na^{+} along with SOS proteins. Then the SOS3 protein interacts and activation of the SOS2 protein occurs by releasing its self-inhibition. The complex SOS2-SOS3 goes down to plasma membrane where SOS1 reacts. The result of phosphorylated SOS1 is improved Na^{+} efflux, reducing Na^{+} toxicity [84].

6. Conclusion

There are a broad variety of abiotic stresses that adversely influence the crops. In crops there are prevalent abiotic stresses such as drought, salinity, elevated temperature, low temperature, and metal toxicity. The symptoms of stress, of course, vary with its severity, from being elusive to disastrous. The crops undergo various kinds of modifications due to abiotic stresses, which can cause antagonistic effects on growth and development of plant. The complexity and type of abiotic stress reactions promote the use of extensive, integrative and multidisciplinary techniques to achieve the various levels of stress response regulation. The crops are undergoing modifications such as decreased relative water content, increased ROS output, enhanced relative stress injury, cell electrolyte leakage, decreased photosynthetic pigment amount, decreased root and shoot length, decreased yield, etc. The crops are undergoing numerous morphological, physiological, biochemical and molecular modifications to overcome the impacts of drought. Lately, there has been a lot of attraction in managing abiotic stress in plants. With the growing growth of high performance genomic instruments, crops have created many new methods to combat abiotic stress.

Author details

Summy Yadav*, Payal Modi, Akanksha Dave, Akdasbanu Vijapura, Disha Patel and Mohini Patel
Division of Biological and Life Sciences, Ahmedabad University, Ahmedabad, India

*Address all correspondence to: summyyadav89@gmail.com

References

[1] Fahad S, Bajwa AA, Nazir U, Anjum SA, Farooq A, Zohaib A, et al. Crop production under drought and heat stress: Plant responses and management options. Frontiers in Plant Science. 2017;**29**:8. Available from: http://journal.frontiersin.org/article/10.3389/fpls.2017.01147/full

[2] Shavrukov Y, Kurishbayev A, Jatayev S, Shvidchenko V, Zotova L, Koekemoer F, et al. Early flowering as a drought escape mechanism in plants: How can it aid wheat production? Frontiers in Plant Science. 2017;**17**:8. DOI: http://journal.frontiersin.org/article/10.3389/fpls.2017.01950/full

[3] Fghire R, Anaya F, Ali OI, Benlhabib O, Ragab R, Wahbi S. Physiological and photosynthetic response of quinoa to drought stress. Chilean Journal of Agricultural Research. 2015;**75**(2):174-183

[4] Shanker AK, Shanker C. Abiotic and Biotic Stress in Plants—Recent Advances and Future Perspectives [Internet]. Rijeka: IntechOpen; 2016. Available from: http://www.intechopen.com/books/abiotic-and-biotic-stress-in-plants-recent-advances-and-future-perspectives

[5] Browse J, Xin Z. Temperature sensing and cold acclimation. Current Opinion in Plant Biology. 2001;**4**(3):241-246

[6] Pearce R. Plant freezing and damage. Annals of Botany. 2001;**87**(4):417-424

[7] Farooq M, Wahid A, Kobayashi N, Fujita D, Basra SMA. Plant drought stress: Effects, mechanisms and management. Agronomy for Sustainable Development. 2009;**29**(1):185-212

[8] Volkov V. Salinity tolerance in plants. Quantitative approach to ion transport starting from halophytes and stepping to genetic and protein engineering for manipulating ion fluxes. Frontiers in Plant Science. 2015;**27**:6. DOI: http://journal.frontiersin.org/Article/10.3389/fpls.2015.00873/abstract

[9] Flowers TJ. Improving crop salt tolerance. Journal of Experimental Botany. 2004;**55**(396):307-319

[10] Hirayama T, Shinozaki K. Research on plant abiotic stress responses in the post-genome era: Past, present and future. The Plant Journal. 2010;**61**(6):1041-1052

[11] Ashkavand P, Zarafshar M, Tabari M, Mirzaie J, Nikpour A, Kazem Bordbar S, et al. Application of SIO2 nanoparticles AS pretreatment alleviates the impact of drought on the physiological performance of *Prunus mahaleb* (Rosaceae). Boletin de la Sociedad Argentina de Botanica. 2018;**53**:207

[12] Bockheim JG, Gennadiyev AN. The role of soil-forming processes in the definition of taxa in soil taxonomy and the world soil reference base. Geoderma. 2000;**95**(1-2):53-72

[13] Carillo P, Grazia M, Pontecorvo G, Fuggi A, Woodrow P. Salinity Stress and Salt Tolerance. In: Shanker A, editor. Abiotic Stress in Plants—Mechanisms and Adaptations [Internet]. Rijeka: IntechOpen; 2011. Available from: http://www.intechopen.com/books/abiotic-stress-in-plants-mechanisms-and-adaptations/salinity-stress-and-salt-tolerance

[14] Akbari G, Modarres Sanavy SAM, Yousefzadeh S. Effect of auxin and salt stress (NaCl) on seed germination of wheat cultivars (*Triticum aestivum* L.). Pakistan Journal of Biological Sciences. 2007;**10**(15):2557-2561

[15] Devasirvatham V, Tan D. Impact of high temperature and drought stresses

on chickpea production. Agronomy. 2018;**8**(8):145

[16] Takahashi D, Li B, Nakayama T, Kawamura Y, Uemura M. Plant plasma membrane proteomics for improving cold tolerance. Frontiers in Plant Science. 2013;**4**:90

[17] Flora SJS, Mittal M, Mehta A. Heavy metal induced oxidative stress & its possible reversal by chelation therapy. The Indian Journal of Medical Research. 2008;**128**(4):501-523

[18] Wuana RA, Okieimen FE. Heavy metals in contaminated soils: A review of sources, chemistry, risks and best available strategies for remediation. ISRN Ecology. 2011;**2011**:1-20

[19] Schutzendubel A. Plant responses to abiotic stresses: Heavy metal-induced oxidative stress and protection by mycorrhization. Journal of Experimental Botany. 2002;**53**(372):1351-1365

[20] Hall JL. Cellular mechanisms for heavy metal detoxification and tolerance. Journal of Experimental Botany. 2002;**53**(366):1-11

[21] Hossain Z, Mustafa G, Komatsu S. Plant responses to nanoparticle stress. International Journal of Molecular Sciences. 2015;**16**(11):26644-26653

[22] Yadav S, Vijapura A, Dave A, Shah S, Memon Z. Genetic diversity analysis of different wheat [*Triticum aestivum* (L.)] varieties using SSR markers. International Journal of Current Microbiology and Applied Sciences. 2019;**8**(02): 839-846

[23] Tandzi LN, Bradley G, Mutengwa C. Morphological responses of maize to drought, heat and combined stresses at seedling stage. Journal of Biological Sciences. 2019;**19**(1):7-16

[24] Cairns JE, Sonder K, Zaidi PH, Verhulst N, Mahuku G, Babu R, et al. Maize production in a changing climate. In: Advances in Agronomy [Internet]. Elsevier; 2012. pp. 1-58. Available from: https://linkinghub.elsevier.com/retrieve/pii/B9780123942753000067

[25] Shaheen T, Rahman M, Shahid Riaz M, Zafar Y, Rahman M. Soybean production and drought stress. In: Abiotic and Biotic Stresses in Soybean Production [Internet]. Elsevier; 2016. pp. 177-196. Available from: https://linkinghub.elsevier.com/retrieve/pii/B9780128015360000086

[26] Oh-e I, Saitoh K, Kuroda T. Effects of high temperature on growth, yield and dry-matter production of rice grown in the paddy field. Plant Production Science. 2007;**10**(4):412-422

[27] Jagadish S, Craufurd P, Wheeler T. High temperature stress and spikelet fertility in rice (*Oryza sativa* L.). Journal of Experimental Botany. 2007;**58**(7):1627-1635

[28] Mukamuhirwa A, Persson Hovmalm H, Bolinsson H, Ortiz R, Nyamangyoku O, Johansson E. Concurrent drought and temperature stress in rice—A possible result of the predicted climate change: Effects on yield attributes, eating characteristics, and health promoting compounds. International Journal of Environmental Research and Public Health. 2019;**16**(6):1043

[29] Cramer GR, Urano K, Delrot S, Pezzotti M, Shinozaki K. Effects of abiotic stress on plants: A systems biology perspective. BMC Plant Biology. 2011;**11**(1):163

[30] Cattivelli L, Rizza F, Badeck F-W, Mazzucotelli E, Mastrangelo AM, Francia E, et al. Drought tolerance improvement in crop plants: An integrated view from breeding to genomics. Field Crops Research. 2008;**105**(1-2):1-14

[31] Samarah NH. Effects of drought stress on growth and yield of barley.

Agronomy for Sustainable Development. 2005;**25**(1):145-149

[32] Fu J, Huang B. Involvement of antioxidants and lipid peroxidation in the adaptation of two cool-season grasses to localized drought stress. Environmental and Experimental Botany. 2001;**45**(2):105-114

[33] Baker NR. Photosynthesis and the Environment [Internet]. Dordrecht: Kluwer Academic Publishers; 2006. Available from: http://accesbib.uqam.ca/cgi-bin/bduqam/transit.pl?&noMan=25127834

[34] Yokota A. Citrulline and DRIP-1 protein (ArgE homologue) in drought tolerance of wild watermelon. Annals of Botany. 2002;**89**(7):825-832

[35] Turner NC, Wright GC, Siddique KHM. Adaptation of grain legumes (pulses) to water-limited environments. In: Advances in Agronomy [Internet]. Elsevier; 2001. pp. 193-231. Available from: https://linkinghub.elsevier.com/retrieve/pii/S0065211301710152

[36] Bota J, Medrano H, Flexas J. Is photosynthesis limited by decreased Rubisco activity and RuBP content under progressive water stress? The New Phytologist. 2004;**162**(3):671-681

[37] Loreto F, Tricoli D, Marco G. On the relationship between electron transport rate and photosynthesis in leaves of the C4 plant sorghum bicolor exposed to water stress, temperature changes and carbon metabolism inhibition. Functional Plant Biology. 1995;**22**(6):885

[38] Hoekstra FA, Golovina EA, Buitink J. Mechanisms of plant desiccation tolerance. Trends in Plant Science. 2001;**6**(9):431-438

[39] Parida AK, Das AB. Salt tolerance and salinity effects on plants: A review. Ecotoxicology and Environmental Safety. 2005;**60**(3):324-349

[40] Grattan SR, Grieve CM. Salinity-mineral nutrient relations in horticultural crops. Scientia Horticulturae. 1998;**78**(1-4):127-157

[41] Barrero-Sicilia C, Silvestre S, Haslam RP, Michaelson LV. Lipid remodelling: Unravelling the response to cold stress in Arabidopsis and its extremophile relative Eutrema salsugineum. Plant Science. 2017;**263**:194-200

[42] Zhu J, Dong C-H, Zhu J-K. Interplay between cold-responsive gene regulation, metabolism and RNA processing during plant cold acclimation. Current Opinion in Plant Biology. 2007;**10**(3):290-295

[43] Nishida I, Murata N. Chilling sensitivity in plants and cyanobacteria: The crucial contribution of membrane lipids. Annual Review of Plant Physiology and Plant Molecular Biology. 1996;**47**(1):541-568

[44] Thomashow MF. Plant cold acclimation: Freezing tolerance genes and regulatory mechanisms. Annual Review of Plant Physiology and Plant Molecular Biology. 1999;**50**(1):571-599

[45] Chen M, Thelen JJ. ACYL-LIPID DESATURASE2 is required for chilling and freezing tolerance in arabidopsis. The Plant Cell. 2013;**25**(4):1430-1444

[46] Miquel M, James D, Dooner H, Browse J. Arabidopsis requires polyunsaturated lipids for low-temperature survival. Proceedings of the National Academy of Sciences. 1993;**90**(13):6208-6212

[47] Janská A, Maršík P, Zelenková S, Ovesná J. Cold stress and acclimation—What is important for metabolic adjustment? Plant Biology. 2010;**12**(3): 395-405

[48] Wahid A, Shabbir A. Induction of heat stress tolerance in barley seedlings by pre-sowing seed treatment with glycinebetaine. Plant Growth Regulation. 2005;**46**(2):133-141

[49] Porter JR. Rising temperatures are likely to reduce crop yields. Nature. 2005;**436**(7048):174-174

[50] Smertenko A, Draber P, Viklicky V, Opatrny Z. Heat stress affects the organization of microtubules and cell division in Nicotiana tabacum cells. Plant, Cell & Environment. 1997;**20**(12):1534-1542

[51] Wahid A, Gelani S, Ashraf M, Foolad M. Heat tolerance in plants: An overview. Environmental and Experimental Botany. 2007;**61**(3):199-223

[52] Huang B, Rachmilevitch S, Xu J. Root carbon and protein metabolism associated with heat tolerance. Journal of Experimental Botany. 2012;**63**(9):3455-3465

[53] Crafts-Brandner SJ. Sensitivity of photosynthesis in a C4 plant, maize, to heat stress. Plant Physiology. 2002;**129**(4):1773-1780

[54] Camejo D, Rodríguez P, Angeles Morales M, Miguel Dell'Amico J, TorrecillasA, AlarcónJJ. Hightemperature effects on photosynthetic activity of two tomato cultivars with different heat susceptibility. Journal of Plant Physiology. 2005;**162**(3):281-289

[55] De Ronde JA, Cress WA, Krüger GHJ, Strasser RJ, Van Staden J. Photosynthetic response of transgenic soybean plants, containing an Arabidopsis P5CR gene, during heat and drought stress. Journal of Plant Physiology. 2004;**161**(11):1211-1224

[56] De Las Rivas J, Barber J. Structure and thermal stability of photosystem II reaction centers studied by infrared spectroscopy. Biochemistry. 1997;**36**(29):8897-8903

[57] Vu JCV, Gesch RW, Pennanen AH, Allen Hartwell L, Boote KJ, Bowes G. Soybean photosynthesis, Rubisco, and carbohydrate enzymes function at supraoptimal temperatures in elevated CO2. Journal of Plant Physiology. 2001;**158**(3):295-307

[58] Morales D, Rodríguez P, Dell'Amico J, Nicolás E, Torrecillas A, Sánchez-Blanco MJ. High-temperature preconditioning and thermal shock imposition affects water relations, gas exchange and root hydraulic conductivity in tomato. Biologia Plantarum. 2004;**47**(2):203-208

[59] Rascio N, Navari-Izzo F. Heavy metal hyperaccumulating plants: How and why do they do it? And what makes them so interesting? Plant Science. 2011;**180**(2):169-181

[60] Uzu G, Sobanska S, Aliouane Y, Pradere P, Dumat C. Study of lead phytoavailability for atmospheric industrial micronic and sub-micronic particles in relation with lead speciation. Environmental Pollution. 2009;**157**(4):1178-1185

[61] Pourrut B, Jean S, Silvestre J, Pinelli E. Lead-induced DNA damage in *Vicia faba* root cells: Potential involvement of oxidative stress. Mutation Research/Genetic Toxicology and Environmental Mutagenesis. 2011;**726**(2):123-128

[62] Saifullah, Meers E, Qadir M, de Caritat P, Tack FMG, Du Laing G, et al. EDTA-assisted Pb phytoextraction. Chemosphere. 2009;**74**(10): 1279-1291

[63] Verbruggen N, Hermans C, Schat H. Molecular mechanisms of metal hyperaccumulation in plants: Tansley review. The New Phytologist. 2009;**181**(4):759-776

[64] Liao YC, Chien SWC, Wang MC, Shen Y, Hung PL, Das B. Effect of transpiration on Pb uptake by lettuce and on water soluble low molecular weight organic acids in rhizosphere. Chemosphere. 2006;**65**(2):343-351

[65] Duarte B, Delgado M, Caçador I. The role of citric acid in cadmium and nickel uptake and translocation, in Halimione portulacoides. Chemosphere. 2007;**69**(5):836-840

[66] Shahid M, Dumat C, Silvestre J, Pinelli E. Effect of fulvic acids on lead-induced oxidative stress to metal sensitive *Vicia faba* L. plant. Biology and Fertility of Soils. 2012;**48**(6):689-697

[67] Maroco JP, Pereira JS, Manuela Chaves M. Growth, photosynthesis and water-use efficiency of two C4Sahelian grasses subjected to water deficits. Journal of Arid Environments. 2000;**45**(2):119-137

[68] Chaves MM, Maroco JP, Pereira JS. Understanding plant responses to drought—From genes to the whole plant. Functional Plant Biology. 2003;**30**(3):239

[69] Wilson J, Ludlow M, Fisher M, Schulze E. Adaptation to water stress of the leaf water relations of four tropical forage species. Functional Plant Biology. 1980;**7**(2):207

[70] Ingram J, Bartels D. The molecular basis of dehydration tolerance in plants. Annual Review of Plant Physiology and Plant Molecular Biology. 1996;**47**(1):377-403

[71] Yadav S, Sharma KD. Molecular and morphophysiological analysis of drought stress in plants. In: Rigobelo EC, editor. Plant Growth [Internet]. Rijeka: IntechOpen; 2016. Available from: http://www.intechopen.com/books/plant-growth/molecular-and-morphophysiological-analysis-of-drought-stress-in-plants

[72] Tangpremsri T, Fukai S, Fischer K. Growth and yield of sorghum lines extracted from a population for differences in osmotic adjustment. Australian Journal of Agricultural Research. 1995;**46**(1):61

[73] El Hafid R. Physiological responses of spring durum wheat cultivars to early-season drought in a mediterranean environment. Annals of Botany. 1998;**81**(2):363-370

[74] Singh V, Singh AK, Raghuvanshi T, Singh MK, Singh V, Singh U. Influence of boron and molybdenum on growth, yield and quality of cauliflower (*Brassica oleracea* L. var. botrytis). International Journal of Current Microbiology and Applied Sciences. 2017;**6**(10):3408-3414

[75] Krasensky J, Jonak C. Drought, salt, and temperature stress-induced metabolic rearrangements and regulatory networks. Journal of Experimental Botany. 2012;**63**(4):1593-1608

[76] Ashraf M, Foolad MR. Roles of glycine betaine and proline in improving plant abiotic stress resistance. Environmental and Experimental Botany. 2007;**59**(2):206-216

[77] Galau GA, Bijaisoradat N, Hughes DW. Accumulation kinetics of cotton late embryogenesis-abundant mRNAs and storage protein mRNAs: Coordinate regulation during embryogenesis and the role of abscisic acid. Developmental Biology. 1987;**123**(1):198-212

[78] Diamant S, Eliahu N, Rosenthal D, Goloubinoff P. Chemical chaperones regulate molecular chaperones *in vitro* and in cells under combined salt and heat stresses. The Journal of Biological Chemistry. 2001;**276**(43):39586-39591

[79] Vierling E. The roles of heat shock proteins in plants. Annual Review of

Plant Physiology and Plant Molecular Biology. 1991;**42**(1):579-620

[80] Shinozaki K, Yamaguchi-Shinozaki K. Gene networks involved in drought stress response and tolerance. Journal of Experimental Botany. 2006;**58**(2):221-227

[81] Shi H, Ishitani M, Kim C, Zhu J-K. The Arabidopsis thaliana salt tolerance gene SOS1 encodes a putative Na+/H+ antiporter. Proceedings of the National Academy of Sciences. 2000;**97**(12):6896-6901

[82] Liu J. The Arabidopsis thaliana SOS2 gene encodes a protein kinase that is required for salt tolerance. Proceedings of the National Academy of Sciences. 2000;**97**(7):3730-3734

[83] Ishitani M. SOS3 function in plant salt tolerance requires N-myristoylation and calcium binding. Plant Cell Online. 2000;**12**(9):1667-1678

[84] Gupta B, Huang B. Mechanism of salinity tolerance in plants: Physiological, biochemical, and molecular characterization. International Journal of Genomics. 2014;**2014**:1-18

Chapter 2

Possibility of No-Input Farming in Lowland Rice Fields in Japan from the Viewpoint of Sustaining Soil Fertility

Naoki Moritsuka

Abstract

In Japan, the area of low-input rice production is gradually increasing with a growing public interest in the quality and safety of our staple food. In an extreme case, rice has been grown over years without using any chemical fertilizers and agrochemicals. However, it is uncertain how much and how long such no-input farming can sustain rice yield and soil fertility. To better understand the sustainability of no-input rice farming in Japan, I briefly review previous results obtained from the long-term field experiments. The topics are (1) rice yield and soil fertility under no-input farming, (2) the environmental factors affecting rice growth and soil fertility under no-input farming, and (3) the dynamics of soil K under continuous rice cropping. The corresponding conclusions are as follows: (1) rice yield and soil fertility under no-input farming in Japan were influenced by various environmental and management factors operating at regional and field scales; (2) the input of K through irrigation and the high-clay content in soil were considered the key environmental factors that enable to sustain no-input farming; and (3) soil K depletion caused by long-term exhaustive cropping should be assessed by monitoring the decrease of soil nonexchangeable K rather than that of exchangeable K.

Keywords: irrigation effect, long-term field experiment, lowland rice, nonexchangeable potassium, soil sustainability

1. Introduction

It is generally considered that crop yield and soil fertility can be maintained by the adequate input of fertilizer elements to soil. However, in the case of irrigated paddy soil, there are several farmers' fields in Japan which have not received any fertilizers for more than a decade but sustained rice yield at around 400 g m^{-2}, i.e., about 80% of the conventionally fertilized fields [1–3].

The wonder of sustaining rice yield without fertilizer input may be explained by the unconscious input of nutrients to lowland fields through irrigation, rainfall, and biological nitrogen (N) fixation [4]. The advantage of lowland rice over upland rice can be found in the nutrient omission trials carried out throughout the country before chemical fertilizer was prevailed [5]. As shown in **Figure 1**, N was the most limiting element for both lowland and upland rice. For lowland rice, however,

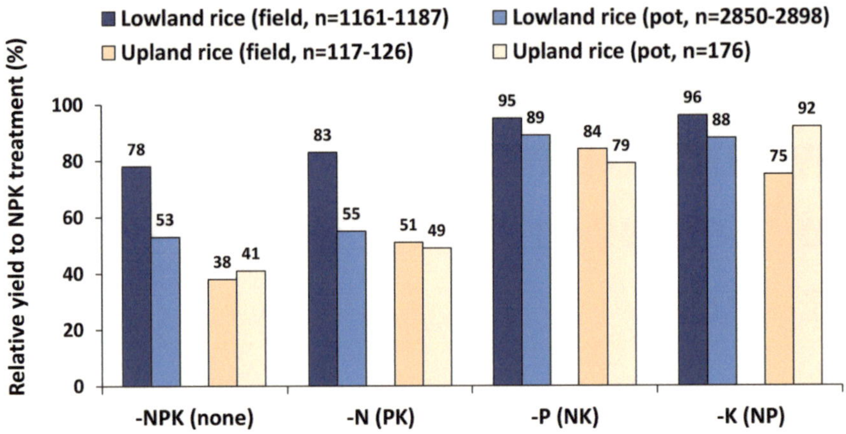

Figure 1.
Response of the yield of lowland rice and upland rice to the omission of N, P, and K fertilizers in Japan (adapted from [5]). Data obtained from the nutrient omission trials conducted under pot and field conditions since 1916 were summarized.

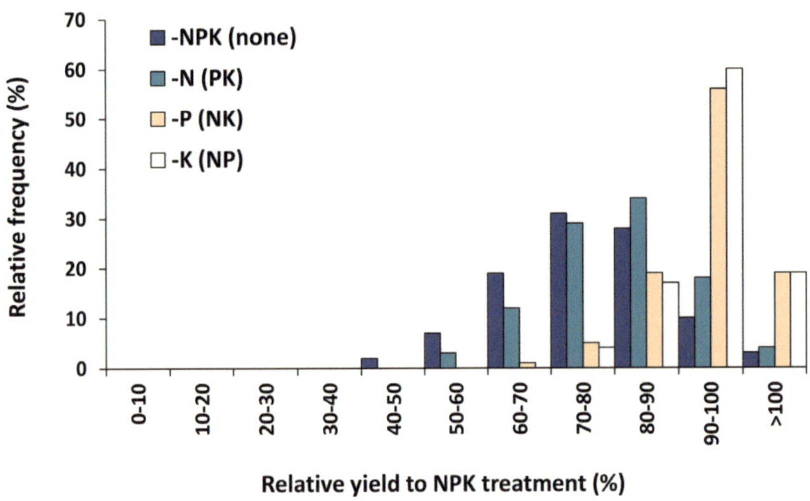

Figure 2.
Response of the yield of lowland rice to the omission of N, P, and K fertilizers in Japan (adapted from [5]). Data obtained from the nutrient omission trials conducted under field conditions (field-grown lowland rice in **Figure 1***) were summarized. The trials were 1097–1138 in number.*

the percentage of yield loss caused by the omission of fertilizers differed with the growth conditions; 22 and 47% under field and pot conditions, respectively. Such a discrepancy was not observed for upland rice. With a closer look at the response of lowland rice to the omission of potassium (K) and phosphorus (P) under field conditions, the omission of these elements also caused more than 10% decrease of the yield in more than 20% of the paddy fields surveyed (**Figure 2**). Accordingly, K or P began to limit rice yield in some of the paddy fields when N limitation was removed by the application of N fertilizer.

These results contributed to predict the necessary amount and type of chemical fertilizers applied to paddy fields. **Figure 3** shows the temporal changes in the average rates of chemical fertilizer applied to paddy fields in Japan [6]. In 1950, N was applied at a higher rate than P and K. With time, the rates of P and K became comparable to the rate of N. This is probably because of the alleviation of N limitation and the use of compound fertilizer containing N and other nutrients. In 1970, more than 60% of

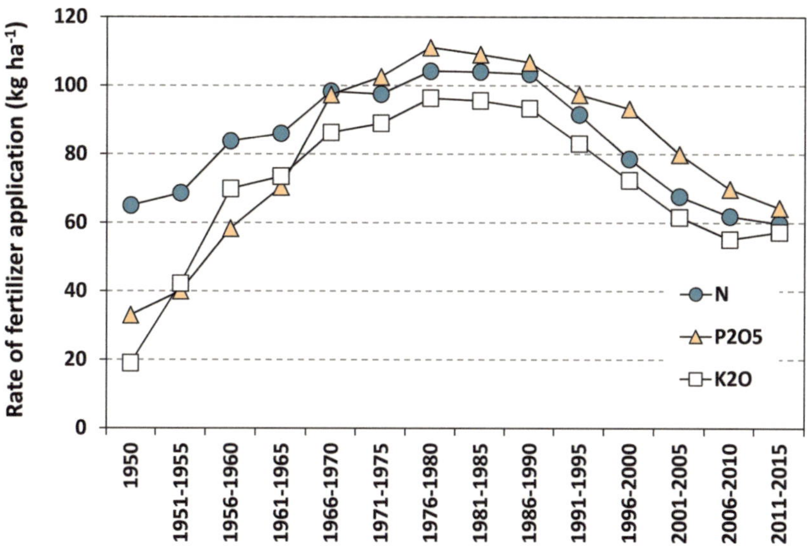

Figure 3.
Rates of chemical fertilizer application to paddy fields in Japan from 1950 to 2015. The values before 1970 were cited from FAO [6], and those after 1970 were cited from several volumes of Pocket Fertilizer Handbook published by the Association of Agriculture and Forestry Statistics, Tokyo.

N, P, and K were applied together in the form of compound fertilizer. The application rates of all nutrients increased rapidly by 1970 and reached a plateau around 1980. Then, the rates decreased from 1990 to 2015. The amount of N applied in 2011–2015 (60 kg ha^{-1}) became smaller than the amount of N applied in 1950 (65 kg ha^{-1}). This would be partly because the percentage of fertilizer N recovered by rice plants was significantly increased by the development of new techniques, e.g., side-dressing of polyolefin-coated urea that can supply N to rice roots according to crop demand. But a more plausible reason is the introduction of the *gentan* policy in 1970 for reducing rice production all over the country and the concomitant shift of the consumer preference to rice from the nutrition to the taste and safety. For example, recent Japanese consumers prefer low-protein rice that is less nutritious than high-protein rice, because cooked rice with high-protein content tends to become hard and nonsticky [7].

From these backgrounds, rice and other crops produced with reduced input of chemical fertilizers and agrochemicals have been attracting more attention by consumers. In 2001, the Japanese government established the guidelines for the certification of crops produced with chemical fertilizers and agrochemicals at less than 50% of the conventional dosage in each region. The area of production of such crops amounted to 0.12 million ha (2.6% of total arable land) in 2017. Organic farming, where chemical fertilizer is fully replaced with organic fertilizer, is also increasing gradually, although the area of organic-farming fields is still 0.5% of the total area of arable land in 2017. The most extreme way of farming is the production of crops without using any chemical fertilizers and agrochemicals. Such no-input farming is called *shizen nouhou* or *shizen saibai* in Japanese, and translated directly as natural farming [1] or nature farming [8]. The amount of rice produced by no-input farming was estimated to be only 0.04% of the national production in 1991 [9].

These histories clearly show that no-input farming in Japan has been developed as a result of the past high-input farming, and it does not represent various types of no-input farming systems in the world. Almost all no-input paddy fields in Japan had received chemical fertilizers and agrochemicals before no-input farming was introduced, and these fields are different from the absolutely no-input fields in other countries that have not received any chemicals since land reclamation.

Recently, no-input rice farming in Japan has been recognized as an economically feasible farming system. Due to the very limited availability, rice produced by no-input farming has been sold at twice or more the price of rice produced by conventional farming [10]. Besides the price of the products, the level of rice yield and its sustainability are also important for farmers [1]. Several researchers have compared rice yield among no-input paddy fields with different periods after introducing no-input farming [1–3]. However, most of the previous studies have used a space-for-time substitution approach instead of monitoring rice yield and soil fertility over years. Thus, it is uncertain how much and how long such no-input farming can sustain rice yield and soil fertility under various environmental conditions.

In order to better understand the sustainability of no-input rice farming in Japan, I briefly review previous results obtained from the long-term field experiments including our no-input trial. The main topics in this review are (1) rice yield and soil fertility under no-input farming, (2) the environmental factors affecting rice growth and soil fertility under no-input farming, and (3) the dynamics of soil K under no-input and high-input rice farming systems.

2. Rice yield and soil fertility under no-input farming

In 1990, Neera et al. [1] surveyed 542 no-input fields in 17 prefectures in Japan and compared rice yield with the average yield by conventional farming according to the corresponding municipal statistics. The sampling of rice plants was performed at one representative site in a paddy field at the rate of 30 hills per field [11]. On average of the surveyed fields, the period of no-input farming was 10.7 years, and the yield of brown rice by no-input farming (445 g m^{-2}) amounted to 87% of the yield by conventional farming (511 g m^{-2}). When the results were compared among different regions, the yield by no-input farming was significantly lower than the yield by conventional farming at six prefectures in Tohoku district located in northern Japan (**Figure 4**). The yield was relatively high in Tohoku district, and the average yield after no-input farming for 28–40 years amounted to 456 g m^{-2} (n = 19). On the other hand, the yield gap was not statistically significant at many prefectures in Kinki and Chugoku districts located in southern Japan except for

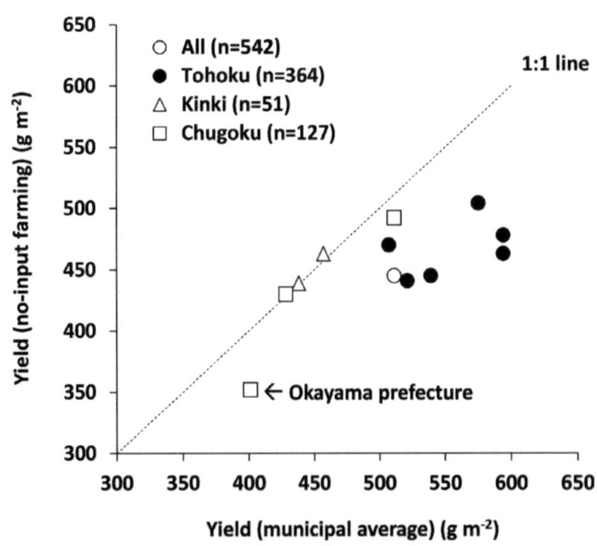

Figure 4.
Relationship between rice yields obtained from municipal statistics and from no-input farming (adapted from [1]).

Okayama prefecture. Furthermore, the coefficients of variation of yield by no-input farming were as large as 12–29% in each region displayed in **Figure** 4, and the number of surveyed fields in each region was relatively large (19–180). These results suggest that rice yield was also influenced by field-specific environmental and management factors. The lack of the yield gap in several prefectures might be due to the arbitrary selection of fertile fields for no-input farming, because it was contrasting to the results obtained from the nutrient omission trials (**Figure** 2). Thus, the authors emphasized that more research is needed to monitor the yield under no-input farming in combination with the yield under conventional farming and to elucidate the factors affecting rice yield under no-input farming.

Following this pioneering work, however, only a few researchers have attempted to identify the factors affecting rice yield under no-input farming. Hosoya and Sugiyama [2] surveyed 16 no-input fields in four prefectures (Aomori, Iwate, Miyagi, and Niigata) in northern Japan. The yield of brown rice in no-input fields was positively correlated with the number of panicles ($r = 0.92$, $p < 0.01$). The panicle number was positively correlated with the air temperature during the vegetative stage ($r = 0.66$, $p < 0.01$) and negatively correlated with the latitude of the location of each field ($r = -0.60$, $p < 0.05$). Tatara et al. [3] also examined 16 no-input fields in three prefectures (Fukui, Shiga, and Kyoto) located in the warmer part of Japan. The yield of rough rice was positively correlated with soil total N content ($r = 0.76$, $p < 0.01$) but was not significantly correlated with the content of mineralizable N in soil. These results imply that, in the case of northern Japan, the yield by no-input farming was limited by the low temperature during the vegetative stage. Thus, the rate of N mineralization from soil rather than soil total N was regarded as an important factor limiting the tiller (panicle) number and yield. In the case of southern Japan, on the other hand, soil total N content was regarded as the most important factor limiting rice yield. These interpretations are based on the assumption that rice growth was not limited by the vigorous growth of weeds in no-input fields, because much more labor is required to remove weeds mechanically without herbicides, and effective weeding is the biggest concern for rice farmers adopting organic farming or no-input farming [12].

Compared to rice yield, much less attention has been paid to soil fertility under no-input farming. When the results in the above two reports [3, 12] were combined, total N content in the surface soil showed a large variation among the fields (**Figure** 5), and the coefficient of variation became 47%. The content was similar to or higher than the national average (2.39 g kg^{-1}) in several fields with a no-input history for more than 5 years. The highest content was recorded in a field with no fertilizer input for 21 years. The soil in this field was classified as one of the Andosols, whereas the soil in all the other fields was classified as non-Andosols according to the digital soil map of Japan [13].

For other soil properties, Kuwamura [8] evaluated the characteristics of soil chemical properties under no-input farming by using a space-for-time substitution approach. An extensive survey was conducted by analyzing 654 soil samples collected from no-input paddy fields throughout Japan from 1992 to 1996. The period of no-input farming ranged from 0 to 49 years. The results were compared with the contemporary national soil inventory data (third survey from 1989 to 1993 in [14]). The average depth of a plow layer in no-input fields (18.5 cm) was larger than that in the conventional paddy fields (14.6 cm). The average content of total N in the surface soil (2.6 g kg^{-1}) was slightly higher than the national average (2.42 g kg^{-1}). On the other hand, the average content of mineralizable N in the surface soil (118 mg kg^{-1}) was slightly lower than the national average (145 mg kg^{-1}). The average content of available P (Truog P) in the surface soil (126 mgP_2O_5 kg^{-1}) was much lower than the national average (298 mgP_2O_5 kg^{-1}). When the soil samples were

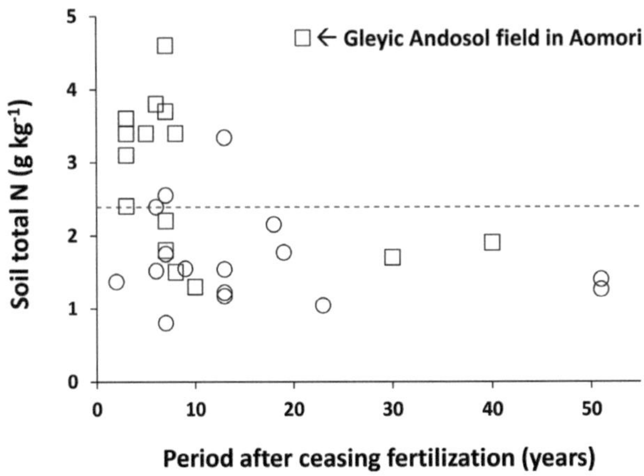

Figure 5.
Relationship between the period after ceasing fertilization to paddy fields and the content of total N in the surface soil (adapted from [3, 12]). The circular and rectangular plots are those reported by Tatara et al. [3] and Hosoya and Sugiyama [12], respectively. A dotted line in the figure indicates the average content in the surface paddy soil of Japan (2.39 g kg^{-1}) reported by MAFF [14].

limited to those classified as non-Andosols (n = 460), the concentration of available P was negatively correlated with the period of no-input farming (Spearman's $r = -0.42$, $p < 0.001$). This negative correlation was partly due to the presence of extremely high P soils in no-input fields with a short history, because the eight outliers with the available P content exceeding 600 mgP_2O_5 kg^{-1} were all sampled from the fields with a history of less than 20 years. Kuwamura [8] interpreted the results as follows: (1) soil available P was depleted by long-term no-input farming; (2) no-input fields with a short history had received more fertilizer-derived P before ceasing fertilization than those with a long history; and (3) no-input farmers with a long experience have managed their fields with lower return of plant residues such as rice straw. In contrast to available P, the concentration of mineralizable N in the non-Andosol samples was positively correlated with the period of no-input farming (Spearman's $r = 0.22$, $p < 0.001$), which implies that soil available N was not depleted by long-term no-input farming.

The above results were obtained from the one-time survey of no-input fields. Due to the lack of long-term monitoring data, it is difficult to make a simple conclusion. Nevertheless, it can be roughly concluded that rice yield and soil fertility in no-input paddy fields were influenced not only by the period of no-input farming but also by various environmental and management factors operating at regional and field scales.

3. Environmental factors affecting rice growth and soil fertility under no-input farming

In this section, I introduce our results obtained from a 5-year no-input trial [15]. To estimate the environmental factors that enable soil fertility to be maintained without fertilization, application of fertilizers to a paddy field at Kyoto University Farm in Takatsuki, Japan, was ceased in 2010. Both planted and unplanted plots were installed in the field (**Figure 6**). Then, changes in rice yield and soil fertility in the field were evaluated until 2015. Surface soil samples were collected from both planted and unplanted plots before transplanting and after harvesting of rice plants. At harvesting, rice straw was also removed from the field. The physico-chemical properties of the samples were monitored. Rice yield and the uptake of N

Figure 6.
Rice plants (cv. Hinohikari) at panicle initiation stage grown without fertilization in Kyoto University farm (right side). In the field with an area of 10 a, two unplanted plots were equipped next to the planted plots. The color of rice leaves in this field was yellower than the color in a fertilized field (left side), suggesting that N was the most limiting nutrient in the unfertilized field. The photograph was taken by the author on July 30, 2012, the third year after ceasing fertilization to this field.

and K by rice plants were also analyzed. The soil in this field was classified as non-Andosol (Gley lowland soil) according to Digital Soil Map of Japan [13]. The surface soil was relatively sandy (sand content higher than 60%) and had the following properties at the start of the experiment: pH(H_2O)—5.95; total C—20.2 g kg^{-1}; total N—1.99 g kg^{-1}; mineralizable N—156 mg kg^{-1}, available P (Bray no.2 P)—484 mgP kg^{-1}; and cation exchange capacity—10.4 $cmol_c$ kg^{-1}. As the soil was relatively rich in available P due to the long-term application of chemical fertilizers, we focused on the dynamics of N and K in this field.

During the experimental period, the yield of unhulled rice was relatively stable; 630, 621, 618, 551, and 639 g m^{-2} from 2010 to 2014, respectively. On the other hand, the levels of mineralizable N, total N, and nonexchangeable K (boiling 1 mol L^{-1} HNO_3-extractable K minus exchangeable K) in the surface soil of both planted and unplanted plots began to significantly decrease after three cropping seasons (**Figure 7**). The amount of total N and boiling HNO_3-extractable K (exchangeable K plus nonexchangeable K) decreased from the surface soil (0–10 cm) of the unplanted plot during the 5 years was estimated to be 55 and 7.2 g m^{-2}, respectively, assuming a bulk density of 1.0 g cm^{-3}. On the other hand, the amount of N and K taken up by a single cropping of rice plants in 2012 was 8.3 and 11.5 g m^{-2}, respectively. Accordingly, N was lost from the unplanted plots with the magnitude comparable to the removal of N by rice plants. The results in **Figure 7** also indicated that the continuous removal of N and K from soil caused the significant depletions of mineralizable N and nonexchangeable K but not of more readily extractable fractions (NH_4^+-N and exchangeable K). By more frequent soil sampling and analysis conducted in 2012, it was revealed that the concentration of exchangeable K in soil decreased from transplanting to the maximum tillering stage and then recovered to the initial level from the booting stage to winter [15]. The reason for the lack of depletion of exchangeable K after continuous removal of K is discussed in the last section.

In 2013, the fourth year after ceasing fertilization, fertilizer trials with N or K application were conducted under both field and pot conditions to identify which element limited rice growth (**Figure 8**). Distilled water was used for irrigation in the pot experiment, whereas river or underground water was used for irrigation in the field experiment; total N and K concentration was measured at each irrigation event. The fertilizer trials demonstrated that the element limiting rice growth was K or N under pot or field conditions, respectively (**Figure 8**). To confirm this result, another nutrient omission trial was conducted in 2016 by using the surface soil collected after six harvests of rice without fertilization. Among the nutrients

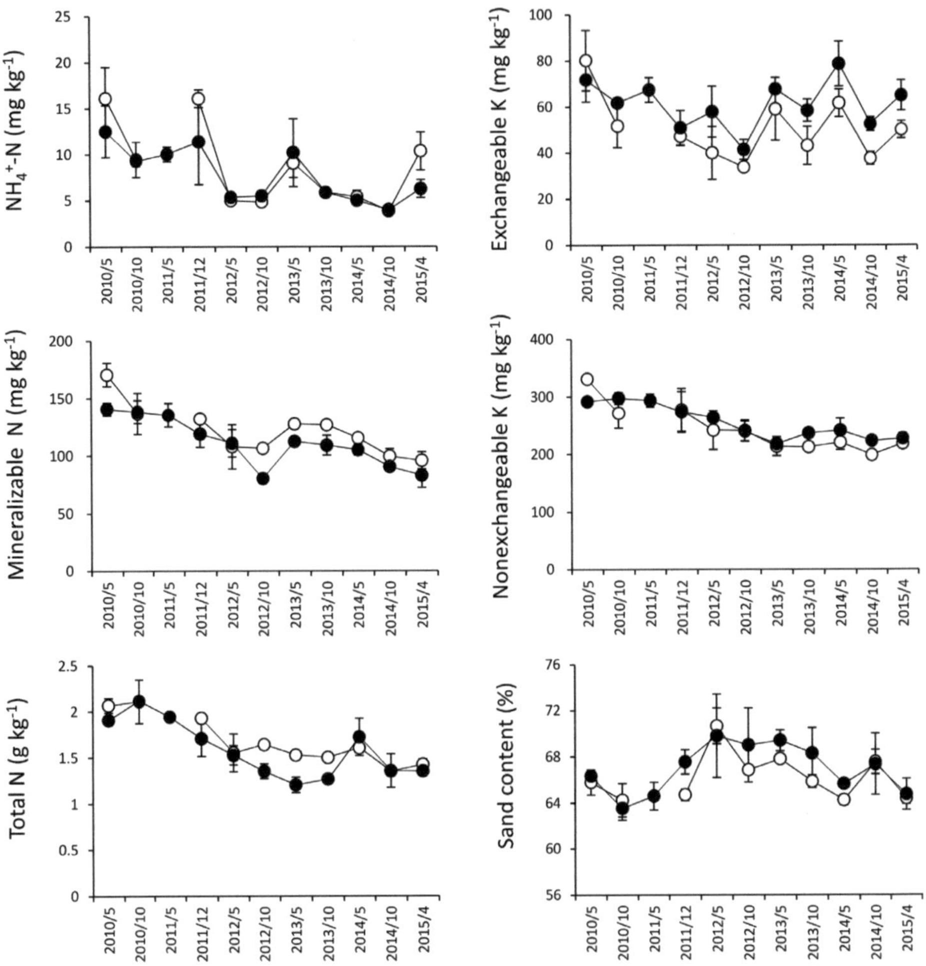

Figure 7.
Temporal changes of surface soil properties at planted (open circle) and unplanted (filled circle) plots in the unfertilized field (adapted from [15]). Error bars indicate the standard error of the mean (n = 2).

omitted (N, K, and Si), K was the most limiting nutrient when distilled water was used for irrigation (**Figure 9**). These results indicate that K, but not N, was the most limiting nutrient in the unfertilized soil and that the amount of K supplied by irrigation was sufficient to overcome the low K status of the unfertilized soil and meet plant demand. This should be the main reason why fertilizer responses were different between pot and field conditions. In other words, previous results on the nutrient omission trials (**Figures 1** and **2**) may have overestimated the ability of soil to supply nutrients to rice plants by allowing the external input of nutrients through irrigation.

In our field, the average concentration of K in irrigation water was 3.8 mg L^{-1} in 2013. If the amount of irrigation was assumed to be 1000 kg m^{-2}, the input of K to the field through irrigation becomes 3.8 g m^{-2}. This amount is slightly higher than the amount of K in rice panicles at maturity stage (3.0 g m^{-2} in our study). Thus, the input of K by irrigation may meet the plant's demand if rice straw is not removed from the fields and the irrigation water is rich in K (>2 mg L^{-1}).

Figure 10 shows the average K concentration in river water sampled from 225 rivers throughout Japan [16]. The sampling was carried out in 1940s and 1950s, when the eutrophication of river water was not a serious problem. The national average of the K concentration in river water was 1.20 mg L^{-1} with a large spatial variation

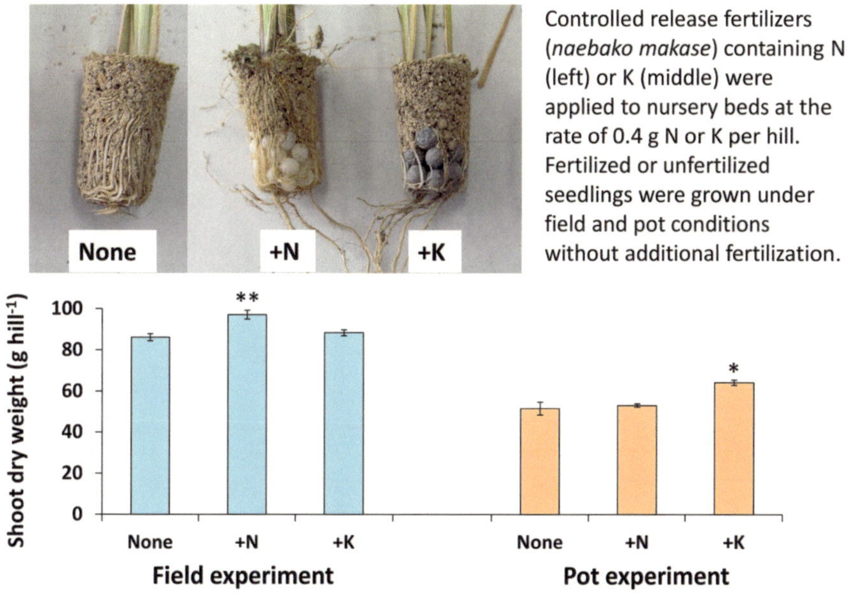

Figure 8.
*Dry matter weight of rice shoot (cv. Hinohikari) at maturity stage as influenced by the fertilizer application and growth conditions (adapted from [15]). Air-dry weight for field experiment and oven-dry weight for pot experiment. Error bars indicate the standard error of the mean (n = 60 for field experiment, n = 3 for pot experiment). In the pot experiment, distilled water was used for irrigation. **and * indicate significant difference from the unfertilized treatment at 1 and 5% (t-test), respectively.*

Figure 9.
Rice plants (cv. Hinohikari) at milk ripe stage grown in Takatsuki soil collected from the field without fertilizer application for 6 years (Moritsuka, unpublished). Distilled water was used for irrigation. The values indicate the average ± standard deviation of the shoot dry matter weight (g pot^{-1}, n = 3) for each fertilizer treatment. The photo was taken by the author on September 11, 2016.

(coefficient of variation = 57%). The concentration of K was higher than 2 mg L^{-1} in 24 rivers (10.7%), and 13 out of 24 rivers were located in Kyushu district in southern Japan. Several rivers originating from the Aso and Kirishima volcanic areas

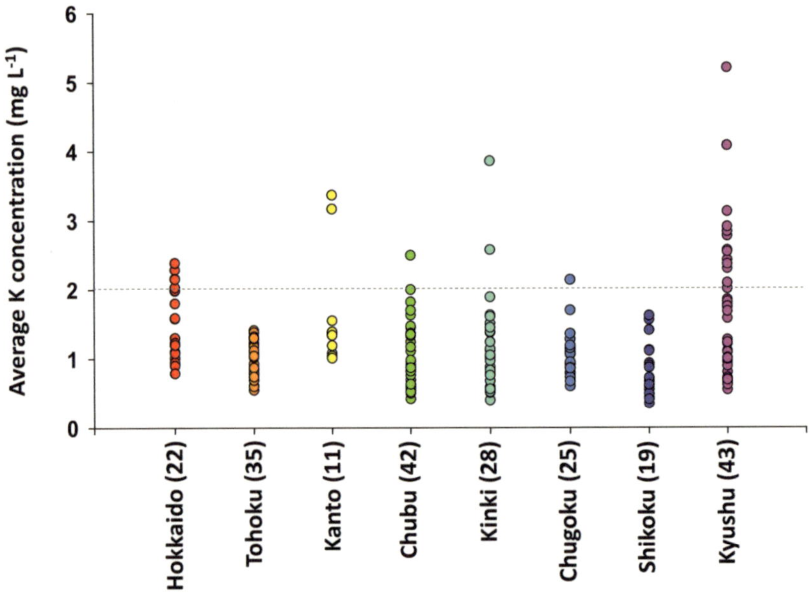

Figure 10.
Average K concentration in river water sampled from 225 rivers in Japan (adapted from [16]). The numbers in the parentheses below the figure indicate the number of rivers surveyed in each region of the country.

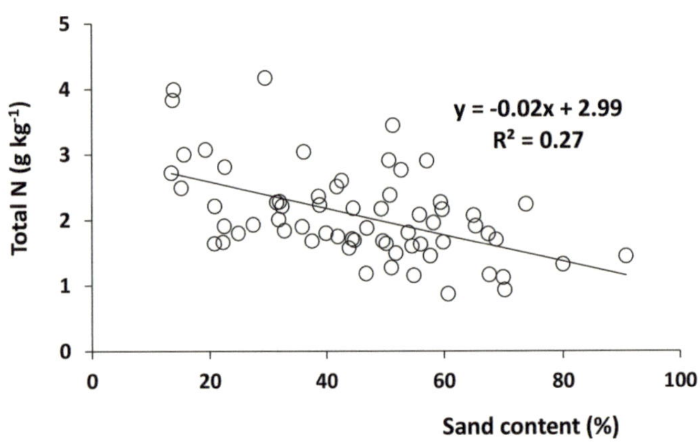

Figure 11.
Relationship between the sand content and the total N content in soils frequently used for paddy fields (adapted from [15]). Among the data of agricultural surface soils in Japan reported by Sano et al. [17], those from soils classified as lowland paddy soils, Gley lowland soils, and Gray lowland soils (n = 65) were displayed.

in Kyushu district showed very high concentrations of both K and Si. These results suggest that, in some of the watersheds in Japan, the input of K by irrigation of river water can meet the plant's demand even without the application of K fertilizer.

In contrast to K, the national average of inorganic N (NH_4^+ and NO_3^-) concentration in river water was 0.28 mgN L^{-1} (coefficient of variation = 99%) [16]. The input of N by irrigation cannot meet the plant's demand unless river water is polluted by eutrophication. As shown in **Figure 5**, there was a large variation in total N content in no-input paddy soils even when an outlier classified as an Andosol was removed. Such a large variation may have originated from the capacity of soil clay particles to accumulate organic matter containing N. This is because a significant negative correlation was observed between the sand content and the total N content in agricultural surface soils frequently used for paddy fields (**Figure 11**) [17].

Summarizing the results of this section, the input of K through irrigation and high-clay content in soil were considered the key environmental factors that enable to continue no-input farming. These factors are indebted to geographical conditions. Furthermore, the accumulation of fertilizer-derived P in the surface soil before ceasing fertilization may be another important factor. As a result of the alleviation of K and P deficiencies in the field, N became the most limiting nutrient in our experimental field. Enhancing the biological N fixation by growing leguminous plants after rice harvest and returning the plant residue to soil before transplanting may help to alleviate the N limitation to rice growth.

4. Dynamics of soil K under no-input and high-input rice farming systems

In the last section, I focus on the dynamics of K in paddy soil. In our experimental field, the concentration of nonexchangeable K in soil decreased significantly by the no-input farming, whereas the concentration of exchangeable K in soil was relatively constant and tended to increase from harvest to next transplanting (**Figure** 7). In this section, these observations are compared with previous results.

Srinivasa Rao et al. [18] evaluated the long-term changes in soil K forms under rice-rice cropping system with different fertilizer management. The experiment was carried out at Hyderabad in India. Surface soils were collected four times over 20 years, and the samples were analyzed for the different forms of K, including nonexchangeable K (boiling 1 mol L^{-1} HNO_3-extractable K minus exchangeable K). **Figure 12** shows the temporal changes in the concentrations of soluble, exchangeable, and nonexchangeable K in surface paddy soil as influenced by different fertilizer treatments. Among the three K forms, nonexchangeable K showed the largest depletion over 20 years (**Figure 12**). The amount of HNO_3-extractable K decreased over 20 years was quantitatively comparable to the net output of K estimated from the total amount of crop removal and fertilization. Thus, continuous cropping of rice caused a significant depletion of nonexchangeable K, while the concentrations

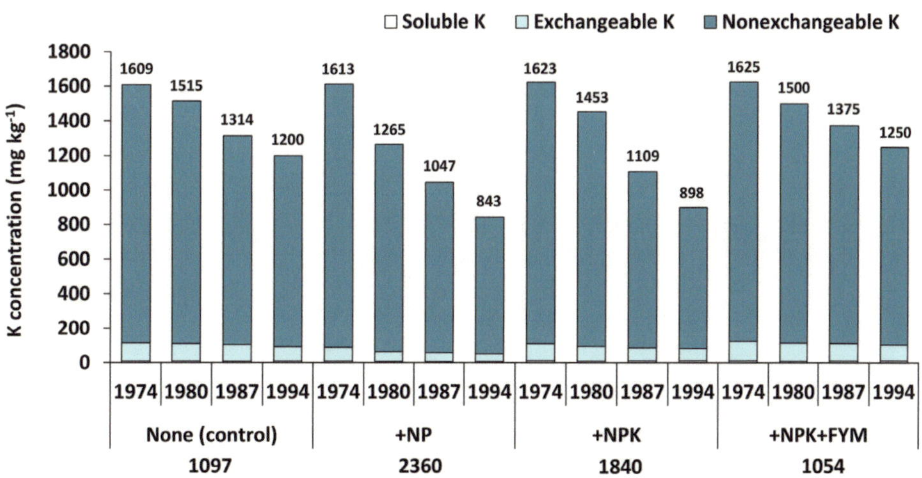

Figure 12.
Effect of 20 years of rice-rice cropping and fertilizer application on the concentrations of soluble, exchangeable, and nonexchangeable K in surface paddy soil (adapted from [18]). The values above each bar indicate the sum of all forms of K. The values in the bottom indicate the net output of K (crop removal minus fertilizer input) over 20 years (kg ha^{-1}). Inorganic fertilizers containing N, P, and K were applied at 115, 9, and 25 kg ha^{-1}, respectively, per cropping, and farmyard manure (FYM) was applied at 15 Mg ha^{-1} per year.

of more readily extractable K fractions were relatively constant. In a soil test, we usually measure the sum of soluble and exchangeable K by extracting soil with neutral 1 mol L^{-1} ammonium acetate. However, compared with nonexchangeable K, these K forms were much less sensitive to the long-term removal of K by rice plants. Based on these results, the authors concluded that the analysis of nonexchangeable K in soil should be added to a conventional soil test for better predicting K fertilizer requirements for long-term operation.

The results of Srinivasa Rao et al. [18] agree well with our results (**Figure 7**) and also with the results from an extensive survey by Khan et al. [19]. By reviewing previous results from the long-term field experiments in the world and comparing the net changes of exchangeable K in the surface soil at the beginning and end of the study period with the net inputs of K due to long-term fertilization and crop removal, Khan et al. [19] revealed that the changes in the soil exchangeable K pool during the study period were much smaller than the net input of K to the field estimated from the total amount of K added and removed (**Figure 13**). In the case of our field, the net decrease of soil exchangeable K during the 5-year experiment amounted to only 5.2% of the cumulative K removed by rice plants (**Figure 13**). From these results, the authors concluded that a one-time measurement of soil exchangeable K cannot account for the highly dynamic interchange of K between exchangeable and nonexchangeable pools. Khan et al. [19]) also reported that the concentration of exchangeable K in soil was increased significantly by air-drying soil samples to the soil moisture content below 50 g kg^{-1}, which is a conventional soil pretreatment required for sample homogenization.

In the case of Srinivasa Rao et al. [18], the application of farmyard manure in combination with chemical fertilizer contributed to recover the concentration of

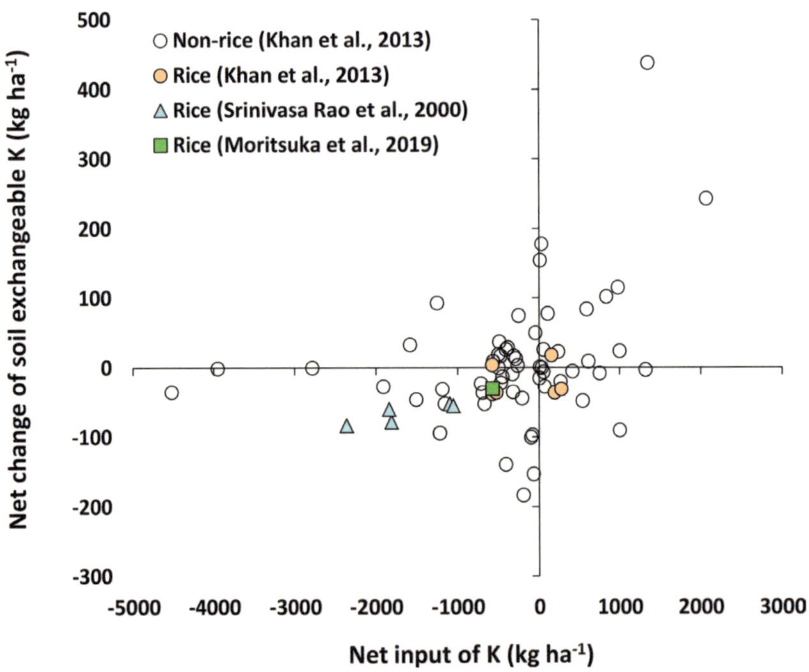

Figure 13.
Relationship between the net input of K due to long-term fertilization and crop removal and the net change of soil exchangeable K (adapted from [19]). Refer to the original papers [18, 19] for calculation methods. For our data, the net input of K was estimated from the extrapolation of crop K removal in 2012, and the net change of exchangeable K was calculated by assuming that soil depth and bulk density were 10 cm and 1.0 g cm^{-3}, respectively.

HNO_3-extractable K to some extent (**Figure 12**). However, a few researchers have reported contrasting results; the co-application of manure over long periods did not necessarily increase the concentration of nonexchangeable K in soil [20, 21]. For example, Kitajima et al. [20] evaluated the effect of long-term co-application of farmyard manure on the K forms in the surface soil at three locations in Japan. As shown in **Figure 14**, the co-application of farmyard manure increased the concentration of exchangeable K (including soluble K) but decreased the concentration of nonexchangeable K (boiling 1 mol L^{-1} HNO_3-extractable K minus exchangeable K) at all the locations. The authors suggested that farmyard manure accelerated the dissolution of K-bearing minerals in soil, by which nonexchangeable K was irreversibly transformed to exchangeable K. Regardless of the processes involved, the results in **Figure 14** cannot be explained by the dynamic equilibrium between the exchangeable K and the nonexchangeable K which operates to minimize the concentration changes of both K forms.

Yamashita et al. [22] recently reported that microbial biomass K in the surface paddy soil is detectable by the conventional fumigation-extraction approach and that the concentration of microbial biomass K was increased by the continuous application of compost to paddy fields. Combining these results with those by Khan et al. [19], it is plausible to consider that exchangeable K pool evaluated by using air-dried soil samples inevitably includes microbial biomass K. The increase of exchangeable K by the long-term application of farmyard manure (**Figure 14**) may be due to the contamination of microbial biomass K in the exchangeable K fraction. From these interpretations, the dynamics of soil K forms in the soil-plant-microbe systems are depicted in **Figure 15** by referring to the concept proposed by Asakawa and Yamashita [23]. The soundness and practical usefulness of this concept need to be evaluated in future experiments.

In summary, our results on soil K dynamics agreed well with previous results from long-term field experiments. Accordingly, it can be generally concluded that soil K depletion caused by long-term exhaustive cropping should be evaluated by monitoring the decrease of soil nonexchangeable K rather than that of exchangeable K. Furthermore, I hypothesized that the dynamics of soil microbial biomass K may cause the fluctuations of soil exchangeable K measured after air-drying pretreatment.

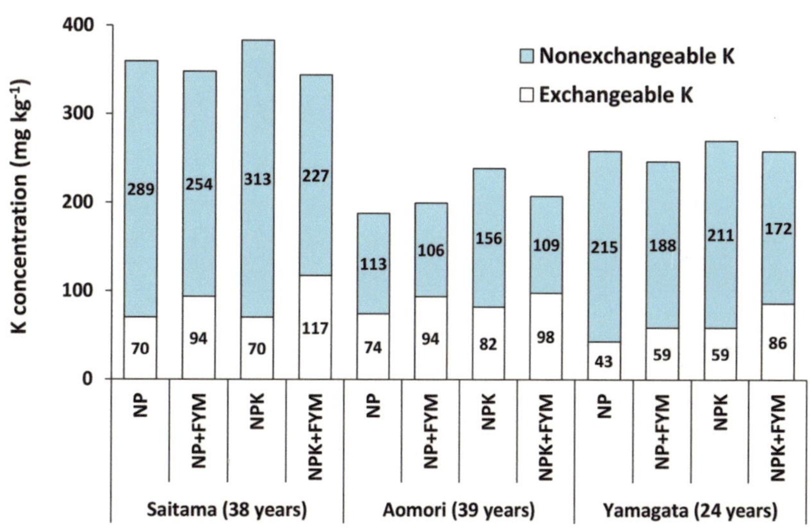

Figure 14.
Effect of long-term co-application of farmyard manure (FYM) on the concentrations of exchangeable and nonexchangeable K in surface paddy soil (adapted from [20]). Animal dung manure had been applied at 11.3 Mg ha^{-1} at all sites. The experiments at Saitama (Konosu) and Aomori sites began in around 1930, and are the earliest fertilizer experiments in Japan. Soil sampling was carried out in December, 1968.

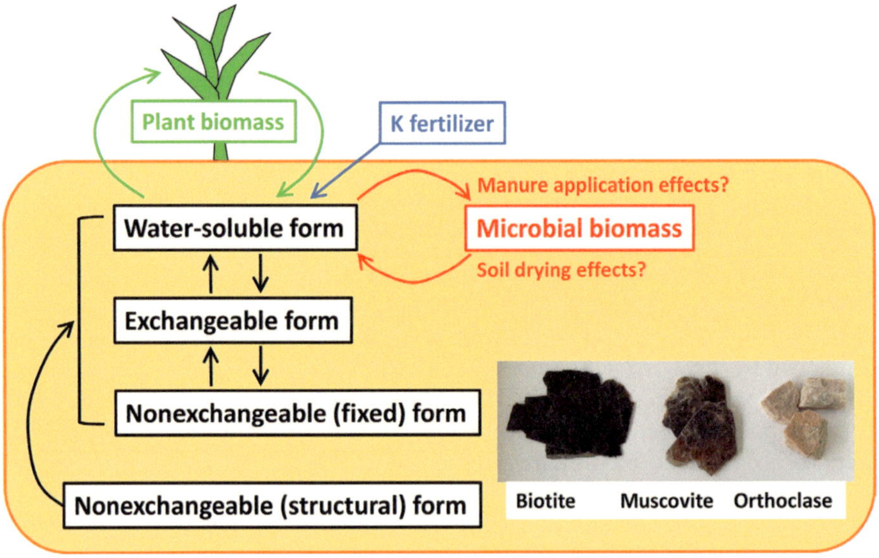

Figure 15.
Dynamics of K in the surface agricultural soil driven by crop plants, soil microbes, and fertilization (adapted from [23]). Both biotite and muscovite contain K in fixed form, and orthoclase contains K in structural form.

5. Conclusions

In this review, no-input rice farming in Japan was evaluated from the viewpoint of soil sustainability. It can be concluded that soil fertility under this farming system has been supported by various environmental and management factors, especially the input of K through irrigation, high-clay content in soil, and accumulation of fertilizer P applied previously to the soil. In the case of our no-input trial for 5 years, a significant depletion of mineralizable N and nonexchangeable K was observed after three cropping seasons, and rice growth was limited by soil K supply when the input of K by irrigation was restricted. These results highlight the importance of monitoring the dynamics of multiple soil nutrients for several years.

Author details

Naoki Moritsuka
Graduate School of Agriculture, Kyoto University, Kyoto, Japan

*Address all correspondence to: morituka@kais.kyoto-u.ac.jp

References

[1] Neera P, Katano M, Hasegawa T. Comparison of rice yield after various years of cultivation by natural farming. Plant Production Science. 1999;**2**:58-64. DOI: 10.1626/pps.2.58

[2] Hosoya K, Sugiyama S. Factors contributing to yield variation among unfertilized paddy fields in northern Japan. Japanese Journal of Crop Science. 2016:266-273(in Japanese with English summary). DOI: 10.1626/jcs.85.266

[3] Tatara S, Homma K, Kuwata M, Kobayashi M, Shiraiwa T. Dynamics of Soil Nutrients and Uptake of Nutrients by Rice Plants in Long-Term no-Input Paddy Fields: Survey Results from 2014 to 2016 (in Japanese). 2017. Available from: http://muhiken.or.jp/wp/entrust/2016/

[4] Kyuma K. Paddy Soil Science. Kyoto: Kyoto University Press; 2004. 280 p

[5] Kawasaki I. Natural Supply of Three Major Elements from Arable Soils in Japan. Tokyo: Nippon Nogyo Kenkyusho; 1953. 19 p. (in Japanese)

[6] FAO. Crop Production Levels and Fertilizer Use. FAO Fertilizer and Plant Nutrition Bulletin No. 2. Rome: FAO; 1981. 69 p

[7] Ohtsubo K, Nakamura S. Evaluation of palatability of cooked rice. In: Li JQ, editor. Advances in International Rice Research. London: IntechOpen; 2017. pp. 91-110. DOI: 10.5772/66398

[8] Kuwamura T. Soil characteristics of nature-farming fields. Research Reports from Institute for Agro-Microbiology. 2005;**9**:7-25. (in Japanese with English summary)

[9] Neera P, Katano M, Hasegawa T. Rice culture under nature farming in Japan. Proceedings of Faculty of Agriculture, Kyushu Tokai University. 1992;**11**:67-74

[10] Sugiyama S. Scientific basis for obtaining rice yield at 480 g m^{-2} without applying fertilizers: Soil conditions suitable for causing biological N fixation. Gendai Nogyo. 2016;**95**(12):134-139. (in Japanese)

[11] Katano M. Rice Production by Natural Farming. Rice Morphology and Farmers' Practices. Rural Culture Association Japan: Tokyo; 1990. 246 p

[12] Hosoya K, Sugiyama S. Weed communities and their negative impact on rice yield in no-input paddy fields in the northern part of Japan. Biological Agriculture and Horticulture. 2017;**33**:215-224. DOI: 10.1080/01448765.2017.1299641

[13] NARO. Digital Soil Map of Japan (in Japanese). 2019. Available from: https://soil-inventory.dc.affrc.go.jp/

[14] MAFF. Summary on the National Soil Survey Programs in Japanese Agricultural Land from 1979 to 2003. Agricultural Production Bureau, Ministry of Agriculture, Forestry and Fisheries; 2008. 485 p (in Japanese)

[15] Moritsuka N, Izawa G, Matsuoka K, Katsura K. Annual changes in soil fertility after ceasing fertilization in an unfertilized paddy field and factors limiting rice growth in the field. Japanese Journal of Soil Science and Plant Nutrition. 2019;**90**:257-267 (in Japanese with English summary). DOI: 10.20710/dojo.90.4_257

[16] Kobayashi J. A chemical study on the average quality and characteristics of river waters of Japan. Nougaku Kenkyu (Research Institute for Bioresources, Okayama University). 1961;**48**:63-106 (in Japanese)

[17] Sano S, Yanai J, Kosaki T. Evaluation of soil nitrogen status in japanese agricultural lands with reference to

land use and soil types. Soil Science and Plant Nutrition. 2004;**50**:501-510. DOI: 10.1080/00380768.2004.10408506

[18] Srinivasa Rao C, Subba Rao A, Swarup A, Bansal SK, Rajagopal V. Monitoring the changes in soil potassium by extraction procedures and electroultrafiltration (EUF) in a Tropaquept under twenty years of rice-rice cropping. Nutrient Cycling in Agroecosystems. 2000;**56**:277-282. DOI: 10.1023/A:1009839704839

[19] Khan SA, Mulvaney RL, Ellsworth TR. The potassium paradox: Implications for soil fertility, crop production and human health. Renewable Agriculture and Food Systems. 2013;**29**:3-27. DOI: 10.1017/S1742170513000318

[20] Kitajima S, Kaneko J, Shiroshita T. The variations of cation status in paddy soils and cation absorption by rice plants with the application of potassic fertilizers. J. Journal of the Central Agricultural Experiment Station. 1975;**22**:105-177 (in Japanese with English summary)

[21] Sharma S, Chander G, Verma TS, Verma S. Soil potassium fractions in rice-wheat cropping system after twelve years of lantana residue incorporation in a northwest Himalayan acid Alfisol. Journal of Plant Nutrition. 2013;**36**:1809-1820. DOI: 10.1080/01904167.2013.815202

[22] Yamashita K, Honjo H, Nishida M, Kimura M, Asakawa S. Estimation of microbial biomass potassium in paddy field soil. Soil Science and Plant Nutrition. 2014;**60**:512-519. DOI: 10.1080/00380768.2014.919237

[23] Asakawa S, Yamashita K. Soil microbial biomass as a source of potassium to plants: Soil microbes bear not only nitrogen and phosphorus but also potassium in their bodies. Kagaku to Seibutsu. 2017;**55**:444-445 (in Japanese). DOI: 10.1271/kagakutoseibutsu.55.444

Chapter 3

Crops Diversification and the Role of Orphan Legumes to Improve the Sub-Saharan Africa Farming Systems

Patricia Vidigal, Maria Manuel Romeiras and Filipa Monteiro

Abstract

Agriculture is the main economic revenue in sub-Saharan African countries, playing a key role on smallholder livelihoods as household incomes and as food. Food insecurity is known to increase with the inevitable climate changes, which already affect the major farming systems, sub-Saharan Africa (SSA) being particularly susceptible, mostly due to the high dependence of rainfall for crop cycles. As such, to promote food security in a long run, new farming systems have to become more sustainable and productive at the same time. In this chapter, a global overview of major farming systems in sub-Saharan Africa is provided, and current and future production scenarios are discussed. Moreover, some of the major pillars under the sustainable land use intensification are highlighted, and the potential of the undervalued African legumes toward a sustainable crop production is debated. Finally, an outline of key opportunities to diversify cropping systems is explored along with the benefits associated to integration of local and "orphan legumes" that are considered. It is argued that the use of these "orphan legumes" and the implementation of appropriated management approaches will promote a sustainable production of more food from the same land area, relying on mutually beneficial ecological relationships and reducing environmental impacts.

Keywords: orphan legumes, sustainable production, farming systems, sub-Saharan Africa

1. Introduction

In the past 50 years, global crop production has expanded, driven largely by higher yields per unit of land and crop intensification, resulting from multiple cropping and/or shortening of fallow periods [1]. The expansion of arable land area allocated to crops has played a less important part in production increases. However, these trends are not uniform across regions. For instance, most of the growth in wheat and rice production in developing countries in the land-scarce regions of Asia and Northern Africa has resulted in yield gains, while expansion of harvested land is a result of rapid production growth of maize in Latin America and the Caribbean and in sub-Saharan Africa (SSA) [2]. Yield growth contributed only one-third of the increase in crop production in the latter region. The arable land area in developed countries peaked in the mid-1980s and has fallen at an accelerating rate ever since. SSA is scientifically known as a rich niche of plant diversity which, in conjunction with local and traditional knowledge, makes the perfect combination to promote a sustainable solution for professional and smallholder farmers while respecting their livelihood needs, traditions, and market demand.

Economic foundations of most SSA are dominated by agriculture, which is recognized to contribute between 15 and 40% of the gross domestic product (GDP). Besides, agriculture sector provides livelihoods for over 70% of SSA's population through family farming [3, 4]. The economically active population in agriculture doubled from 100 million people in 1980 to 212 million in 2013. Considering that 75% of the SSA population is involved directly or indirectly in farming and related employment, the strategic role of family farms, mainly by women, has been recognized by key actors [5]. Over the last 40 years, the SSA population has been increasing from 279 to 826 million people, both in rural and urban populations. It is expected that due to the climate changes, there will be an increase in rural-urban migration as a consequence of agriculture abandonment and toward the search for better opportunities for both livelihoods and work, which will also cause an expansion and reclassification of urban boundaries [6]. As a result, by 2050 about 50% of SSA's population will be living in towns and cities [7]. In fact, a migration from rural-to-urban areas has been increasing at a fast pace (**Figure 1**).

To answer the increasing growth in consumers, production growths have stemmed mostly from area expansion at the expense of biodiversity, cultural value, and the rise in greenhouse gas emission (GGE). To respond to both market needs and the feeding of continuously growing population, crop production has been marked by extensive growth of staple crops, namely, in SSA. Over the last 20 years, crop staple production has risen at the cost of more land for agriculture. By 2014, most of African arable land was occupied by staple crops with more than 80 million hectares (ha), and the major contributors are maize, sorghum, and millet,

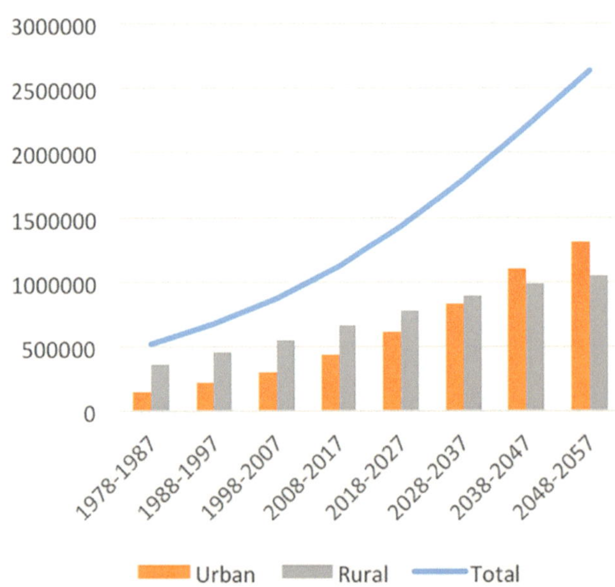

Figure 1.
Annual growth of population in rural and urban regions in sub-Saharan Africa within the period of 1950s and to future projections until the 2050s [7].

accounting for 80% of total food production. From the 1960s to the 1970s, there was an increase of 1 million ha dedicated to maize that increases by the 1980s with more 3 million ha, and from them on, there has been an increase of 4 million ha for every decade [7]. When restricting to the top six agriculture commodities in SSA region on FAOSTAT data [7] (**Figure 2**), major crops were analyzed in terms of production (A), area harvested (B), and yield (C), and key staple crops were highlighted, namely, rice, cassava, sorghum, and grain legumes/pulses, along with maize.

Maize is the crop that occupies the largest portion of agriculture land use, with an increasing area harvested devoted to its production that does not translate to an increment on crop production and thus yield. However, its production has been in an increasingly trend due to maize being Africa's most important food crop, and it is held up as a model food crop to meet Africa's growing urban demand for convenient food products [8–10]. Maize production, however, is risky because of unpredictable rainfall. On the other hand, cassava is known as Africa's second most important food staple in terms of per capita calories consumed, as a major source of calories. Accordingly, cassava production is among the higher number in SSA, occupying less agriculture land but with increasing steady production, translated in high yields. For instance, investment from the Bill & Melinda Gates Foundation in projects such as accelerated varietal improvement and seed delivery of legumes and cereals in Africa (AVISA) has contributed to more efficient cassava varieties. Yet, cassava has several other advantages over rice, maize, and other grains as a food staple in areas where there is a degraded resource base, uncertain rainfall, and weak market infrastructure. It is drought tolerant; this attribute makes it the most suitable food crop during periods of drought and famine. Cassava has historically played an important famine prevention role in Eastern and Southern Africa where maize is the preferred food staple and drought is a recurrent problem. While rice is produced in vast areas of the world, the physical requirements for growing it are limited to certain zones. Economically viable cultivation typically requires high average temperatures during the growing season, abundant supplies of water applied in a timely manner, smooth land surfaces to facilitate uniform flooding and drainage, and a subsoil stratum that inhibits the percolation of water. The bulk of world rice production is destined for food use and is the primary staple for more than half of the world's population. In recent years, rice has also become an important staple throughout Africa as part of the changing dietary habits. However, rice production requires high workforce and has limitations due to low mechanization of major SSA countries, which makes rice a crop usually bought at higher prices, without increasing its production.

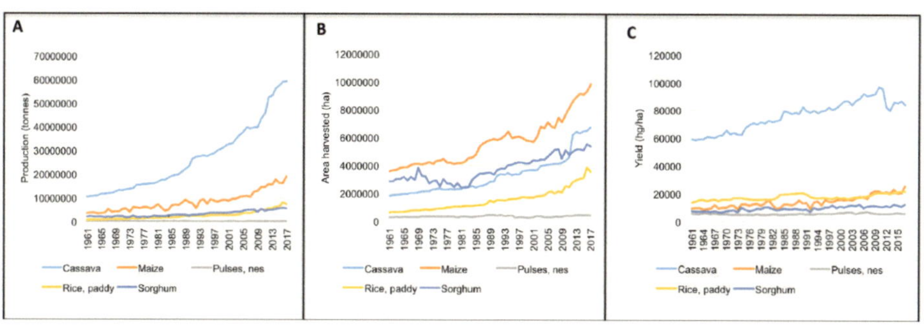

Figure 2.
Top six agriculture commodities in the sub-Saharan Africa region, in terms of production (A), area harvested (B), and yield (C), from the period 1961 to 2011 [7].

Considering this overall trend of major staple crop production in SSA region, it is undeniable that agricultural growth will contribute to poverty reduction, within a sustainable crop production scenario. A great diversity of farming systems across SSA shapes the current agriculture production in the region. Thus, in this chapter, we first provide an overview analysis of the major farming systems in SSA along with agroecological zoning, which delivers clear evidences on the sustainability of current agriculture production. After, we pinpoint how to ally sustainable intensification to integrated land use in SSA farming systems, by recurring to intercropping systems focusing on pulse crops (grain legumes, which are grown primarily for their edible seeds) and more particularly on legumes that have been named orphan legumes. Orphan, or underutilized, legumes are domesticated legumes with useful properties but with less importance than major world crops due to use and supply constraints. However, they play a significant role in many developing countries, providing food security and nutrition to consumers, as well as income to resource-poor farmers. Being legumes, these plants have the advantage of fixing atmospheric nitrogen for their own needs and for soil enrichment, thereby reducing the cost of fertilizer inputs in crop farming [11].

2. Farming systems in sub-Saharan Africa: an overview toward sustainable intensification

The diversity of agroecological zones (AEZs) across SSA (**Figure 3A**) results in the wide range of farming systems. According to the availability of natural resources (land, water, grazing areas, and forest) and climate, especially length of growing period and altitude, as well as the pattern of farm activities and household livelihood, African farming systems can be classified in 15 farming classes (**Figure 3B**). AEZs are climate-based and are a useful basis for determining the general suitability and production potential of crops and livestock in any given area. Thus, by matching AEZs with SSA farming systems, one can disclose potential or constraints toward SSA farming system (**Figure 3B**), by using a correlation analysis on agroecological zones and farming systems area based on HarvestChoice data (https://harvestchoice.org/).

From the 15 farming systems in SSA, there are 5 that occupy a higher percentage of the SSA region: (1) maize mixed, (2) arid pastoral oases, (3) pastoral, (4)

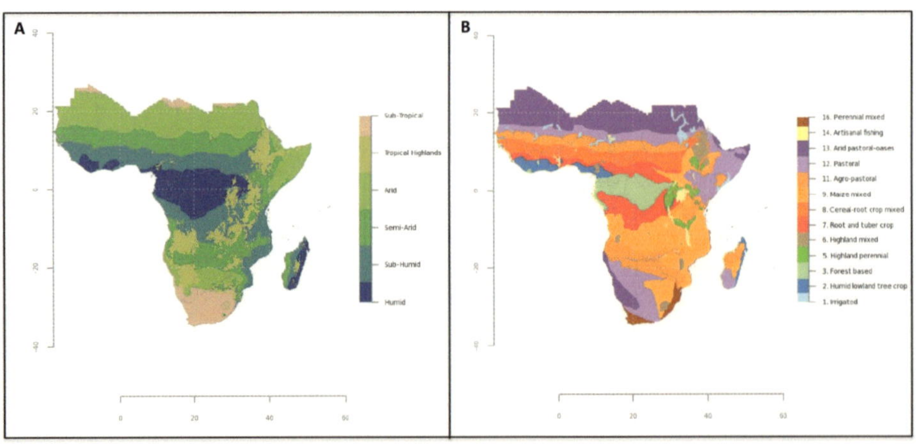

Figure 3.
Farming systems and agroecological zones in sub-Saharan Africa. (A) Agroecological zones 5-class [12]; (B) farming system classes [13].

Farming systems	SSA (%)	SSA regions (%)				Agroecological zones (%)					
		EA	MA	SA	WA	Arid	Humid	Semiarid	Subhumid	Subtropical	Tropical highlands
Irrigated	0.9	0.8	0.1	0.1	2.1	1.7	0.0	2.2	0.0	0.0	0.1
Agropastoral	**15.3**	20.2	10.7	16.5	14.4	6.0	1.7	47.9	3.8	0.3	18.7
Pastoral	**15.5**	21.0	3.3	40.4	10.9	27.5	1.9	19.4	1.7	36.1	12.8
Arid pastoral oases	**17.1**	3.1	10.8	13.7	40.1	62.5	0.0	0.0	0.0	32.8	2.4
Artisanal fishing	2.0	3.9	1.3	0.0	1.9	0.3	3.6	1.3	5.2	0.0	0.6
Perennial mixed	1.6	0.3	0.0	11.3	0.0	0.0	0.0	1.3	0.5	15.2	0.5
Humid lowland tree crop	2.9	1.3	1.9	0.0	6.8	0.0	14.2	0.0	4.1	0.0	0.0
Forest based	6.0	0.0	20.1	0.0	0.2	0.0	41.6	0.0	0.9	0.0	0.1
Highland perennial	1.9	5.6	0.8	0.0	0.0	0.0	1.1	0.1	1.2	0.0	11.5
Highland mixed	2.2	5.3	1.1	2.1	0.2	0.0	0.7	0.4	1.1	3.1	12.2
Root and tuber crop	9.8	0.8	27.0	0.0	5.4	0.0	30.0	0.0	25.0	0.0	2.4
Cereal root crop mixed	7.2	0.6	6.7	0.0	18.0	0.0	0.0	8.8	24.6	0.0	0.6
Maize mixed	**17.7**	36.9	16.4	15.8	0.0	2.0	5.3	18.6	31.9	12.5	38.0

Table 1.
Percentage of prevalence of each farming system in all SSA, in each SSA regions, and in different agroecological zones [12, 13].

agropastoral, and (5) root and tuber crop (**Table 1**). The most prominent farming system in SSA is maize mixed, occupying 18% of SSA, especially in East Africa (37%) with the prevalence of the AEZ subhumid, semiarid, and tropical highlands (**Table 1**). East Africa tropical highlands and subhumid highlands have a bimodal rainfall pattern offering farmers two cropping seasons, but in drier areas such as semiarid AEZ, farmers usually harvest only once a year. This farming systems is one of the most important food production system in East Africa, with only 6% of the irrigated area in SSA [14], thus depending mostly on rainfall (**Figure 4**).

Considering a projection of increased number of drying days over East Africa [17] and a 0.96% annually increasing temperature (**Figure 5A**), the sustainability of this farming system is of great concern, and there is an urgent need of capacity building in crop management technologies, such as nitrogen efficiency in rainfed systems. The main staple crop in the maize mixed farming system is maize, with the main income being migrant allowances, cattle, small ruminants, tobacco, coffee, and cotton, plus the sale of food crops such as maize and pulses [14]. In the past, most of the production has been boosted by a subsidized combination of high doses of inorganic fertilizers and hybrid maize varieties. Once subsidies were removed, the use of high-cost inputs on maize became unprofitable, and the majority of smallholders reverted to traditional varieties with low to no market value, resulting in low household income. Although maize is the main crop, the intercropping system exists with pulses, oil seeds, cotton, sorghum, and millet. Intercropping with pulses, such as common bean, cowpeas, and soybeans, is common where landholdings are small and there is less pressure on the land. Most of the area occupied today by the maize mixed system was heavily afforested as farmers have pushed arable land into the forests, decreasing biodiversity to increase area devoted to commercial species. Pressure on the land to respond market needs led to problems related to declining soil fertility in combination with long dry seasons resulting in lower crop yields, food insecurity, hunger, and poverty [18]. Nevertheless, maize mixed is one of the farming systems that has a good long-term agricultural growth prospects with high potential for poverty reduction [14], which is reflected in East African lowest annual percentage of prevalence of severe food insecurity (**Figure 5A**).

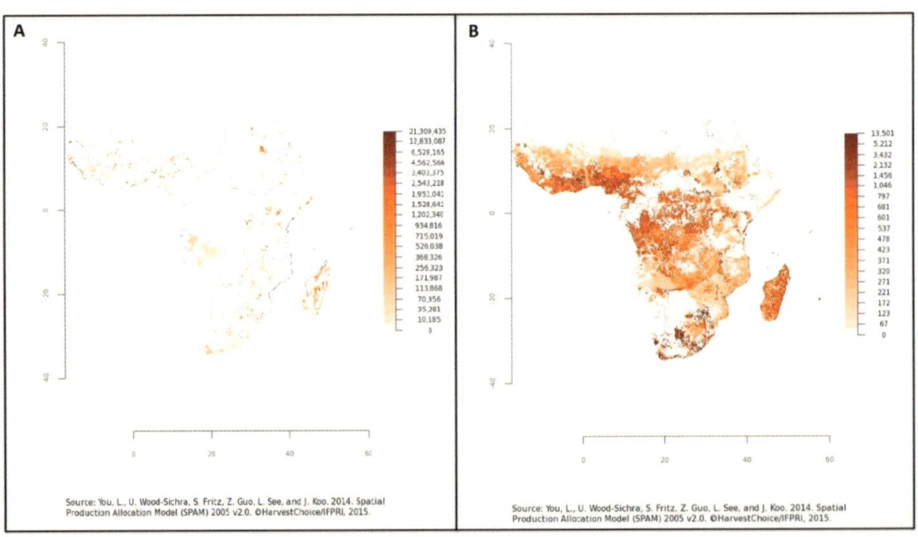

Figure 4.
Food crops irrigated (A) and rainfed (B) value production (Int$, 2005) [15, 16].

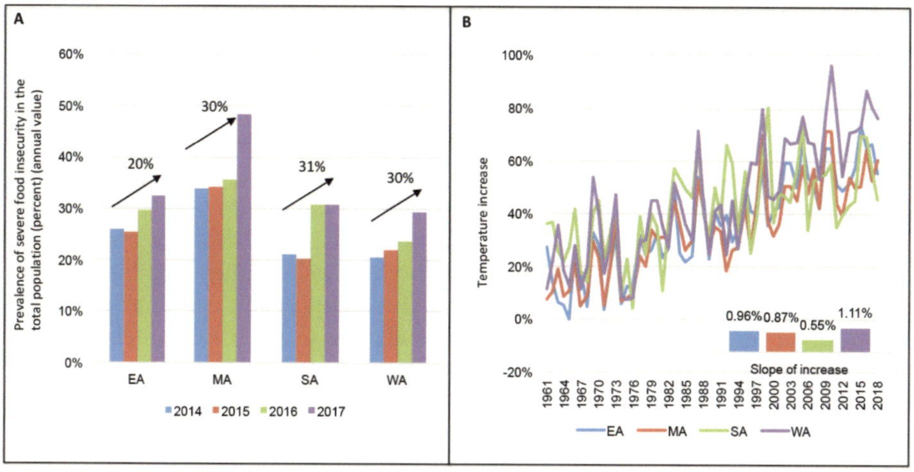

Figure 5.
Temperature increase in all regions of SSA and the annual percentage of prevalence of severe food insecurity in the total population of each SSA region. Abbreviations: East Africa (EA), Middle Africa (MA), Southern Africa (SA), West Africa (WA) [7].

The second most relevant farming system is the arid pastoral oasis farming system covering 62% of the arid AEZ and 40% of the West Africa region (**Table 1**). This farming system contains some oasis farming and a number of irrigation schemes, producing date palm (*Phoenix dactylifera* L.) and other palms, vegetables, and stable crops such as maize and rice [14].

This farming system is the most dependent on rainfall, and although West Africa has an annual temperature increase of 1.1% (**Figure 5B**), there is also a projection of 30–70% increased precipitation within semiarid and subhumid AEZs in West Africa [17] which account for 25 and 22% of the area, respectively, thus presenting a minor prevalence of severe food insecurity (**Figure 5A**). In the third place, there are two farming systems that have relevance in SSA, agropastoral and pastoral. Agropastoral farming system, generally in the semiarid and tropical highlands of East and South Africa (**Table 1**), is characterized by producing both crops and livestock. Approximately, 22 million ha are used for crops, mainly rainfed sorghum and pearl milted for family subsistence, whereas sesame and pulses are for household income. Livestock are also kept for subsistence (milk and milk products), offspring, transportation (camels, donkeys), land preparation (oxen, camels), income revenue, exchange, savings, bride wealth, and/or insurance against crop failure [14]. One of the major concerns and fragilities of this area is its vulnerability to drought, leading to crop failure and consequently to weaker animals due to a decrease in crop biomass production [19]. As animals are insurance to crop failure, severe drought leads to decapitalizations of herds and therefore lack of animals to exchange for grain. In addition, the search for more land, to mitigate the decrease in millet and sorghum yields for subsistence, along the investment in other crops used for trading (e.g., pulses) promotes a decline in soil fertility and weed infestation, mainly by *striga* in cereals. Prevalence of severe food insecurity has increased greatly in South Africa from 2016 to 2017 (**Figure 5A**), which is of great concern considering the alarming prediction of drought severity for South Africa [20].

The pastoral farming system, generally in the arid and subtropical AEZs, occupying 40.4% in South Africa region and 21% in East Africa, is dominated by livestock, where livelihoods depend mainly from cattle, camels, sheep, goats, some cereal crops, and off-farm work [21]. Being mostly present in arid regions and in

South Africa, the main source of vulnerability is the great climatic variability and consequently high incidence of drought, similarly to the agropastoral farming systems.

Overall, regardless of the farming system, the major concern of SSA food security is connected to drought, due to the dependency on the rainfall periods in most of the farming systems. As such, SSA farming system sustainability has been largely affected by climate changes, such as increases in temperature (**Figure 5A**) and the occurrence of 291 events of extreme drought [22], posing a clear threat to the maintenance of current and future crop production, affecting smallholder's livelihoods and food security in the long run. Increasing food production by expansion of agricultural land is fragile, as population grows, thus demanding more land through deforestation. FAO Special Programme for Food Security considers intensification of existing production patterns and diversification of production and processing, as the two main strategies to eradicate poverty and hunger. These two strategies meet the objective of sustainable intensification (SI) concepts, in combination with site-specific factors and agroecological conditions. SI is defined as the process of "producing more food from the same area of land while reducing the environmental impacts" [23]. Pretty et al. [24] stated that from 40 projects over 20 countries involving over 10 million farmers, SI increased farm productivity over twofold. Moreover, an adequate implementation of SI worldwide could respond to 2050 food demand while supporting land conservation from 1 to 0.2 billion ha and decreasing gasoline gallon equivalent (GGE) from 3 to 1 Gt per year [25]. Thus, it is imperative to emphasize and implement efficiently SI practices and agricultural technologies in SSA to ensure both food security and profitability. To sustainably increase yields of smallholders in their farming systems, it is essential to adopt an effective land management and implement strategies that aid farmers to face climate uncertainties.

2.1 Promoting sustainable crop production: the potential of multipurpose pulse crops

A sustainable crop production needs an efficient soil fertility management, in order to prospect future high yield production. Most African soils are poor compared to most other parts of the world, due to the lack of volcanic rejuvenation. This has caused African soils to undergo various cycles of weathering, erosion, and leaching, resulting in poor nutrient soils [26]. As the population continues to grow at a fasting rate leading to an increased demand for food, Africa's agricultural land is becoming increasingly degraded (**Figure 6A**), due to ill management practices and of external inputs. In East Africa the rate of depletion is so high that even drastic measures, such as doubling the application of fertilizer (**Figure 6B**) or manure or halving erosion losses, would not be enough to offset nutrient deficits. In African soil, there is higher depletion of nitrogen and potassium than phosphorus due to leaching and soil erosion. These soil problems are the result of continuous cropping of cereals without rotation with legumes, inappropriate soil conservation practices, and inadequate amounts of fertilizer use [28]. These problems are aggravated by short growing seasons together with limited water availability from rainfall resulting in restricted crop diversification contributing for additional pressure on the land.

Among all the plant nutrients essential for crop production, nitrogen is the key nutrient [29]. African farmers to fulfill this large nitrogen requirement for crop production in an increasing depleted soil are using 16 metric tons of nitrogen each year (**Figure 6B**) [7]. Pulse crops and soil microorganisms have potential to convert nitrogen into plant-usable forms, contributing significant amounts of nitrogen to

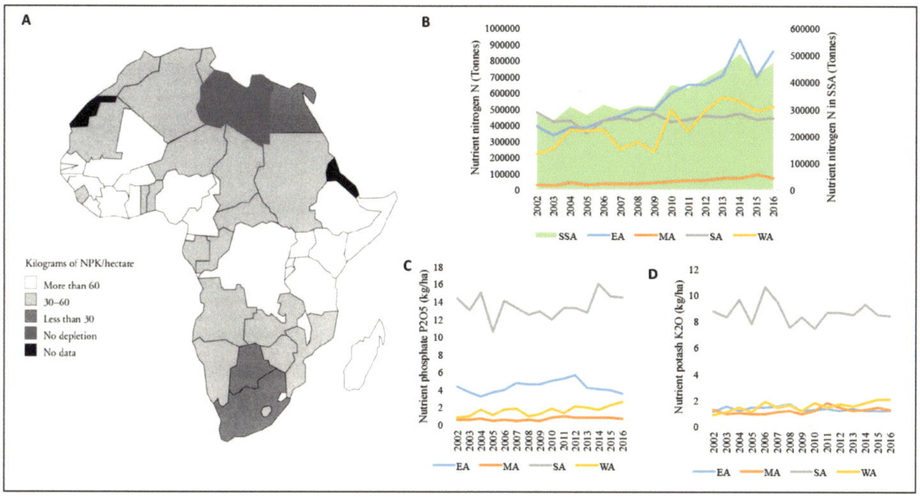

Figure 6.
(A) Average annual nutrient depletion (NPK) in Africa, 1993–1995 [27]; (B) average of total nitrogen (N) from all fertilizer products; (C) average of total phosphate (P_2O_5) from all fertilizer products; (D) average of total potassium (K_2O) from all fertilizer products [7].

satisfy crop needs. To respond population food needs, natural sources of nitrogen are not sufficient to achieve required yields; thus, there is a need to complement with chemical nitrogen but in an efficient, eco-friendly, and environmental management manner. Under current scenario, there is an urgent need for improving nitrogen use efficiency and balance use of natural resources which is essential for sustainable agricultural production [30]. Pulses, and especially multipurpose pulses, are part of Africa history due to their multiple benefits in agriculture and society. Multipurpose pulses serve and are needed for different functions and in general are best to respond to the diverse needs of farmers, including food, fuel, and fodder, and ecosystem services such as pollination and improving soil fertility and organic matter content. By increasing soil organic matter content, an improvement in soil structure is obtained, promoting an increase in water-holding capacity [31]. Moreover, pulses and legumes in general have the natural ability to biologically fix atmospheric nitrogen and to enhance the biological turnover of phosphorous [32]. However, over time, consumers' preferences have changed with traditional crops which have been replaced by staple crops (e.g., rice, cassava, and maize) and which have been subject to intensive research and political support worldwide. The quantity of arable land used for pulses is much less than the area cultivated with important cereals (**Figure 2B**), thus negatively affecting the nutrient balance in African soils [32]. Multipurpose pulse crops offer smallholder farmers a multifaceted way to improve food security, diet, and soil health as well as economic returns and income stability. SSA smallholder's farmers have been incentivized to produce common bean and cowpea (**Figure 7**), but with climate change and most of SSA agriculture being rainfall dependent, future is compromised. Although the production of cowpea (**Figure 7E**) and bean (**Figure 7F**) is far greater than other pulses (**Figure 7G**), the area distribution of pulses is more comprehensive within AEZ and farming systems, especially in the major SSA farming systems as maize mixed and agropastoral (**Figure 3B**). Thus, it is important to recover and enhance agriculture productivity of local crops, known as orphan legumes, known to local farmers and communities. Moreover, these orphan legumes are a likely source of important traits for introduction into major crops to aid in combating the stresses associated with global climate change.

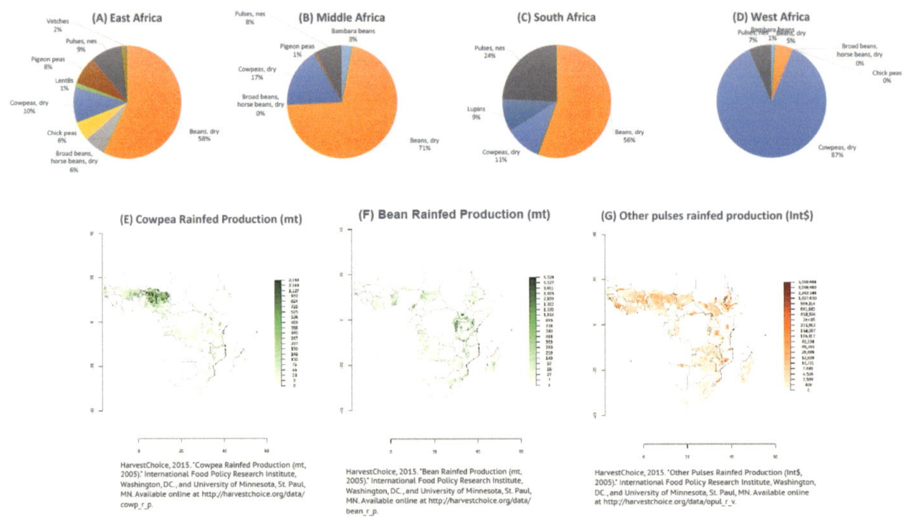

Figure 7.
Area harvested of pulses in East Africa (A), Middle Africa (B), South Africa (C), and West Africa (D). Data retrieved from FAOSTAT (accessed March 2019). Rainfed production of cowpea (E) and beans (F) in metric tons (mt) for SSA. Rainfed production of other pulses in international dollars (Int$) for SSA [33–35].

Thus, the potential use of multipurpose pulse crops as a sustainable strategy to overcome the present problems associated with the agricultural intensification is undeniable to cope soil depletion and decreasing crop yields annually, as well as inevitable environmental changes that will occur in the next 50 years. The promotion of neglected and underutilized species (NUS) African legumes adapted to rainfed and drought conditions will contribute not only to the diversity of cropping systems but also to decrease food insecurity. However, there is the need to address critical knowledge gaps that will allow the full use and advantages to introduce successfully pulse crops within agricultural and food systems. Part of this includes promoting pulse farming and implementing different farm management practices in order to contribute to the resilience of SSA farming systems. As the world celebrated the International Year of Pulses in 2016, there is a continuous need to establish the potential and invest in the innovation of undervalued role that pulses can play and that have to play in the post-2016 agenda.

3. Climate change and crop production in SSA: the key role of pulse crops

Changes in temperature and rainfall regime may have considerable impacts on agricultural productivity and on the ecosystem on which many people depend [36]. Rainfall amounts, distribution, and intensity are already producing floods, droughts, and changes in large-scale hydrological cycle [37, 38] which will affect the duration of crop growing seasons. Changes in temperatures affect plant growth and animal feed intake. Increases in maximum temperatures can lead to yield reductions and reproductive failure in maize, and animals reduce their feed intake. Maize being the most produced staple crops in SSA is particularly sensitive to temperatures above 30°C [17]. Also, wheat growing temperature is already above optimal, and it is expected to increase [39]. Increase in nighttime temperatures can also lead to decrease in rice yields, especially during the dry season. Another concerning factor is the increasing carbon dioxide concentration in the atmosphere that is beneficial for C3 plants such as wheat, but not for C4 plants such as maize and sorghum, and it may also decrease protein concentration in wheat grain, reducing nutrient availability for animals. Climate projections indicate losses of 27–32% for maize, millet, and groundnut [40] and 71% for beans [41] especially soybean that is the most common legumes produced.

Rainfed farming system in SSA produces 90% of staple food in SSA [42] and in the face of long periods of drought or dry spells in the growing season causes an unbalancing of the cycle of by-products in the mix farming system [43]. In order to find more suitable agriculture conditions, population migration takes place. This strategy, together with the increase in population, is leading to tropical rainforest destruction to conquer agriculture land, plus the general land degradation due to inappropriate land use, which in turn causes desertification, salinization, sodification, and soil and water erosion, increasing atmospheric carbon dioxide and creating a spiraling decline in the productivity of the land in terms of both food and other natural resources [14]. Projections are alarming; showing climate variability on agricultural production will have substantial effects in mixed smallholder systems, resulting in reduced food security that potentially increases the risk of hunger and undernutrition. However, it is the mixed system that presents the best capacity to tackle the inevitable change in climate. For that, farmers may have to respond by increasing the system resilience, diversification, and risk management [36]. To increase system resilience, farmers have to improve soil and nutrient management, through manure and crop residues, using, for example, legumes for natural nitrogen fixation and suitable for livestock feed. Also, they need to improve ecosystem management and biodiversity, by considering the substantial genetic variability in domestic crops and livestock that have the ability to withstand extreme temperatures, drought, and other environmental constrains, as well as pests and diseases. The combination of different crops and livestock breeds with their wild relatives is fundamental in developing a sustainable resilience [36].

Of the 400,000 plants species in existence today, only actinorhizal plants and legumes have evolved nitrogen-fixing nodules [44]. The primary role that legumes play is to fix atmospheric nitrogen through their symbiotic relationship with *Rhizobium* spp., contributing with nitrogenous compounds to the soil, either directly, by nodule excretion, or indirectly, by decomposition of root nodules and tissues [45]. The ability to fix atmospheric nitrogen makes legumes excellent partners within various farming systems, as they provide nitrogen and therefore reduce the needs for mineral nitrogen fertilizers by associated non-legumes [46].

The use of inorganic nitrogenous fertilizers has increased exponentially over the last 50 years, but just 30–50% of crop yields are sustained by inorganic fertilizers, although between 1960 and 2000, the efficiency of nitrogen for global cereal production decreased from 80 to 30%. Moreover, more than 50% of nitrogen fertilizer applied was lost from cereal crops between 1961 and 2010, and in some cases more than 80% is lost [46]. As a result there has been a 5% increase year by year of carbon dioxide equivalent emission [47]. These data show an unsustainable trend for African farmers. Increases of atmospheric carbon dioxide benefit cereal growth, and it decreases protein content in the grain, as opposite to what has been observed in cereal grains produced followed by legume crops. Therefore, intercropping or rotation of grain legumes with cereals or other non-leguminous crops increases nitrogen-use efficiency, reducing greenhouse gas emissions. It is estimated that grain legumes can offer 20–40% wheat nitrogen needs [48]; thus, intercropping is important for the development of sustainable systems, particularly in systems with limited external inputs [49]. About 21 Mt. of nitrogen is fixed annually by legume-rhizobia symbiosis, returning 5–7 Mt. of nitrogen to soils from about 190 million ha of grain legumes [48]. Without a doubt, cultivation of grain legumes is a very promising approach to increase farmers' income, especially when compared to cereal monoculture that was boosted by the "Green Revolution" [50]. Grain legumes are a very important food crop in many parts of Africa, as they are a source of nitrogen-rich edible seeds, providing high-protein products. Grain legume yields vary more than staple crops, mostly due to environmental constrains such as drought that limits symbiotic nitrogen fixation [51, 52], which in turn diminishes nutrient grain quality [53]. Soybean has clearly dominated yields, with increases of 2.9% year by year, whereas cowpea yield is stable but occupying 4.3% more land every year, trying to minimize the loss to diseases as well as insect pests and drought, low soil fertility, other abiotic stresses, and low availability of seed of improved varieties [47].

There is fast evidence that intercropping and rotation with grain legumes are beneficial as legumes improve soil structure, increase organic matter [54, 55], and provide food and feed to the most widespread farming system in Africa, the mixed-crop-livestock farming system. Moreover, intercropping or rotation with grain legumes improves water efficiency by saving water for subsequent crops or by providing soil coverage, minimizing soil evaporation, erosion, and weeds, which makes feasible the production of grain legumes in dry, drought-vulnerable, and low-labor availability areas. Residue from grain legumes provides an excellent source of high-quality feed to livestock especially during the dry season, when animal feeds are in short supply. Synergies between crops and livestock offer various opportunities for raising productivity and increasing efficiency of resources, thus increasing household incomes and securing availability and access to food [36]. Moreover, the residues from grain legume cultivation will preserve soil moisture, prevent soil erosion, and increase yields in the same piece of land, which are all big constraints of SSA farming systems that are constantly facing anthropic pressure.

Farmers have been neglecting these native grain legumes, as they are incentivized to produce common bean and soybean. However, with climate change and most of SSA agriculture being rainfall dependent, future is compromised. Many grain legume breeding programs are suffering from low genetic diversity and several bottlenecks that occurred during and after domestication. Thus, it is vital to consider the considerable large genetic variability in native crops that have the ability to withstand extreme temperatures, drought, and other environmental constrains. In agricultural statistics *Lablab*, jack, or sword bean (*Canavalia* spp.), winged bean (*Psophocarpus tetragonolobus* DC), guar bean (*Cyamopsis tetragonoloba* Taub.), velvet bean (*Stizolobium* spp.), yam bean (*Pachyrhizus erosus* Urb.), and others are recognized worldwide as "minor crops," pooled in "Pulses, nes" (pulses that are not identified separately, according to FAO). However, these pulses have been showing a steady but modest yield increase over the last 50 years, without occupying more land [7] (**Figure 1**).

3.1 The role of orphan legumes for the crop production sustainability of SSA farming systems

There is a lack of consensus in the definition and what orphan or neglected and underutilized species (NUS) should be referred to. These crops have been referred by different names, such as orphan crops, neglected crops, underutilized crops, forgotten crops, and minor crops. In this study we will refer this group as NUSs, under the definition of plants with prospective value as crops but which have been paid limited attention by agricultural researchers, plant breeders, seed companies, and policymakers [56]. However, due to the potential that these crops hold as food, nutritional content, and economic security of the developing and undeveloped parts of the world, they are appropriately referred to as crops for the future [57]. As such, these crops represent an opportunity for innovation in research, capacity building, social empowerment, and food value chains (i.e., production, processing, consumption, marketing, and product development). Understanding the importance that these crops hold, the African Orphan Crop Consortium (AOCC) was established with the full support of the African Union in 2011, assigned to work on 101 selected crops originated or naturalized in Africa (http://africanorphancrops.org) by investing in training, products, tools, services, practices, and processes to mainstream them into the African agro-food system [58]. In conjunction with this initiative, the FAO builds the database named International Network of Food Data Systems (INFOODS) listing more than 1000 unique NUS (http://www.fao.org/fileadmin/templates/food_composition/documents/Copy_of_INFOODS-List-of-underutilized-species-2_0_Jan15.xls). The AOCC partnership works to make high-nutritional-value crops grown by African farmers available to rural and urban consumers by promoting the adoption of modern breeding methods for crop improvement purposes. Under these pillars, genomic resources through next-generation sequencing from the collection of 101 African NUS are being generated (see http://africanorphancrops.org/meet-the-crops/), which included important annual and perennial (tree) species, e.g., *Moringa oleifera* L. known as the tree of life and the iconic boabab tree (*Adansonia digitata* L.). Through the high-throughput genomic resources gathered, the AOCC is also engaged to develop tools to assess genetic diversity in crops and to support breeding programs. Among such NUSs, there are several pulse crops (**Table 2**), and with the high genomic data generated, it will enable to promote research and breeding studies on the crops that will open a new venue toward understanding its suitability on several farming systems.

Lablab [*Lablab purpureus* (L.) Sweet] and velvet [*Mucuna pruriens* (L.) DC] beans are among the selected NUSs, which are known to display agronomic, nutritional, and versatile characteristics, and thus can be faced as important examples of multipurpose legumes that could and should be integrated in the most representative African farming systems, maize mixed and the agropastoral farming systems. Studies have reported that crude protein concentration in maize silage increases when intercropping with *Lablab* without compromising forage yield or milk production [59] and dry matter yield [60]. *Lablab*, locally called *njahe*, has a special significance and is intimately associated with women fertility [61] probably due to its abundance in palmitic acid [62], a fatty acid that has a structural and functional role in utero [63]. Being women the cornerstone of African economic development, contributing with approximately 70% of agricultural labor and produce about 90% of all food [47], the interest in boosting *Lablab* and other NUC legumes development is key for diversifying agriculture sector [64]. It is estimated that grain legumes increase wheat productivity by 77% and of maize by 25–33%. An intercrop system of maize with common bean resulted in maize with higher biomass yield, plant total nitrogen concentration, and crude protein concentration [65]. Intercropping maize and soybean was shown to be beneficial, as there was an increase of 19–36% in crude protein over monoculture corn [66]. Intercrop of maize with cowpea increased 9% in crude protein compared with monoculture corn [67], adding that cowpea nitrogen fixation under drought conditions is highly tolerant [68]. Also, pigeon pea (*Cajanus cajan*) is another legume mainly grown by poor farmers and is known as the poor people's meat because of its high protein content. It is among the most drought-tolerant and nutritious orphan legume crops and withstands drought because of its deep roots and osmotic adjustment in the leaves [69].

Finally, the characterization of orphan legumes on the "omics" level is still starting, and these legumes remain unexplored on the genomic, transcriptomic, and proteomic level, despite the efforts such as the African Orphan Crops Initiative (http://africanorphancrops.org), which are starting to fill the genomic information gap.

Scientific name	Common name	Assembled	Stages of assembly	Ref.
Artocarpus heterophyllus	Jack tree		Reference genome	[58]
Artocarpus altilis	Breadfruit		Reference genome	[58]
Faidherbia albida	Acacia (apple ring)	Reference genome		[69]
Moringa oleifera	Drumstick tree	Reference genome		[69]
Sclerocarya birrea	Marula	Reference genome		[69]
Digitaria exilis	Fonio		Reference genome	[58]
Eleusine coracana	Finger millet		Reference genome	[58]
Lablab purpureus	*Lablab* bean	Reference genome		[70]
		RNAseq		[71]
Solanum aethiopicum	African eggplant	Reference genome		[72]
Vigna subterranea	Bambara groundnut	Reference genome		[70]
		RNAseq		[71]

In the pipeline or soon: Abelmoschus caillei; Adansonia digitata; Allanblackia floribunda; Allanblackia stulhmannii; Allium cepa; Amaranthus cruentus; Amaranthus tricolor; Anacardium occidentale; Annona reticulata; Annona senegalensis; Balanites aegyptiaca; Basella alba; Boscia senegalensis; Brassica carinata; Canarium madagascariense; Carica papaya; Carissa spinarum; Casimiroa edulis; Cassia obtusifólia; Celosia argentea; Chrysophyllum cainito; Citrullus lanatus; Cleome gynandra; Cocos nucífera; Colocasia esculenta; Corchorus olitorius; Crassocephalum rubens; Crotalaria juncea; Crotalaria ochroleuca; Cucumis metuliferus; Cucurbita maxima; Cyphomandra betacea; Dacryodes edulis; Detarium microcarpum; Detarium senegalense; Dioscorea alata; Dioscorea dumetorum; Dioscorea rotundata; Diospyros mespiliformis; Dovyalis caffra; Ensete ventricosum; Eragrostis tef; Garcinia livingstonii; Garcinia mangostana; Gnetum africanum; Hibiscus sabdariffa; Icacina oliviformis; Ipomoea batatas; Irvingia gabonensis; Landolphia spp.; Lannea microcarpa; Lens culinaris; Macadamia ternifólia; Macrotyloma geocarpum; Mangifera indica; Momordica charantia; Morus alba; Musa acuminata AAA Group; Musa balbisiana; Opuntia monacantha; Parinari curatellifolia; Parkia biglobosa; Passiflora edulis; Persea americana; Phaseolus vulgaris; Plectranthus esculentus; Plectranthus rotundifolius; Psidium guajava; Ricinodendron heudelotii; Saba comorensis; Saba senegalensis; Solanum scabrum; Sphenostylis stenocarpa; Strychnos cocculoides; Strychnos spinosa; Syzygium guineense; Talinum fruticosum; Tamarindus indica; Telfairia occidentalis; Tylosema esculentum; Uapaca kirkiana; Vangueria infausta; Vangueria madagascariensis; Vicia faba; Vigna radiata; Vitellaria paradoxa; Vitex doniana; Xanthosoma sagittifolium; Xanthosoma spp.; Ximenia caffra; Ziziphus mauritiana

Table 2.
Present status and progress of AOCC developing genomic resources—reference genome sequencing of 100 accessions/species for 101 crops [58].

4. Final remarks

In sub-Saharan Africa, countries rely mostly on agriculture as economic revenue and as a base for smallholder farmers, for both household income and food. Considering the diversity of the farming systems along the different agroecological zonings, evaluating its performance under climate changes is key to determine its future sustainability for alleviating poverty and food security. Overall, major farming systems in SSA are under threat since they are rainfall-dependent and thus pose a scenario of food insecurity if no proper agriculture management and solutions are taken. In this chapter, the potential of pulse crops as a viable and sustainable strategy for upholding farming systems' intercropping and production indices was highlighted. The promotion of legumes adapted to semi- and arid conditions will contribute to the diversity of cropping systems and diets of African people living in rural areas. However, there is a need to address critical knowledge gaps that will allow the full use and advantages to introduce successfully the so-called neglected and underutilized crops, native to Africa, within agricultural and food systems. By exploring native legumes adapted to arid conditions, namely, low rainfall periods, it will be a key tool for adaptation to climate change. This will also contribute to improve soil fertility and enhance food, forage, and mulching quality, which is of main importance particularly for the developing countries. Therefore, promoting its cultivation and implementing different farm management practices will contribute to the resilience of SSA farming systems. As the world celebrated the International Year of Pulses in 2016, there is a need to establish the potential and invest in the innovation of undervalued role that pulses can play in the post-2016 agenda. In spite of their recognized importance, some African native legumes are still underutilized or overlooked crops, and its use is a viable option to raise farming productivity.

Acknowledgements

The work was funded by the Portuguese Rural Development Program (PDR2020) for the Operational Group STEnCIL, Initiative 27 [PDR2020–1.0.1-FEADER-031465], within the European Innovation Partnership (EIP-AGRI) supported by the European Agricultural Fund for Rural Development and undertaken in the scope of project CajOmics [PTDC/AGR-PRO/5727/2014] funded by the Fundação para a Ciência e Tecnologia (FCT)-FCT/MCTES/PIDDAC, Portugal, and the project CVAgrobiodiversity/333111699 funded by the Fundação para a Ciência e Tecnologia (FCT) and Aga Khan Development Network (AKDN). The work was supported by FCT funds, to the following research units: LEAF [UID/AGR/04129/2019] and cE3c [UID/BIA/00329/2019]. FM was individually funded by FCT-awarded postdoctoral fellowship SFRH/BPD/115162/2016.

Conflict of interest

The authors declare no conflict of interest.

Author details

Patricia Vidigal[1], Maria Manuel Romeiras[1,2] and Filipa Monteiro[1,2]*

1 Linking Landscape, Environment, Agriculture and Food (LEAF), Instituto Superior de Agronomia, Universidade de Lisboa, Lisboa, Portugal

2 Faculdade de Ciências, Centre for Ecology, Evolution and Environmental Changes (cE3c), Universidade de Lisboa, Lisboa, Portugal

*Address all correspondence to: fimonteiro@fc.ul.pt; fmonteiro@isa.ulisboa.pt

References

[1] Grönemeyer JL, Reinhold-Hurek B. Diversity of bradyrhizobia in subsahara Africa: A rich resource. Frontiers in Microbiology. 2018;**9**. DOI: 10.3389/fmicb.2018.02194. Epub ahead of print 20 September 2018

[2] FAO. World Food and Agriculture—Statistical Pocketbook 2018. Rome, Italy: FAO; 2018. p. 255

[3] Monteiro F, Catarino L, Batista D, Indjai B, Duarte MC, Romeiras MM. Cashew as a High Agricultural Commodity in West Africa: Insights towards Sustainable Production in Guinea-Bissau. Sustainability. 2017;**9**:1666. https://doi.org/10.3390/su9091666

[4] Havik PJ, Monteiro F, Catarino S, Correia AM, Catarino L, Romeiras MM. Agro-Economic Transitions in Guinea-Bissau (West Africa): Historical Trends and Current Insights. Sustainability. 2018;**10**:3408. https://doi.org/10.3390/su10103408

[5] FAO. State of Food and Agriculture in the Region, Including Future Prospects and Emerging Issues. In: Regional Conference for Europe. Food and Agriculture Organization of the United Nations; 2014. p. 15

[6] Djurfeldt AA, Jirström M. Urbanization and Changes in Farm Size in Sub-Saharan Africa and Asia from a Geographical Perspective: A Review of the Literature. CGIAR's Independent Science and Partnership Council; 2013. pp. 1-40

[7] FAO. Food and Agriculture Organization of the United Nations. 2019. Available from: http://www.fao.org/faostat/en/ [Accessed: 12 April 2019]

[8] Blackie M, Malcom J. Maize, food self-sufficiency and policy in East and Southern Africa. Food Policy. 1990;**15**:383-394. DOI: 10.1016/0306-9192(90)90055-5

[9] Byerlee D, Eicher CK, editors. Africa's Emerging Maize Revolution. Boulder, London: Lynne Rienner; 1997. 301 p. ISBN: 1-55,587-776-1

[10] Mellor, John W, Delgado C, et al. Accelerating food production in sub-Saharan Africa. In: Agricultural Development. Baltimore: Johns Hopkins; 1987. 417 p

[11] Catarino S, Duarte MC, Costa E, et al. Conservation and sustainable use of the medicinal leguminosae plants from Angola. Peer Journal. 2019;7:e6736. DOI: 10.7717/peerj.6736

[12] HarvestChoice. AEZ Tropical (5-class, 2009). Washington, DC, and University of Minnesota, St. Paul, MN: International Food Policy Research Institute; 2015. Available from: http://harvestchoice.org/data/aez5_clas. [Accessed: 21 April 2019]

[13] HarvestChoice. Farming System Name. International Washington, DC, and University of Minnesota, St. Paul, MN: Food Policy Research Institute; 2015. Available from: http://harvestchoice.org/data/fs_2012_tx [Accessed: 21 April 2019]

[14] Dixon J, Gulliver A, Gibbon D. Sub-Saharan Africa. In: Farming Systems and Poverty: Improving Farmers' Livelihoods in a Changing World. Rome and Washington DC: FAO and World Bank; 2001. 412 p

[15] HarvestChoice. Food Crops Irrigated Value Production (Int$, 2005). Washington, DC, and University of Minnesota, St. Paul, MN: International Food Policy Research Institute; 2015. Available from: http://harvestchoice.org/data/vp_food_i [Accessed: 21 April 2019]

[16] HarvestChoice. Food Crops Rainfed Value Production (Int$, 2005). Washington, DC, and University of Minnesota, St. Paul, MN: International Food Policy Research Institute; 2015. Available from: http://harvestchoice.org/data/vp_food_r [Accessed: 21 April 2019]

[17] Serdeczny O, Adams S, Baarsch F, et al. Climate change impacts in sub-Saharan Africa: From physical changes to their social repercussions. Regional Environmental Change. 2017;**17**:1585-1600. DOI: 10.1007/s10113-015-0910-2

[18] Shiferaw B, Prasanna BM, Hellin J, et al. Crops that feed the world 6. Past successes and future challenges to the role played by maize in global food security. Food Security. 2011;**3**:307-327. DOI: 10.1007/s12571-011-0140-5

[19] Masike S, Ulrich P. Vulnerability of traditional beef sector to drought and the challenges of climate change: The case of Kgatleng district, Botswana. Journal of Geography and Regional Planning. 2008;**1**:12-18

[20] Tian-Jun Z, Tao H. Projected changes of palmer drought severity index under an RCP8.5 scenario. Atmospheric and Oceanic Science Letters. 2013;**6**:273-278. DOI: 10.3878/j.issn.1674-2834.13.0032

[21] Garrity D, Dixon J, Boffa JM. Understanding African Farming Systems: Science and Policy Implications. 2012. Available from: http://aciar. gov. au/aifsc/sites/default/files/images/understanding_african_farming_systems_11_dec_update. pdf

[22] Masih I, Maskey S, Mussá FEF, et al. A review of droughts on the African continent: A geospatial and long-term perspective. Hydrology and Earth System Sciences. 2014;**18**:3635-3649. DOI: 10.5194/hess-18-3635-2014

[23] Godfray HCJ, Beddington JR, Crute IR, et al. The challenge of food security. Science. 2012;**327**:812. DOI: 10.4337/9780857939388

[24] Pretty J, Toulmin C, Williams S. Sustainable intensification in African agriculture. International Journal of Agricultural Sustainability. 2011;**9**:5-24. DOI: 10.3763/ijas.2010.0583

[25] Tilman D, Balzer C, Hill J, et al. Global food demand and the sustainable intensification of agriculture. Proceedings of the National Academy of Sciences. 2011;**108**:20260-20264. DOI: 10.1073/pnas.1116437108

[26] Smaling EM. The balance may look fine when there is nothing you can mine: Nutrient stocks and flows in West African soils. In: Gerner H, Mokwunye AU, editors. Use of Phosphate Rock for Sustainable Agriculture in West Africa; 21-23 November 1995; IFDC-Africa; 1995

[27] Henao J, Baanante C. Nutrient depletion in the agricultural soils of Africa. In: Pinstrup-Andersen P, Pandya-Lorch R, editors. The Unfinished Agenda: Perspectives on Overcoming Hunger, Poverty, and Environmental Degradation. Washington, D.C: IFPRI; 2001. p. 301

[28] Bationo A, Hartemink A, Lungu O, et al. Knowing the African soils to improve fertilizer recommendations. In: Improving Soil Fertility Recommendations in Africa using the Decision Support System for Agrotechnology Transfer (DSSAT). Dordrecht: Springer Netherlands; 2012. pp. 19-42

[29] Mosier AR, Bleken MA, Chaiwanakupt P, et al. Policy implications of human-accelerated nitrogen cycling. Biogeochemistry. 2001;**52**:281-320. DOI: 10.1023/A:1015798424363

[30] Yadav MR, Kumar R, Parihar CM, et al. Strategies for improving nitrogen

use efficiency: A review. Agricultural Reviews. DOI: 10.18805/ag.v0iOF.7306. Epub ahead of print 16 February 2017

[31] Pimentel D, Hepperly P, Hanson J, et al. Environmental, energetic, and economic comparisons of organic and conventional farming systems. Bioscience. 2005;**55**:573-582. DOI: 10.1641/0006-3568(2005)055[0573:EEAECO]2.0.CO;2

[32] Snapp S, Rahmanian M, Batello C. Pulse Crops for Sustainable Farms in Sub-Saharan Africa. Rome: FAO; 2018. Available from: http://www.fao.org/3/i8300en/I8300EN.pdf

[33] HarvestChoice. Cowpea Rainfed Production (mt, 2005). Washington, DC, and University of Minnesota, St. Paul, MN: International Food Policy Research Institute; 2015. Available from: http://harvestchoice.org/data/cowp_r_p [Accessed: 21 April 2019]

[34] HarvestChoice. Bean Rainfed Production (mt, 2005). Washington, DC, and University of Minnesota, St. Paul, MN: International Food Policy Research Institute; 2015. Available from: http://harvestchoice.org/data/bean_r_p. [Accessed: 21 April 2019]

[35] HarvestChoice. Other Pulses Rainfed Production (Int$, 2005). Washington, DC, and University of Minnesota, St. Paul, MN: International Food Policy Research Institute; 2015. Available from: http://harvestchoice.org/data/opul_r_v. [Accessed: 21 April 2019]

[36] Thornton PK, Herrero M. Adapting to climate change in the mixed crop and livestock farming systems in sub-Saharan Africa. Nature Climate Change. 2015;**5**:830-836. DOI: 10.1016/j.gfs.2014.02.002

[37] Dai A. Drought under global warming: A review. Wiley Interdisciplinary Reviews: Climate Change. 2011;**2**:45-65. DOI: 10.1002/wcc.81

[38] IPCC. Summary for policymakers. In: Climate Change 2014, Mitigation of Climate Change. Climate Change 2014 Synthesis Report Summary for Policymakers. 2014. Available from: https://www.ipcc.ch/site/assets/uploads/2018/02/AR5_SYR_FINAL_SPM.pdf

[39] Liu J, Fritz S, van Wesenbeeck CFA, et al. A spatially explicit assessment of current and future hotspots of hunger in sub-Saharan Africa in the context of global change. Global and Planetary Change. 2008;**64**:222-235. DOI: 10.1016/j.gloplacha.2008.09.007

[40] Schlenker W, Lobell DB. Robust negative impacts of climate change on African agriculture. Environmental Research Letters. 2010;**5**:014010. DOI: 10.1088/1748-9326/5/1/014010

[41] Thornton PK, Jones PG, Ericksen PJ, et al. Agriculture and food systems in sub-Saharan Africa in a 4 C+ world. Philosophical Transactions of the Royal Society A—Mathematical Physical and Engineering Sciences. 2011;**369**:117-136. DOI: 10.1098/rsta.2010.0246

[42] Rosegrant M, Cai X, Cline S. World Water and Food for 2025: Dealing with Scarcity. IFPRI: Washington, D.C; 2002. Available from: http://www.ifpri.org/themes/impact/impactwater.pdf

[43] Mohammed-Saleem MA. Mixed farming systems in sub-Saharan Africa. In: Wilson RT, Ehui S, Mack S, editors. Livestock Development Strategies for Low Income Countries—Proceedings of the Joint FAO/ILRI Roundtable on Livestock Development Strategies, for Low Income Countries, ILRI, Addis Ababa, Ethiopia. Nairobi, Kenya: Food and Agriculture Organization/International Livestock Research Institute; 1995

[44] Ribeiro-Barros AI, Catarino S, Moura I, et al. Actinorhizal trees and shrubs from Africa: Distribution, conservation and uses. Antonie Van Leeuwenhoek. 2019;**112**:31-46. DOI: 10.1007/s10482-018-1174-x

[45] Tothill JC. The role of legumes in farming systems of sub-Saharan Africa. In: Potential of Forage Legumes in Farming Systems of sub-Saharan Africa. Proceedings of a workshop held at ILCA; 16-19 September 1985. Ethiopia: Addis Ababa; 1986

[46] Foyer CH, Lam HM, Nguyen HT, et al. Neglecting legumes has compromised human health and sustainable food production. Nature Plants. 2016;**2**:16112. DOI: 10.1038/nplants.2016.112

[47] FAO. The Future of Food and Agriculture—Trends and Challenges. 2017. Available from: http://www.fao.org/3/a-i6583e.pdf

[48] Reeves TG, Thomas G, Ramsay G. Save and Grow in Practice Maize, Rice, Wheat. A Guide to Sustainable Cereal Production. Rome: Food and Agriculture Organization of the United Nations; 2016. Available from: www.fao.org/3/a-i4009e.pdf

[49] Adesogan AT, Salawu MB, Deaville ER. The effect on voluntary feed intake, in vivo digestibility and nitrogen balance in sheep of feeding grass silage or pea–wheat intercrops differing in pea to wheat ratio and maturity. Animal Feed Science and Technology. 2002;**96**:161-173. DOI: 10.1016/S0377-8401(01)00336-4

[50] Chibarabada T, Modi A, Mabhaudhi T. Expounding the value of grain legumes in the semi- and arid tropics. Sustainability. 2017;**9**:60. DOI: 10.3390/su9010060

[51] Devi MJ, Sinclair TR, Beebe SE, et al. Comparison of common bean (*Phaseolus vulgaris* L.) genotypes for nitrogen fixation tolerance to soil drying. Plant and Soil. 2013;**364**:29-37. DOI: 10.1007/s11104-012-1330-4

[52] Sinclair TR, Messina CD, Beatty A, et al. Assessment across the United States of the benefits of altered soybean drought traits. Agronomy Journal. 2010;**102**:475. DOI: 10.2134/agronj2009.0195

[53] Farooq M, Gogoi N, Barthakur S, Baroowa B, Bharadwaj N, Alghamdi SS, et al. Drought stress in grain legumes during reproduction and grain filling. Journal of Agronomy and Crop Science. 2017;**203**:81-102. DOI: 10.1111/jac.12169

[54] Courty PE, Smith P, Koegel S, et al. Inorganic nitrogen uptake and transport in beneficial plant root-microbe interactions. CRC Critical Reviews in Plant Sciences. 2015;**34**:4-16. DOI: 10.1080/07352689.2014.897897

[55] Peix A, Ramírez-Bahena MH, Velázquez E, et al. Bacterial associations with legumes. CRC Critical Reviews in Plant Sciences. 2015;**34**:17-42. DOI: 10.1080/07352689.2014.897899

[56] Padulosi S, Thompson J, Rudebjer P. Fighting Poverty, Hunger and Malnutrition with Neglected and Underutilized Species (NUS): Needs, Challenges and the Way Forward. Rome: Bioversity International; 2013

[57] Baldermann S, Blagojević L, Frede K, et al. Are neglected plants the food for the future? CRC Critical Reviews in Plant Sciences. 2016;**35**:106-119. 56p

[58] Hendre PS, Muthemba S, Kariba R, et al. African Orphan Crops Consortium (AOCC): Status of developing genomic resources for African orphan crops. Planta. DOI: 10.1007/s00425-019-03156-9. Epub ahead of print 9 May 2019

[59] Armstrong KL, Albrecht KA, Lauer JG, et al. Intercropping corn

with lablab bean, velvet bean, and Scarlet Runner bean for forage. Crop Science. 2008;**48**:371. DOI: 10.2135/cropsci2007.04.0244

[60] Contreras-Govea FE, Muck RE, Armstrong KL, et al. Fermentability of corn-lablab bean mixtures from different planting densities. Animal Feed Science and Technology. 2009;**149**:298-306. DOI: 10.1016/j.anifeedsci.2008.05.009

[61] Robertson CC. Black, white, and red all over: Beans, women, and agricultural imperialism in twentieth-century Kenya. Agricultural History. 1997;**71**:259-299. Available from: http://www.jstor.org/stable/3744315

[62] Vidigal P, Duarte B, Cavaco AR, et al. Preliminary diversity assessment of an undervalued tropical bean (*Lablab purpureus* (L.) Sweet) through fatty acid profiling. Plant Physiology and Biochemistry. 2018;**132**:508-514. DOI: 10.1016/j.plaphy.2018.10.001

[63] Agostoni C, Moreno L, Shamir R. Palmitic acid and health: Introduction. Critical Reviews in Food Science and Nutrition. 2016;**56**:1941-1942. DOI: 10.1080/10408398.2015.1017435

[64] Miller NR, Mariki W, Nord A, et al. Cultivar selection and management strategies for *Lablab purpureus* (L.) Sweet in Africa. In: Handbook of Climate Change Resilience. Cham: Springer International Publishing; 2018. pp. 1-14

[65] Dawo MI, Wilkinson JM, Sanders FE, et al. The yield and quality of fresh and ensiled plant material from intercropped maize (*Zea mays*) and beans (*Phaseolus vulgaris*). Journal of the Science of Food and Agriculture. 2007;**87**:1391-1399. DOI: 10.1002/jsfa.2879

[66] Herbert SJ, Putnam DH, Poos-Floyd MI, et al. Forage yield of intercropped corn and soybean in various planting patterns 1. Agronomy Journal. 1984;**76**:507. DOI: 10.2134/agronj1984.00021962007600040001x

[67] Bryan WB, Materu MB. Intercropping maize with climbing beans, cowpeas and velvet beans. Journal of Agronomy and Crop Science. 1987;**159**:245-250. DOI: 10.1111/j.1439-037X.1987.tb00096.x

[68] Sinclair TR, Manandhar A, Belko N, et al. Variation among cowpea genotypes in sensitivity of transpiration rate and symbiotic nitrogen fixation to soil drying. Crop Science. 2015;**55**:2270. DOI: 10.2135/cropsci2014.12.0816

[69] Odeny DA. The potential of pigeon pea (*Cajanus cajan* (L.) Millsp.) in Africa. Natural Resources Forum. 2007;**31**:297-305

[70] Chang Y, Liu H, Liu M, et al. The draft genomes of five agriculturally important African orphan crops. Gigascience. DOI: 10.1093/gigascience/giy152. Epub ahead of print 7 December 2018

[71] Chapman M. Transcriptome sequencing and marker development for four underutilized legumes. Applications in Plant Sciences. 2015;**3**:1400111. DOI: 10.3732/apps.1400111

[72] Song B, Song Y, Fu Y, et al. Draft genome sequence of the *Solanum aethiopicum* provides insights into disease resistance, drought tolerance and the evolution of the genome. bioRxiv. DOI: 10.1101/532077. Epub ahead of print 2019

Chapter 4

Review on the Role of Salicylic Acid in Plants

Ali S. Hassoon and Inas Abdulsattar Abduljabbar

Abstract

Salicylic acid and its derivatives as one of the plant hormones produced by the plant naturally belong to the group of phenolic acids and consist of a ring linked to the group of hydroxyl and carboxyl group, and the starting ingredient to form the cinnamic acid. It is mainly manufactured within the plant in cytoplasmic cell. This acid was first discovered in *Salix* spp., which contains the Salicin compound by 9.5–11% and is present in the plant in the form of free phenolic acids or associated with amino compounds. Symbolized by the symbol SA called chemical ortha hydroxyl benzoic acid chemical formula is $C_7H_6O_3$.

Keywords: salicylic acid, plants

1. Introduction

Salicylic acid and its derivatives as one of the plant hormones produced by the plant naturally belongs to the group of phenolic acids and consists of a ring linked to the group of hydroxyl and carboxyl group, and the starting ingredient to form is the cinnamic acid (**Figure 1**). It is mainly manufactured within the plant in cytoplasmic cell. This acid was first discovered in *Salix* spp., which contains the salicin compound by 9.5–11% and is present in the plant in the form of free phenolic acids or associated with amino compounds [1]. Symbolized by the symbol SA called chemical ortha hydroxyl benzoic acid chemical formula is $C_7H_6O_3$ [2].

Salicylic acid plays an important role in the growth and development of the plant for important physiological roles such as increasing the plant's response to stress conditions (biotic and abiotic) by increasing the resistance of the plant to System Acquired Resistance (SAR) by stimulating or changing the internal paper dissection endogenous signaling to withstand a large number of stresses. Salicylic acid acts as a stimulant or transmitter of the cell to withstand environmental stress conditions such as dryness, coldness, heat, stress of heavy elements, and conditions of ammonia tension and also increases the plant's ability to withstand salt stress salt particularly harmful sodium chloride compound NaCl [3].

It also has the ability to bind conjugate with some amino acids such as proline and arginine, which increase the plant's effectiveness in resisting environmental stresses and at the same time maintain systemic acquired resistance [4].

The most important effects of salicylic acid are to stimulate the production of antioxidants. Antioxidant against the effect of free radicals from the group Reactive Oxygen species (ROS) when exposed to heat stress and stress Drought stress and prevents the oxidation of algebraic and oxytin and cytokinein and also has a role at

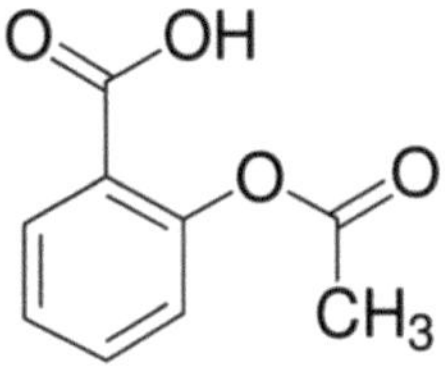

Figure 1.
Chemical structure of SA acid.

the genetic level. It stimulates the genes of antioxidant enzymes such as manganese superoxide dismutase (Masud) [5].

Salicylic acid increases the plant's response to tolerance and resistance to various diseases affecting plants as it is found that increasing its internal concentration activates the protective role of pathogenic pathogens [6]. The SA also has many important physiological roles, such as stimulating the flowering, ion absorption, nutrient transfer, increasing the representation of CO_2 gas, controlling the movement of stomata, photo materials, gas exchange, and protein synthesis. It also contributes to increasing the percentage of nucleic acids and amino acids and the accumulation of dry matter and speeds up the formation of various plant dyes and increasing their levels such as chlorophyll and carotene and prevents the representation of ethylene gas, and it is contrary to the work of ABA responsible for the fall of leaves. It also plays an important role in increasing metabolic rates, which contributes to the energy saving of the plant through alternative pathways accompanied by a change in the level of nucleic and amino acids within the plant [7].

2. Effect of salicylic acid in growth and yield

De Kock et al. [8] were the first to talk about the role of salicylic acid as a growth regulator during the past two decades, after which the interest in this compound has increased, and many studies have been conducted that showed a relationship between salicylic acid and the growth and development of plants. Among these studies is the finding of the cotton plant *Gossypium hirsutum* L., which belongs to the Malvaceae family in three levels of salicylic acid (50, 100, and 150) mg/l had it. The highest rate of the studied traits was the plant height (143.80) cm, the number of branches (34.28 branches), and the total cotton yield (3371.9) kg/Ha in relation to other concentrations used [9].

Najafian [10] concluded that *Rosmarinus officinalis* L. spraying with three levels of salicylic acid (450, 300, and 150) mg/l resulted in a significant increase in growth rates and photosynthesis compared to untreated plants. The increase was more pronounced when spraying plants with a concentration of 300 Mg.

Najafian [11] found that spraying SA acid at three levels (150, 300, and 450) mMol on *Thymus vulgaris* L. had a significant effect on the studied traits. Spraying at a concentration of 150 mM gave an increase in the dry weight of the vegetative total and photosynthesis and increased plant tolerance for salt stress conditions.

In a study on the response of the Indian mustard *Brassica juncea* L. to spraying with two levels of salicylic acid (35 and 70 mg), there was a significant superiority in all vegetative traits studied (plant height, number of branches, and leaf area). In addition, there was a significant increase in all the parameters of the crop (the weight of one mustard, the total yield of the seed, and the seed yield), when spraying the plants at a concentration of 70 mg/l in comparison with the concentration of 35 mg/l and spraying with distilled water only [12].

In a study conducted in Pakistan on the *Abelmoschus esculentus* L., which belongs to the marsh family, [13] found that salicylic acid spraying with concentrations of 50 and 75 mg/l had a significant effect on most studied traits. The effect of spraying was 50 mg/l is more pronounced in increasing vegetative growth rates and leaf content than chlorophyll.

Abbas and Ibrahim reported [14] that the growth regulator SA was sprayed on *Niggella sativa* L. at several levels (50, 100, and 200) mg/l with significant effect on the studied traits. And 200 mg/l. Spraying at a concentration of 50 mg/l was the best in increasing growth, yield, and oil ratio indices.

Al-Mohammadi and Al-Rawi also [15] observed a study on the effect of spraying on some of the growth catalysts on *Datura stramonium* L. The spraying with acetylsalicylic acid at 200 ppm gave the highest rate of all vegetative and studied traits (plant height, dry and vegetative content of the leaves, nitrogen and potassium, number of fruits, plant and the total yield kg/hectare) compared to non-treated plants.

3. Effect of salicylic acid in qualitative and medical qualities

The significant phylogenetic effects reflected by the salicylic acid act towards the growth and development of the plant and the improvement of its health made it a popular vehicle for those interested in agricultural production. This has already been shown to improve the qualities of many plants that occupy a high economic position. It also activates the roles of many enzymes and also has an important action towards syphilis and the bio-synthesis of ethylene gas (the maturation hormone and aging) and the movement of stomata and contributes to plant metabolism and transfer of ions [16, 17].

Through research and studies on the effect of salicylic acid treatment on the specific qualities of plants, Gharib [18] noted that the spray of the basil plants *Ocimum basilicum* L. and the *Majorana hortensi* L. were planted in a 40 cm pot with three concentrations of salicylic acid [10^{-3}, 10^{-4}, 10^{-5}] mole resulted in a significant increase in the ratio of the active ingredient of both plants compared to the comparison treatment. The spraying of two varieties of *Cymbopogon flexuous* L. with a concentration of 5–10 m of salicylic acid developing in the plants gave a significant increase in the specific qualities and active substances of plants and antioxidants compared to non-treated plants [19].

Khandaker et al. showed [20] that spray of red *Amaranthus tricolor* L. plants with three concentrations of salicylic acid (10^{-3}, 10^{-4}, and 10^{-5} mm) had the most significant effect on plant active compounds compared to untreated plants and significant increase in properties (total phenols, antioxidant, and plant pigments). Salicylic acid spraying with concentrations (25 and 50 mg/l) on the vegetative group of *Cumin cyminum L.* resulted in a significant increase in the percentage of plant pigments at a concentration of 50 mg/l compared to comparison plants [21].

The addition of salicylic acid with three concentrations (30, 60, and 90 mg/l) resulted in a significant increase in the production of some plant antioxidants from blackwheat leaves when treated with concentrations of 60 and 90 mg/l compared with non-treated plants [22].

Majoul showed [23] a significant increase in the percentage of nutrients P, N, and the leaf content of chlorophyll when spraying the okra plants were measured at two levels of salicylic acid (78 and 155 mg/l) and in two steps compared to the comparison treatment.

The medicinal seeds of the *Digitalis trojanaivanina* collected from the Turkish Aida mountains with three concentrations of salicylic acid resulted in significant superiority in the studied active substances (pigment content, total phenols, phenols, and flavonoids) compared to non-treated plants [24].

Author details

Ali S. Hassoon* and Inas Abdulsattar Abduljabbar
Al-Musaib Technical College/Al-Furat Al-Awsat Technical University, Iraq

*Address all correspondence to: alisalealtaie2015@gmail.com

References

[1] Janda M, Ruelland E. Magical mystery tour: Salicylic acid signaling. Environmental and Experimental Botany. 2014. In Press

[2] Pál M, Szalai G, Kovács VG, Janda OK. Salicylic acid-mediated abiotic stress tolerance. In: Hayat S, Ahmad A, Alyemeni MN, editors. Salicylic Acid. Netherlands: Springer; 2013. pp. 183-247

[3] Simaei M, Khavari-Nejad RA, Bernard F. Exogenous application of salicylic acid and nitric oxide on the ionic contents and enzymatic activities in NaCl-stressed soybean plants. American Journal of Plant Sciences. 2012;**3**:1495-1503

[4] Baghizadeh A, Hajmo-hammadrezaei M. Effect of drought stress and its interaction with as corbate and salicylic acid on okra (*Hibiscus esculents* L.) ruination and seedling growth. Journal of Stress Physiology & Biochemistry. 2011;**7**(1):55-65

[5] Horvath E, Szalai G, Janda T. Induction of a biotic stress tolerance by salicylic acid signaling. Journal of Plant Growth Regulation. 2007;**26**(3):290-300

[6] Kumar D. Salicylic acid signaling in disease resistance. Plant Science. 2014;**228**:127-134

[7] Davies PJ. The plant hormones: Their nature, occurrence and function. In: Davies PJ, editor. Plant Hormones: Biosynthesis, Signal Transduction, Action. 3rd ed. Dordrecht: Kluwer Academic Publishers; 2004. pp. 1-15

[8] De Kock PC, Grabowsky FB, Innes AM. The effect of salicylic acid on the growth of *Lemna gibba* L. Annals of Botany. 1974;**38**:903-908

[9] Al-Rawi ANT, Al-Ani MH, Al-Saad TM. Response of cotton *Gossypium hirsutum* L. for different irrigation periods and salicylic acid. Anbar Journal of Agricultural Sciences. 2014;**12**(2):283. The Republic of Iraq

[10] Najafian S, Khoshkhui M, Vahid T. Effect of salicylic acid and salinity in rosemary (*Rosmarinus officinalis* L.): Investigation on changes in gas exchange, water relations, and membrane stabilization. Advances in Environmental Biology. 2009;**3**(3):322-328. Iran

[11] Najafian S, Khoshkhui M, Tavallali V, Saharkhiz MJ. Effect of salicylic acid and salinity in thyme (*Thymus vulgaris* L.): Investigation on changes in gas exchange, water relations, and membrane stabilization and biomass accumulation. Australian Journal of Basic and Applied Sciences. 2009;**3**(3):2620-2626

[12] Dugogi EH, Mahdi NK, Matroud SAK. Response of Indian mustard (*Brassica juncea* L.) to the distance of planting and spraying with salicylic acid and their effect on growth, seed yield and hard oil. In: Second Scientific Conference of the Faculty of Agriculture, University of Karbala; 2012. pp. 173-181. The Republic of Iraq

[13] Raza SH, Shafiq F, Khan I. Seed invigoration with water, ascorbic and salicylic acid stimulates development and biochemical characters of okra (*Ablemoschus esculentus*) under normal and saline conditions. International Journal of Agriculture and Biology. Faisalabad University. 2013;**15**:486-492. Pakistan

[14] Abbas J, Ibrahim BA. Effect of salicylic acid in some vegetative and fruit traits of the black bean. Iraqi Journal of Agricultural Sciences. 2014;**45**(8):845-853. (Special number) The Republic of Iraq

[15] Al-Mohammadi AN, Al-Rawi AR. Effect of planting date, growth promoters, type of organic manure on growth and yield, and active ingredient of *Datura stramonium* L. Tikrit Journal for Agricultural Sciences. 2016;**16**(2):26-50. The Republic of Iraq

[16] Krasavina MS. Effect of Salicylic Acid on Solute Transport in Plants. Timiryazev Institute of Plant Physiology. Moscow, Russia: Springer; 2007. pp. 25-68

[17] Hayat S, Ali B, Ahmed A. Salicylic acid: Biosynthesis, metabolism and physiological role in plants. In: Salicylic Acid—A Plant Hormone. Netherlands: Springer; 2007. pp. 1-14

[18] Gharib FAE. Effect of salicylic acid on the growth metabolic, activities and oil content bail and major. International Journal of Agriculture and Biology. 2006;**8**(4):485-492

[19] Idrees MM, Khan MA, Aftab T, Naeem M, Hashmi N. Salicylic acid-induced physiological and biochemical changes in lemongrass varieties under water stress. Journal of Plant Interactions. Aligarh Muslim University. 2010;**5**(4):293-303. India

[20] Khandaker L, Akond AS, Oba S. Foliar application of salicylic acid improved the growth, yield and leaf's bioactive compounds in red amaranth (*Amaranthus tricolor* L.). Vegetable Crops Research Bulletin. 2011;**74**:77-86. Japan

[21] Shouaily MS. A physiological study of the effect of planting method, the number of plants in Joura and the spraying of salicylic acid concentrations and their interactions in the vegetative and syphilis growth of *Cuminum cyminum* L. Basrah Journal of Agricultural Sciences. 2012;**25**(2):37-46. The Republic of Iraq

[22] Nasrallah AY, Saadallah HSE, Ne'ma SI. Effect of some plant growth regulators on field properties and antioxidant production of blackwheat leaves. Iraqi Journal of Agricultural Sciences. 2015;**46**(5):682-694. The Republic of Iraq

[23] Majoul AK. Effect of salicylic acid and boron in the growth and yield of the plant *Abameoschus esculentus* L. Btira. Al-Furat Journal of Agricultural Sciences. 2015;7(3):27-35. The Republic of Iraq

[24] Cingoz GS, Ekrem G. Effect of salicylic acid on thermo tolerance and cardenolide accumulation under high temperature stress in *Digitalis trojana* Ivanina. Journal of Plant Physiology and Biochemistry. Abnat Izzet Baysal University. 2016;**105**:145-149. Turkey

Chapter 5

Development of Androgenesis Studies on Eggplant (*Solanum melongena* L.) in Turkey from Past to Present

Gülsün Elif Vural, Esin Ari, Sinan Zengin and Sekure Sebnem Ellialtioglu

Abstract

Eggplant is one of the most widely cultivated vegetable species in the world and Turkey. The breeding of eggplant with high yields and quality is one of the important efforts in the seed sector today. Traditional breeding activities cannot respond quickly to market mobility. With the integration of dihaploidization methods into the breeding cycles, breeding programs have gained significant momentum. The most used haploidy technique in eggplant is the anther culture based on androgenesis, and its use in public and private sectors has become widespread in recent years. To date, the use of the isolated microspore culture technique as another androgenesis technique is limited; however, the studies are in progress in particular for indirect microspore embryogenesis. Genotype effect is one of the most decisive factors determining the success of androgenesis in eggplant. Also, the other factors such as nutrient medium content, types and concentrations of plant growth regulators, age and growing conditions of donor plants, determination of the appropriate microspore developmental stages, different pre-treatments, temperature shocks and incubation conditions are also effective on androgenesis success. In this review, it is aimed to provide information about the in vitro eggplant androgenesis studies, which have been carried out and are currently being conducted in Turkey.

Keywords: anther culture, aubergine, haploid, doubled haploid, microspore culture, microspore embryogenesis

1. Introduction

The eggplant also known as *Solanum melongena* L. (2n = 2x = 24) from Solanaceae family is one of the important vegetable species in the world and in Turkey, and thus, great importance is given to the breeding studies of this plant. Although it is a perennial plant in tropical climates and annual plant in cool climates, its economic production is done annually. The primary gene center of eggplant is considered to be the Indo-Burmese region of Asia and while India and

IntechOpen

China are the secondary gene centers [1]. It was first cultivated in Asia [2, 3]. While the wild forms of the species were small and prickly, and with bitter fruits, current forms of the plant have been achieved through the selections, during the period of cultivation, in the direction of large, thornless and tasty fruits [2, 4, 5]. Of the 51.3 million tons of eggplants produced in the world, 93% is grown in Asia and only 2.5% (about 875.000 tons) across Europe. As for Turkey, with 854.049 tons of eggplant production, it performs almost as much as the eggplant production of all European countries. In addition, Turkey ranks fourth in the eggplant production of the world after China, India and Egypt [6].

Plants are affected by the positive or negative conditions of their environments and try to adapt to those regions. In this adaptation process, non-adaptable plants perish, and adapters hide or develop some properties in order to survive, and in this way, different variations occur in their regions. The different climatic features in Turkey have contributed to the formation of our local cultivars, which have been cultivated by the people living in Anatolia for centuries, and have caused Turkey to be the second diversity region of eggplant. Different researchers have classified the eggplants according to their being early-late, adaptation to the environmental conditions, origin, shape, color and other characteristics [7]. Zhukovsky collected eggplant samples while traveling in Turkey between the years 1925 and 1926 and identified five different eggplant varieties on the Anatolian soils over those samples [8]. However, *S. melongena* L., which is cultivated as a culture plant in different countries, is generally considered to have four subspecies. These are (1) spp. *esculentum*, (2) spp. *insanum*, (3) spp. *serpentium*, and (4) spp. *depressum*, and they differ in color, shape, habitus, efficiency, usage area, etc. [9].

There are many different uses of eggplant in Turkish cuisine. Besides being consumed as fresh, dried and frozen, it is used in jams, pickles, sauces and salads. In addition, it has recently attracted attention with its use in diet lists due to its fibrous structure and low-calorie value. Eggplant, in contrast to general belief, is rich in vitamins and minerals. It has important nutritional value and contains very valuable antioxidants and phenolic substances for human health. For this reason, it has been used frequently in drug production and alternative medicine in the world since ancient times. While the fruit and leaves of eggplant have a lowering effect on cholesterol levels in the blood, eggplant extract is used in different treatments such as diabetes, asthma, bronchitis and digestive disorders [10, 11].

Eggplant cultivation in Turkey was being conducted in the field conditions before. Then, with the development of greenhouse and seed technologies, eggplant production was started in greenhouse and became one of the vegetables produced throughout the year. The yield of the cultivars cultivated in greenhouses has increased by means of the F_1 hybrid cultivars, because they have higher yield values than the standard cultivars and do not show phenotypic variations. Thus, hybrid cultivars have entered the market with great speed. This has led to a rapid decrease in the standard cultivars with phenotypic variations at high rate and even caused some local species to become near extinct. Since the 1970s, genetic erosion has started all over the world, and it is spreading rapidly in Turkey, too. It is compulsory to take serious steps toward the preservation of resources against this genetic erosion and to protect the local populations. It is possible to produce new eggplant hybrids from the local genotypes in the characteristics desired by the market by taking them into breeding programs, selfing and crossing them with the other parent lines. For this purpose, the combined use of classical and modern breeding methods will lead to speed up the progress. The use of haploidy techniques among modern breeding techniques provides tremendous advantages for breeders, especially since it allows the production of 100% homozygous pure lines only in one generation.

2. Haploidization and doubled haploid (DH) technique

If the number of chromosomes in the somatic cells of the plant species is as much as the number of chromosomes found in their generative cells, these plants are called haploid (with n chromosome, single chromosome set), and the process of obtaining haploid plants is called haploidization. The implementation of chromosome doubling of haploid plants with spontaneous doubling or certain antimitotic chemicals brings the chromosome numbers to the normal chromosome numbers and 100% homozygous plants are produced. This stage is called doubled haploidy (DH) or dihaploidization [12]. With the use of DH lines, providing the desired properties, as a male or female parent in hybrid (F_1) seed technology, the importance of DH technology has become even better understood, and its use has become increasingly widespread.

Haploidy studies, which began with the discovery of spontaneous haploid plants in 1922 [13], gained momentum in the laboratory conditions after the 1960s, and they are becoming increasingly popular nowadays. As far as it is known, spontaneous parthenogenetic haploid plant formation in eggplant has not been encountered so far. In laboratory conditions, the first study to obtain this type of haploid plant is on anther culture and reported by Raina and Iyer [14]. It was later reported by the Chinese Haploid Research Group that developed first healthy haploid and doubled haploid (DH) eggplant plants [15]. In other anther culture studies [16–21] following them, the production of haploid and DH eggplant plants has been successfully carried out.

In the eggplant haploidy studies to date, anther culture among the androgenesis techniques has been mostly used. Protocols used in eggplant anther cultures are based on different versions of the protocol used by Dumas de Vaulx et al. [22] for pepper anther culture. Dumas de Vaulx and Chambonet [17] developed a protocol similar to pepper anther culture and started using it in eggplant and reported successful results. The success of this protocol is based on the treatment of anthers with high-temperature (+35 °C) when they are first introduced into the medium. The positive effects of high-temperature applications on haploid embryogenesis have also been reported in the studies conducted on Brassicaceae family [23, 24].

In eggplant microspore culture studies, the anthers are either cultured after being pre-treated and isolated [25], or different pre-treatments were applied to the isolated microspores [11, 26, 27]. Although the use of microspore culture was quite limited in eggplant breeding until today, the studies on this subject have been continued because of its different advantages compared to anther culture. In particular, the practical microspore culture protocols to be developed for direct embryogenesis have the potential to give great momentum to eggplant breeding studies.

While DH technique is used commercially in vegetable breeding for certain species (pepper, eggplant, melon, cucumber, squash) in Turkey, it is in the process of being improved to an effective level for a variety of vegetables (onion, leek, gherkin, tomato, watermelon, cabbage, carrot, spinach). The use of DH technique is restricted due to the response of haploid plant formation varying on the species. This is because haploid response is under the effect of various factors. For instance, each of the different genotypes within a species reacts differently to the technique used, or several genotypes respond positively, while some genotypes do not respond at all. The basis of these differences is due to genotype. In addition to the genotype, the growing conditions of plants used as donor parents, climatic effect, season, temperature, light intensity, the age of the plant, irrigation and fertilization regimes, type of stress factors and severity of exposure, and the chemicals used for plant protection have also a great effect on success. The healthier the donor plant is grown, the greater the chance of success in obtaining haploid plants. Another important issue is the nutrient medium consisting of macro- and micro-elements, vitamins, plant growth regulators (PGRs), carbohydrate sources and other unidentified substances. On the other hand, the correct determination of the period of taking the buds and the culture of the microspores at the appropriate stage also directly affects the success. Although different staining techniques are used to determine the bud stage, the most practical method used today is the 4′-6-diamidino-2-phenylindole (DAPI) staining method. As the pretreatments prior to culture, different pre-temperature applications (such as 4, 10, 28, 35°C) are applied to the anthers or buds. However, incubation in the dark at +35°C for 8 days is generally used successfully in eggplant anther cultures. To date, a number of factors affecting microspore embryogenesis have been investigated, and their effects have been demonstrated in several eggplant haploidy studies [11, 17–19, 28–34] conducted in the world and Turkey; however, this article will mainly focus on studies in Turkey.

3. Eggplant androgenesis studies conducted in Turkey

When the public and private sectors' vegetable breeding studies in Turkey were examined, it can be seen that the use of biotechnology, and in particular the use of DH technology based on embryo formation from male gamete cells, namely, androgenesis, is not going far back. The first DH study in vegetables was made in the 1980s in pepper, and then, studies on other species followed. The haploidy studies in eggplant consist of anther, isolated microspore or shed-microspore culture applications in the world. The first haploidization study in eggplant began in 1991 in Turkey with Karakullukçu's [19] anther culture study. This was followed by other anther culture [20, 35–46], isolated microspore culture [32, 34, 47] and shed-microspore culture [19] studies. Although the number of studies on this subject is small at the beginning, the efforts to obtain DH plants through anther culture have been accelerated in recent years thanks to the increase in the state-supported projects and the engagement of private sector. Androgenesis studies on eggplant have a history of 45 years worldwide and 30 years in Turkey.

4. Factors affecting success of eggplant androgenesis studies conducted in Turkey

4.1 Genotype

'Genotype' as being one of the most important factors affecting androgenesis in eggplant was also revealed with the studies conducted in Turkey [19, 20, 31, 33, 38, 43, 48, 49].

Karakullukçu [19], in various anther culture trials, reported the differences between genotypes in regard to embryo and haploid plant formations. The anthers of 13 genotypes of eggplant were cultured in the nutrient medium containing 5 mg L^{-1} kinetin and 5 mg L^{-1} 2–4 D, but only four genotypes have shown androgenesis response. While only embryoid occurred in Kemer and Prelane F_1, both embryo and haploid plants were achieved from Halep Karası and Baluroi F_1 genotypes. In this study, no embryos could be obtained from the other nine genotypes (Dourga, Pala, Şeytan, Birecik Yerlisi, Adana Topağı, Fabina F_1, Galine F_1, Black Beauty, Marfa F_1).

In the anther cultures of Alpsoy [38], conducted with 15 genotypes consisting of Pala, Kemer, Topan, Aydın Siyahı, Manisa, Adana, Urfa Yerlisi, Munica,

Baluroi, Mileda, Ancha, Leila, Barbentane, Bellissima and Purpurea, haploid embryo and plants were obtained from Kemer, Urfa Yerlisi, Adana, Barbentane and Leila genotypes and he reported that the genotype influenced the success of anther cultures.

In the microspore culture study of Özdemir [34], Faselis, Amadeo and Aydın Siyahı were used as the genotypes. Microspores were subjected to the same pretreatment and cultivated in the same nutrient medium. In Aydın Siyahı, no multinucleated structure happened, and only symmetric nucleus divisions were observed; however, symmetrical nucleus divisions and also multinucleated structures formed in Phaselis and Amadeo genotypes. It was stated that these differences are caused by genotype. It has also been reported in previous anther culture studies [38, 41, 43] that the genotype Aydın Siyahı has the low ability to form embryos.

Using the anthers of Yamula, Karabaş F_1, Malkara F_1, Çantalı F_1, and Tatlıcan F_1 genotypes, as explants, Doksöz [20] obtained 136 and 25 embryos and 77 and 15 plants only from Yamula (9.4%) and Karabaş F_1 (1.73%) genotypes, respectively.

According to the literature, there are big differences in embryo formation and haploid plant yield responses of genotypes, even though all the procedures starting from the growing conditions of donor plants until obtaining haploid plants are the same. The basis of genotype-related differences was associated with different internal amino acid contents of different genotypes by Dunwell [50]. As a result of different androgenesis studies, it has been demonstrated that the genotype effect is related to the genetics of the plant and that it is not possible to alter the androgenetic response caused by the genotype even if the conditions of donor plant growth, medium, culture and other factors affecting androgenesis were optimized [12, 51].

Thanks to genetic studies conducted in different plants, it is known that haploid formation is under the effects of genes, and in some species, certain genes initiating haploid formation were identified. For example, in some in vivo studies, *ig* gene in the corn [52] and *hap* gene in the barley [53] were defined as the genes responsible for haploid formation [54]. Thus, in vitro androgenesis is also under genetic control, and this property can be transferred to F_1 progeny by crossing androgenic genotypes with non-androgenic ones. In this respect, Tuberosa et al. [55] cultured the anthers of 8 different eggplant cultivars collected from different countries and 16 hybrid genotypes obtained from their crosses. While parents formed embryos at 17.3%, hybrids generated 42% embryo formation. A similar study was conducted by Başay and Ellialtioglu [44] in Turkey. The researchers examined the androgenesis response of the F_1 hybrid plants, which were obtained from the crosses among responsive genotypes Topan and Halep Karası and unresponsive three different genotypes. From the crosses, haploid plants were obtained from Topan × Vd-1, while embryo and haploid plant occurred from Topan × Teorem F_1 and Teorem F_1 × Topan crosses. These and similar studies show that the success of androgenesis in eggplant is highly dependent on genotype.

One of the most commonly used genotypes as the donor plant in eggplant androgenesis studies in Turkey is Aydın Siyahı. The first embryo formation from this genotype (1.25%) was reported by Başay et al. [41]. However, in other studies [34, 38, 43], the androgenic response of Aydin Siyahı is reported to be lower.

From the anther culture studies of Yücel [56] on *Solanum torvum*, only callus was obtained from Aydın Siyahı and Kemer cultivars in 2012. Haploid plants were produced by using DDV protocol at 4.6% from Aydın Siyahı in 2014. The highest haploid plant yield (36.4%) was obtained in the fall season of 2015 from the anthers of Aydın Siyahı cultured in the medium containing 1 mg L^{-1} 2,4-D + 1 mg L^{-1} kinetin. This yield was followed by Kemer cultivar (33.8%)

cultured in the medium supplemented with 5 mg L^{-1} 2,4-D + 5 mg L^{-1} kinetin. Of the 79 plants obtained from above anther cultures and examined in terms of ploidy level, 60 were identified as haploid, 13 were diploid, 4 were triploid, 1 was tetraploid and 1 was mixoploid.

In the most recent study, Vural [46] used A117 F_1, Anamur F_1 and Darko F_1 in the anther culture studies. The most embryo response was obtained from A117 F_1 up to 320 embryos/100 anthers.

4.2 Growing conditions of donor plants

Even though the androgenic response of the donor genotype is high, if the conditions of the growing environment are not suitable, the chance of success decreases. The climatic conditions applied during the in vitro haploid culture of a plant should generally be consistent with the environmental conditions required for the cultivation of that plant, except for special temperature shocks. Dunwell [57] stated that the growing conditions of donor plants affect the development of microspores and hence the embryo yield and that successful results will only be achieved when the appropriate temperature, light intensity and lighting period for the plant are optimized. Therefore, although the optimum environmental conditions in donor plant growth vary depending on plant species, various parameters such as temperature in the growth season, amount of daily lighting and light intensity reaching to plant, amount of CO_2 in the environment, fertilization, irrigation and other cultural practices must be met at the right time and in the right amount [12]. Suitable environmental conditions for eggplant cultivation are stated as places where the temperature is 15–20°C at night and 21–30°C at daytime and where the lighting time and light intensity are high [58]. It should also be considered that artificial lighting is not as efficient as sunlight [59]. As another important issue, pesticide applications of donor plants should be ceased at least 3-4 days before the culture to prevent microspores from getting stressed during their development [38, 60].

The effect of growing season of the donor plants is important for the androgenic response of anthers. In the last anther culture study conducted in Turkey, Vural [46] compared the spring- and autumn-season anther cultures in her thesis. Interestingly, the performances of anthers grown in the autumn gave much more successful results. Furthermore, the same genotype may show different responses under different conditions. The best example of this is the eggplant genotype called 'Dourga'. Although this genotype showed high success in forming embryos and haploid plants in the study of Dumas de Vaulx and Chambonnet [17], it showed low success in embryo formation in studies of Tuberosa et al. [55] and Rotino et al. [61]. Also, Karakullukçu [19] reported that it did not generate embryos at all. This study with Dourga and other genotypes grown at two different locations (Adana and Ankara) revealed once again the importance of the growth conditions of donor plants and climate differences. Different results were obtained from this study even though the anthers were cultured in the same laboratory conditions by the same person and the same practices were carried out. Karakullukçu [19] also reported that the anthers taken from the buds of plants grown in short days and low-temperature conditions (greenhouse conditions in winter, especially in December and January) did not produce embryos.

Alpsoy [38] could not receive any androgenesis response from the plants cultivated in the greenhouse in 1994 and 1995. However, the embryo and haploid plants were able to be obtained from the plants grown in field conditions in 1996 and 1998. The researcher obtained embryo and haploid plants in 5 genotypes (Kemer, Urfa Yerlisi, Adana, Barbentane, Leila) out of 15 genotypes by optimizing cultivation conditions with different applications in Bursa and Ankara locations. This study emphasizes the importance of cultivation conditions of the donor plant and the need to optimize conditions.

The first embryo formation in Aydın Siyahı (1.25%), which is one of the most focused genotypes in eggplant androgenesis studies in Turkey, was reported by Basay et al. [41] who cultured the anthers of the donor plants grown in humid and temperate greenhouse conditions in Yalova province. With the same genotype, no success was obtained in Ankara and Yalova locations at the same time, which shows again that donor plant growth is affected by the environmental growing period conditions. In the same study in 2015, it was shown that androgenetic embryos at 36.4% ratio were obtained in warm and humid Adana province conditions from genotype of Aydın Siyahı.

The effect of growing season of the donor plants is important for the androgenic response of anthers. In the last anther culture study conducted in Turkey, Vural [46] compared the spring- and autumn-season anther cultures. Interestingly, the performances of anthers grown in the autumn gave much more successful results.

The ratio of androgenesis response of the donor plants grown in natural growth conditions is higher than that of the plants grown in artificial conditions such as climatic chambers or aquacultures. However, compared to the plants grown in natural conditions in open fields, higher anther culture success is obtained from the plants grown in the soil in greenhouses under optimized conditions [12].

In the microspore and anther culture studies conducted by Özdemir Çelik [47] in two F1 cultivars (A117 and Amadeo F1), DH plants and lines were produced. In this first successful microspore culture study conducted in Turkey, appropriate culture season has been determined, and haploid plants were produced indirectly from callus regeneration. Significant differences were found among the genotypes in androgenic response not only in anther culture but also in microspore culture. These developments would allow for new prospects for the future eggplant androgenesis studies.

4.3 Developmental stages of buds and microspores

The other important factor affecting the androgenesis success is determination of the suitable microspore development stage for the culture. There will be no embryogenic induction from the microspores unless they are cultured in appropriate development stage. In different eggplant androgenesis studies, it has been reported that no progress took place in microspores cultured in the early or late stages, and also necrosis formed starting from the first days of culture [18, 19].

The suitable microspore stage for androgenesis studies varies to species, genotype, growing conditions of donor plants and the androgenesis technique used [62], which requires a cytological examination step prior to culture on the basis of each species and even genotypes within each species. In this step, one of the classic staining methods such as acetocarmine [60], Feulgen [63] or taking sections with paraffin [64, 65] can be used. However, if the laboratory infrastructure is appropriate, the easiest, fastest and most reliable staining method using DAPI (4′,6-diamidino-2-phenylindole) [66] specific to DNA should be preferred [67]. In various studies carried out in different species, it has been shown that the appropriate microspore covers the time period beginning from the formation of the tetrad's after meiosis to the starch accumulation after the first mitotic division [68, 69].

In order to detect the appropriate microspore stage and bud morphology of four different eggplant genotypes, Karakullukçu [19] primarily divided buds into eight different groups according to their morphological characteristics and, then, determined the microspore development stages of these buds with paraffin and acetocarmine methods. Four eggplant genotypes (Baluroi F_1, Prelane F_1, Pala and Kemer) were used in this study where anthers of the suitable bud groups are cultured on equal conditions and pre-treated at 35°C for 8 days in dark conditions. While all anthers in the first, second, third, seventh and eight groups among the anther groups turned to dark color and did not show any development, the anthers in groups 5 and 6 were found to be in the correct stage for anther culture. The morphological appearance of buds in these stages was as follows: The sepals and petals were equal for the fifth group, while the sepals were slightly opened, and the petals were extended for 1–2 mm for the sixth group [35]. The appropriate microspore stages in these groups were determined as the uninucleate microspores before the first pollen mitosis or binucleated microspores at the beginning of cytokinesis.

In addition to these indicators in eggplant, anther color can also be a determining feature for the buds in the appropriate stage. Considering the fact that bud selection based on only bud morphologies may be inaccurate with the aging of the plants, it is also recommended to look at the color of the anthers. It is stated that the color of the anthers at the right stage for eggplant anther culture is greenish yellow, while yellow and dark yellow anthers are in late stage, and yellowish green anthers are in the early stage [43]. These criteria and phenotypic markers determined by Karakullukçu [19] constitute the basis for the bud selection phase of all androgenesis studies made in eggplant in Turkey.

In Şeker's study [70], in which DAPI staining technique was used to determine the appropriate microspore stage, DAPI stain and phosphate-buffered saline (PBS) buffer solutions were prepared, and PBS buffer solution was autoclaved. Anthers of each of different sized buds were crushed with PBS buffer in the petri dish. After the second addition of PBS buffer together with DAPI stain onto the crushed anthers, they were incubated in dark conditions for 5 min. Then, 1–2 drops of incubated microspores were placed on a slide and monitored under a fluorescent microscope. As the result of the study, the most appropriate microspore stage for anther culture was determined to be the late uninucleate microspore development stage. Bud morphology of this stage was confirmed as that the calyx and the corolla in the flower buds should be in the same length, or the corolla may be a little longer.

4.4 The nutrient media and additives

In the culture media prepared for androgenesis studies, several components in different proportions are added to induce microspores to divide and form embryos. These components cover macro- and microelements, vitamins, amino acids, carbohydrates as the energy source, PGRs, solidifying agents and substances called unidentified substances such as activated carbon, coconut milk, etc. In addition, other factors like how to prepare the nutrient medium (solid, biphasic or liquid) and the pH of the medium should be handled with care. In the eggplant androgenesis studies in Turkey, C and R media of Dumas de Valux et al. [22] were used as the basal nutrient media in all anther cultures. Differently, Özzambak and Atasayar [71] used MS [72] and NN [73] media, while Alpsoy [38] and Doksöz [20] investigated the effect of MS medium as well as DDV-C and DDV-R media. In the microspore culture studies [32, 34, 47], NLN [74] medium was used.

It is mandatory to use PGRs to convert gametophytic development to sporophytic development in eggplant microspore embryogenesis. For this purpose, initially an auxin (such as 2,4-D, NAA, IAA, IBA, etc.) is needed, while the cytokinin (such as kinetin, BAP, zeatin, etc.)-type PGR is required in the regeneration phase. In eggplant androgenesis studies, sucrose was generally preferred as the carbon source, whereas maltose and glucose were used in some studies. In the initial stage of culture, adding of sucrose at 12% encourages the formation of embryos and haploid plantlets in eggplant [17, 19].

In the first eggplant anther culture study in Turkey, Karakullukçu [19] examined the effects of sucrose and glucose, different types and concentrations of PGRs and activated charcoal in different ratios on 13 genotypes. Among the different medium trials, the nutrient medium consisting of 120 g L^{-1} sucrose, 5 mg L^{-1} 2–4 D and 5 mg L^{-1} kinetin was determined as the most successful as in the original DDV medium. In this, embryos were obtained at a rate of 12.1% from Balouri F_1, 3.8% from Halep Karası and 1.5% from Kemer. No positive response was received from trials with activated charcoal. In this study [19] a total of 22 embryos and 13 haploid plants were achieved from all genotypes.

Özzambak and Atasayar [71] investigated the effects of BAP (0.1, 4, 8, 10, 15 mg L^{-1}), kinetin (1, 2, 4 mg L^{-1}), 2,4-D (2 mg L^{-1}) and NAA (0.1 and 2 mg L^{-1}) added to the MS and NN basic nutrient media on callus formation in an anther culture study. The highest callus ratio (8%) was obtained from the medium consisting of NN + NAA 2 mg L^{-1} + kinetin 1 mg L^{-1} + 40 g L^{-1} sucrose.

Alpsoy [38] tested different combinations of NAA (0.3, 1, 2, 4 mg L^{-1}), BA (0.7, 1, 3 mg L^{-1}), kinetin (0.1, 1, 5 mg L^{-1}) and 2,4-D (5 mg L^{-1}) as PGR in his anther culture study in which MS and DDV-C were used as the basic nutrient medium. As the result, DDV-C medium with 5 mg L^{-1} 2,4-D and 5 mg L^{-1} kinetin was found to be the most successful medium for haploid plant production as in Karakullukçu's study [19], which was followed by MS medium with 4 mg L^{-1} NAA and 1 mg L^{-1} kinetin.

Ellialtioglu et al. [75] used a total of 36 eggplant genotypes consisting of local accessions and commercial F_1 hybrid cultivars in an anther culture to increase haploid plant yield to be used as parents in the hybrid breeding. The protocol developed by Dumas de Vaulx and Chambonnet [17] was implemented in this study which is carried out in the Serene Laboratories of Dikmen Agriculture Co. According to the DDV protocol. The anthers containing uninucleate microspores were first cultured in C medium and kept in darkness at +35°C for 8 days and in photoperiod at +25°C for 4 days and then transferred to R medium. Haploid embryos were seen after 30–50 days and transferred to MS medium. The haploid embryo formation rates ranged from 0 to 45% depending on the genotypes. There was no embryo development in the genotypes with foreign origin, whereas more or fewer haploid embryos were formed in local genotypes.

In the anther culture study of Doksöz [20], MS and DDV-C media with vitamin B12 (0.03 mg L^{-1}) were compared. 2,4-D (0.01 mg L-1) + kinetin (0.01 mg L^{-1}) was used in the DDV-C medium, while only kinetin (0.1 mg L^{-1}) was added in the DDV-R medium. In the result, DDV-C medium was found to be more successful in terms of embryo yield than MS medium.

Ellialtioglu et al. [76] cultured the anthers of Tombak, Malkara and Mabel eggplant cultivars on DDV-C medium. Maltose and sucrose were used as the carbon source, and different concentrations of kinetin and 2,4-D or NAA and BAP combinations were tested as PGRs. The ratio of haploid embryo formation ranging between 0 and 59.6% differed according to the 'cultivar × carbon source × PGR applications' basis. The highest haploid embryo formation frequency occurred in Mabel cultivar, cultured in C medium with 120 g L^{-1} sucrose and 5 mg L^{-1} 2,4-D and 5 mg L^{-1} kinetin.

Geboloğlu et al. [45] compared the effects of different types of carbohydrate sources and PGR concentrations on anthers cultured in DDV-C medium supplemented with 0.03 mg L^{-1} vitamin B12. Different combinations and concentrations of sucrose (30, 60, 90, 120 and 150 g L^{-1}), honey (30, 60, 90, 120 and 150 g L^{-1}), kinetin and 2,4-D (1, 3 and 5 mg L^{-1}) were tested. Then, anthers were transferred to DDV-R medium supplemented with 30 g L^{-1} sucrose and 0.1 mg L^{-1} kinetin and subcultured at the fourth week. The highest embryo yield was obtained from the application of 120 g L^{-1} sucrose +1 mg L^{-1} kinetin +3 mg L^{-1} 2,4-D in Yamula genotype (10.7 embryos/10 anthers). In this study, the effect of 'honey' was investigated for the first time in eggplant androgenesis in the world. Although the results obtained from the honey are lower than the sucrose applications, it has been reported that the honey concentrations can be optimized and the protocol can be improved.

In the studies of Ellialtioglu et al. [77] in which anthers of 12 eggplant genotypes are cultured in different laboratories in Ankara, Antalya and Tokat, the anthers with uninucleate microspores, determined by DAPI, are placed on DDV-C medium supplemented with different PGR concentrations. The cultures were first subjected to the heat shock at 35°C in dark conditions for 8 days and then moved to photoperiod conditions at 25°C for 4 days. At the end of fourth day, cultures were transferred to DDV-R medium. In the result, the highest embryo formation (38.4%) was determined in the anthers of Anamur F1 cultured in DDV-C medium containing 120 g L^{-1} sucrose, 1 mg L^{-1} 2,4-D and 1 mg L^{-1} kinetin. Also, an interaction was found between the most suitable hormone combinations in compliance with the genotypes.

Vural [46] compared the effect of several culture media modified from DDV-C and DDV-R medium with the addition of certain carbohydrate sources and unidentified substances.

As to the media used in the microspore cultures conducted in Turkey, Bal et al. [32] who made the first study modified a protocol used for tobacco microspore culture and tested this protocol on Bambino eggplant cultivar. According to the modified protocol, microspores were pre-cultured in the B medium and then transferred to the AT3 medium containing 0.25 M maltose. No embryo was formed, but symmetrical nucleus divisions and multinucleated structures were detected in the study.

In the ovary co-cultured microspore culture study of Özdemir [34], the effects of different concentrations of 2,4-D, kinetin, NAA and BAP in NLN medium were investigated on microspores of three different genotypes (Phaselis, Amadeo and Aydın Siyahı) cultured together with wheat ovaries. As for the pretreatment, eggplant anthers were incubated in 0.3 M mannitol solution under dark conditions at +35°C for 8 days, and then microspores were isolated from the anthers at the end of the eighth day and cultured in NLN medium supplemented with 5 mg L^{-1} 2,4-D + 5 mg L^{-1} kinetin or 5 mg L^{-1} NAA + 5 mg L^{-1} BAP for embryogenic stimulation. After that, microspores were co-cultured with wheat ovaries in NLN medium containing 0.5 mg L^{-1} 2,4-D + 0.5 mg L^{-1} kinetin or 0.5 mg L^{-1} NAA + 0.5 mg L^{-1} BAP. In the study, a combination of kinetin and 2,4-D was observed to be more effective in inducing eggplant androgenesis. Multinucleated structures were obtained only in the medium containing 0.5 mg L^{-1} 2,4-D + 0.5 mg L^{-1} kinetin + ovary, while no embryo and plant were formed.

4.5 Pretreatment shocks applied to cultures

In order to encourage embryo formation in androgenesis studies, the buds prior to culturing or the anthers or microspores after their transfer to the nutrient medium are subjected to different pre-temperature shocks. In addition to the commonly used temperature shocks, keeping under dark–light conditions at different times, using different rates of different PGRs, starvation applications, high osmotic pressure, centrifugation, ethanol treatment, low atmospheric pressure, the use of radiation sources such as UV and Co_{60} and the use of various chemicals are among the other applications. The most common shock application used in eggplant androgenesis is the incubation of cultured anthers in high temperatures (35°C) in the first days.

Dumas de Valux et al. [22] who observed the positive effects of 35°C temperature applied in the first days of pepper anther cultures established a similar experiment in eggplant according to their previous results. The anthers cultivated at 35°C in dark conditions for the first 8 days of culture gave higher success rate than the control anthers cultivated at 25°C [17].

In the first eggplant anther culture study in Turkey, after the application of 12, 24 or 48 h of cold shock at 4°C to the buds, the dissected anthers were transferred to the culture medium and subjected to heat shock at 35°C for 8 days [19]. All of the cold-pretreated anthers turned to black and could not further develop. In another experiment, the anthers were exposed to different heat shock applications at 25, 30 and 35°C for 4 and 8 days after culture in DDV-C medium, and then they were kept in the climate room with 16-h light/8-h dark photoperiod regime at 25°C for 12 days without cold application to the buds. After 12 days, the cultures were transferred to DDV-R medium. In the trials, there were regular increases in the rate of embryogenesis with the increase in temperature. Based on these results, it was reported that cold pretreatment was not suitable for eggplant anther cultures, and the application of 35°C in 8 days under dark conditions was more successful to encourage embryogenic development than other applications.

Alpsoy [38] could not obtain any embryos from the anthers cultivated without shock pretreatment, during 1994–1995. In the experiments in 1996–1998, he achieved embryo and haploid plant formation from anthers pretreated at 35°C in dark for 8 days and reported that high-temperature application to anther cultures in dark had a direct effect on the success of androgenesis.

Ellialtioglu and Tıpırdamaz [37] applied cold shock to the flower buds of Kemer cultivar at 4°C for 80 h or 9°C for 5 days, besides the control group to which any shock was not applied. In addition, they investigated the effects of activated charcoal, added to the nutrient medium, on the amount of internal abscisic acid (ABA), in the anthers. Cold shocks and activated charcoal decreased the amount of ABA in eggplant anthers but did not have a positive effect on embryo formation. The formation of the embryos was provided by only the pre-treatments of anthers with heat shock at 35°C in dark conditions for 8 days and the control group (7.75%).

In anther culture studies of Doksöz [20], 24 h of cold shock at 4°C were applied to the flower buds as pre-shock application. The post-culture incubation of the anthers was carried out at 9 or 35°C for 8 days in the dark. As in the previous study [37], the anthers coming from control group buds that were not pretreated were found to be more successful. Generally, no result was obtained from the anthers subjected to 9°C for 8 days in the dark. However, the embryo and regenerated plant yield of the anthers pre-shocked at 35°C in dark conditions for 8 days has the highest value consisting of 161 embryos and 88 plants.

In microspore culture study of Bal et al. [32], the isolated microspores of Bambino cultivar were first subjected to 4, 25 or 33°C for 2 days in R medium, then transferred to the AT3 medium and cultured at 25°C in the dark. No embryo was formed, but the symmetrical divisions and the formation of multinucleated structures (19.4%) were observed only in the microspores pretreated at 32°C for 2 days. As the result, it was stated that the modified tobacco protocol was effective and the high-temperature shock as the pre-treatment had an inducing effect for eggplant microspore embryogenesis.

According to the studies conducted in Turkey and the world, the cold shocks applied to the eggplant buds taken at the appropriate stage did not generally resulted in positive response for microspore embryogenesis. On the contrary, it is generally accepted that high-temperature shocks such as 35°C for 8 days at dark conditions have positive effects on induction of microspore embryogenesis and regeneration.

4.6 Culture conditions

Ellialtioglu et al. [77], of which details were mentioned above previously, compared the growth performance of haploid embryos cultured under fluorescent lamps or LED lighting conditions. The ratio of haploid embryo formation ranging between 0 and 38.4% was reported to differ under the interaction of 'cultivar × light source × PGR application'.

After the anthers or microspores are isolated and cultured in a nutrient medium and a heat shock treatment is applied, the culture conditions based on climate data of the growth environment greatly affect the androgenesis success. The temperature and light to be two important variables of environmental conditions should be optimized. In in vitro conditions, light intensity can be used between 300 and 10,000 lux depending on the plant species, explant type, nutrient medium [78] and culture stage. It is recommended that light intensity should be low for anther cultures or even in the dark during the early days of culture. Thus, anthers are usually cultivated in the dark during the first period of culture, and then the 300–1,500 lux light intensity is applied to the cultures for embryo development. The embryos are germinated at 2,000–3,000 lux [12].

As a result of the large number of studies made in both the world and Turkey, the climatic conditions preferred for eggplant anther cultures are:

- Firstly, to exposure the cultures in DDV-C medium to a heat shock pretreatment consisting of 35°C in the dark for 8 days temperature shock

- As additional pretreatment, to keep the cultures in the same nutrient medium for 4 more days but under photoperiod conditions consisting of 16/8 h of light/darkness at 25°C

- At the end of 12 (8 + 4) days, to transfer the cultures to DDV-R medium and culture them under photoperiod conditions consisting of 16/8 h of light/darkness at 25°C

5. Ploidy detection and chromosome doubling

The use of haploid plants, which are very valuable for breeding, depends on chromosome doubling of these plants, thus bringing their chromosome number to the number before haploidization and making them 100% homozygous. Following haploidization procedures, chromosomal set numbers of the regenerated plants are determined by using different ploidy analysis techniques. Although various methods have been used to determine the ploidy in eggplant, flow cytometry analysis is becoming more widespread, since it is more practical and faster.

For chromosome doubling, haploid plants are exposed especially to colchicine or oryzalin, trifluralin and other chemicals with antimitotic effect in either in vitro or in vivo conditions. It has been reported that the lanolin treatment with 0.5%

colchicine for 48 h in darkness to the buds starting from the secondary axillary buds in in vivo conditions resulted in 50–70% doubling [21].

In Turkey, Ellialtioglu et al. [40] compared in vitro and in vivo colchicine treatments. In in vitro treatment, micro shoots were incubated in the colchicine solution containing 0.5 or 1%, for 1 or 2 h. In in vivo treatment, the acclimatized haploid plants transferred to greenhouse were pruned, and then the cotton pieces absorbed with same concentrated colchicine (0.5 or 1%) were placed in their axillary buds for 1 or 2 h. One hundred percent chromosome doubling and dihaploid shoots can be achieved in in vivo by using both 0.5% colchicine treatment for 2 h and 1% colchicine for 1 h from the well-grown haploid plants, of which leaves were pruned after the development of four to five nodes. However, this way takes longer time than in vitro method. Each of the waiting steps in in vivo method consisting of the plant growth in greenhouse, shooting of the buds, formation of the first flower and finally selfing requires some time. In addition, all the procedures may need to be repeated if the doubling did not happen. The in vitro method has been identified as a way that can be an alternative to in vivo and may even be seen as advantageous for saving time. Although there have been a few losses in in vitro method during the applications and acclimatization, these losses will remain unimportant when working with a large number of materials. It has been shown by Ellialtioglu et al. [40] that especially the application of 0.5% colchicine for 2 h in in vitro can be used for dihaploidization of haploid shoots in the eggplant. The most important advantage of this method is that the regenerated plants developed from this application thrive as the complete diploid plants, which offers earlier flowering and thus the possibility of earlier selfing.

In the study conducted by Özdemir Çelik [47], the in vitro method described by Ellialtioglu et al. [40] was used to double the chromosome numbers of haploid plants obtained from anther culture, and chromosome doubling was successfully performed.

6. Microspore embryogenesis efficiency

Among the eggplant androgenesis studies conducted in Turkey so far, anther culture technique has been used predominantly, and feasible androgenetic results are generally derived from these cultures. The highest microspore embryogenesis rate in all the anther culture studies conducted in Turkey was recorded as 320 embryos/100 anthers from A117 F_1 [46] cultivar. This was followed by 59.6% from Mabel cultivar [76], 38.4% from Anamur F_1 [77] and 36.4% [56] from open pollinated cultivar Aydın Siyahı.

As for the efficiency of microspore embryogenesis in eggplant androgenesis studies in the world, as far as we know, the highest embryogenesis rates obtained from anther culture studies were 237.5% (237.5 embryos/100 anthers) from DH36 line developed from Bandera F1, 146.5% from Bandera F1, the parent of DH36 line [79], 60% from Ecavi F1 [33] and 53% from Cristal F1 [80].

In the studies of isolated microspore culture, the first remarkable improvement was achieved by Corral-Martinez et al. [27]. After the refinement studies, the highest success was reported by Rivas-Sendra et al. [81] from the first generation of DH population lines (DHS1 lines) developed from Bandera F1. The callus yields were obtained to be 65.08 callus mL^{-1}, 76.86 callus mL^{-1}, 92 callus mL^{-1}, 149.11 callus mL^{-1} and 267.36 callus mL^{-1} from Bandera F1 (control) and DH15, DH41, DH40 and DH36 lines, respectively. According to the literature, in order to obtain DH eggplant plants via isolated microspore culture nowadays, it is attempted to get callogenesis at first and then to develop regenerated plants through organogenesis [47, 81]. However, an efficient direct embryogenesis protocol have not yet been developed in the isolated microspore culture studies in eggplant so far.

7. Overview of an efficient anther culture protocol used in eggplant

The stages of an eggplant anther culture protocol (applied in TUBITAK TEYDEB Project No. 68989 conducted by Antalya Tarim Co. R&D Center) which can be used effectively and practically in breeding studies were briefly summarized in **Figures 1–9**.

Figure 1.
Donor plant growing under biotic and abiotic stress-free conditions.

Figure 2.
Suitable bud morphology for eggplant anther culture.

Figure 3.
Culture of anthers under aseptic conditions after appropriate surface sterilization treatment and nutrient medium preparation.

Figure 4.
Pretreatments (temperature applications) and incubation of cultures.

Figure 5.
Direct androgenesis and haploid embryo formation.

Figure 6.
In vitro regeneration and development of haploid plantlets.

Figure 7.
Acclimatization of in vitro regenerated plantlets.

Figure 8.
Plants obtained from anther culture at the different ploidy levels. Application of chromosome doubling in the greenhouse.

Figure 9.
Colchicine treatment with lanolin. Fruits on the DH branches.

8. Conclusion and recommendations

Each of the androgenesis methods including anther, shed-microspore and isolated microspore culture has advantages or disadvantages from each other. Among the eggplant androgenesis studies in the world, there are applicable protocols for anther culture, the most commonly used method, while the use of shed-microspore culture is not widespread and has not been given enough attention. As for the isolated microspore culture, the studies on indirect microspore embryogenesis have been tried to develop for a long time, and successful protocols have been improved in recent years. However, a practical direct embryogenesis protocol in isolated microspore culture is still missing.

When the recent eggplant androgenesis studies were evaluated, in the latest anther culture study, Vural informed the highest embryo yield to be 320 embryos/100 anthers [46]. Another striking improvement was performed by a Spanish research group who recently made major advances in eggplant microspore culture, that the microspore embryogenesis response is generally caused by genotype rather than by the culture protocol. This group has developed a superior DH eggplant line with very high androgenic response from a DH population improved from a hybrid cultivar with high androgenic response and their inbred lines. The haploidy performance of this superior line has folded the performance of its hybrid parent 1.6 times in anther culture and 4.1 times in microspore culture. These important improvements have shown the significance of population development and the use of DH lines to obtain haploid plant in eggplant, which is still considered to be recalcitrant between tomato and tobacco in Solanaceae [82].

It has long been known that hybrids and DH lines exhibit higher haploid performance than their parents. Therefore, at the beginning of the recommendations to increase the efficacy of microspore embryogenesis in eggplant, the development of populations, carrying the genes responsible from high androgenic performance, comes first by means of the crosses between the high androgenic hybrids, genotypes and in particular DH lines and non-androgenic elite genotypes. Other suggestions are to focus more on the isolated microspore culture technique which has a much higher embryo or callus yield potential than the anther culture and to improve more practical protocols in particular for direct embryogenesis. Therefore, it may be dwelt on the development of shed-microspore culture protocols since it has an application between anther culture and isolated microspore culture. However, shed-microspore culture is more practical than isolated microspore culture and has the potential for higher embryo yield than anther culture. In addition, the culture media and also the culture conditions can be optimized to increase embryo yield, embryo quality and plant regeneration rate in eggplant androgenesis by different chemical, biochemical, PGR or especially phytohormones that were tested and proved in other plant species.

Finally, it is considered beneficial to increase in vitro androgenesis studies also in wild *Solanum* species to develop rootstock.

Icons and abbreviations

Icons

%	percent
°C	degrees Celsius
Co_{60}	cobalt 60
CO_2	carbon dioxide
$g\ L^{-1}$	gram/liter
M	molar
$mg\ L^{-1}$	milligrams/liter
L	liter
UV	ultraviolet

Abbreviations

ABA	abscisic acid
AT3 medium	consisted of 13 mM KNO_3, 8.6 mM $(NH_4)_2\ SO_4$, 2.9 mM $KH_2\ PO_4$, 1.1 mM $CaCl_2\ .2H_2O$, 0.7 mM $MgSO_4\ .7H_2O$, 10 mM MES buffer, 8.6 mM glutamine, 0.25 M maltose and Fe-EDTA, vitamins and microelements according to Murashige and Skoog [72]
$AgNO_3$	silver nitrate
AVG	aminoethoxy-vinylglycine
BA	N6-benzyladenine
BAP	6-benzylaminopurine
B medium	Kyo and Harada (1986) nutrient medium (consisted of KCl, 1.49 g L^{-1}; $MgSO_4.7H_2O$, 0.25 g L^{-1}; $CaCl_2$, 0.11 g L^{-1} and Mannitol (0.3 M) 54.63 g L^{-1} and 1 mM phosphate buffer of pH 7) [83]
DAPI	4′-6-diamidino-2-phenylindole
DDV-C	Dumas de Vaulx—C medium [17]
DDV-R	Dumas de Vaulx—R medium [17]
DH	doubled haploid
IAA	indole-3-acetic acid
IBA	indole-3-butyric acid
LED	light-emitting diode
NAA	α-naphthalene acetic acid
MS	Murashige and Skoog [72] nutrient medium

NLN	The nutrient medium, whose origin is from Nitsch and Nitsch [73] medium, modified by Lichter (1982) [74]
NN	Nitsch and Nitsch [73] nutrient medium
PBS buffer	phosphate-buffered saline
PGRs	plant growth regulators
2,4-D	2,4-dichlorophenoxyacetic acid

Author details

Gülsün Elif Vural[1,2], Esin Ari[2], Sinan Zengin[1] and Sekure Sebnem Ellialtioglu[3*]

1 Antalya Tarim Co. R&D Center, Antalya, Turkey

2 Department of Agricultural Biotechnology, Faculty of Agriculture, Akdeniz University, Antalya, Turkey

3 Department of Horticulture, Faculty of Agriculture, Ankara University, Ankara, Turkey

*Address all correspondence to: sebnemellialti@gmail.com

References

[1] Kalloo G. Eggplant (*Solanum melongena* L.). In: Genetic Improvement of Vegetable Crops. New York: Pergamon Press; 1993. pp. 587-604

[2] Daunay MC, Lester RN, Ano G. Eggplant. In: Charrier A, Jacquot M, Hamon S, Nicolas D, editors. Tropical Plant Breeding. Enfield, NH, USA: Science Publishers Inc.; 2001. pp. 199-222

[3] Boyacı HF. Resistance resources and its inheritance against to fusarium wilt in eggplants [PhD thesis]. Adana, Turkey: Cukurova University, Institute of Natural and Applied Sciences, Department of Horticulture; 2007. 97 p (in Turkish)

[4] Choudhury B, Simmonds NW, Smartt J. Eggplant. In: Simmonds NW, editor. Evolution in Crop Plants. Smartt J, New York: John Wiley & Sons; 1995. pp. 464-465

[5] Doğanlar S, Farry A, Daunay MC, Lester RN, Tanksley SD. Conservation of gene function in the Solanaceae as revealed by comparative mapping of domestication traits in eggplant. Genetics. 2002;**161**:1713-1726

[6] FAOSTAT. 2016. Available from: http://www.fao.org/faostat/en/ [Accessed: 01 May 2018]

[7] Günay A. Special Vegetable Growing (Özel Sebze Yetiştiriciliği). Vol. 4. Ankara, Turkey: Çağ Publish, Ankara University, Faculty of Agriculture Department of Horticulture; 1992. p. 6

[8] Galenbus VL. Anatolian eggplants (Anadolu Patlıcanları). Türkiye'nin Zirai Bünyesi (Anadolu). Türkiye Şeker Fabrikaları A.Ş Neşriyatı. 1951;**20**:738-745 (in Turkish)

[9] Güvenç İ. Vegetable Growing: Fundamental Knowledges, Storage and Production (Sebzecilik: Temel Bilgiler Muhafaza ve Yetiştiricilik). Avşar Campus, Kahramanmaraş, Turkey: Kahramanmaraş Sütçü İmam University, Faculty of Agriculture, Department of Horticulture; 2016. 206 p (in Turkish)

[10] Porcelli S. Problemi e prospettive delia melanzana. Cenni introduttivi. Agricoltura Ricerca. 1986;**60**:1-2

[11] Rotino GL. Haploidy in eggplant. In: Jain SM, Sopory SK, Veilleux RE, editors. *In vitro* Production in Higher Plants. Vol. 3. Dordrecht, Netherlands: Kluwer Academic Publishers; 1996. pp. 115-141

[12] Ellialtioglu Ş, Sarı N, Abak K. In: Babaoğlu M, Gürel E, Özcan S, editors. Plant Biotechnology-I Tissue Culture and its Applications. Haploid Plant Production (Bitki Biyoteknolojisi-I Doku Kültürü ve Uygulamaları. Haploid Bitki Üretimi). Konya, Turkey: S.Ü.Vakfı Yayınları; 2001. pp. 137-189

[13] Blakeslee AF, Belling J, Farnham ME, Bergner AD. A haploid mutant in the Jimson weed, *Datura stramonium*. Science. 1922;**55**:646-647

[14] Raina SK, Iyer RD. Differentiation of diploid plants from pollen callus in anther cultures of *Solanum melongena* L. Zeitschrift für Pflanzenzüchtung. 1973;**70**:275-280

[15] Research Group of Haploid Breeding. Induction of haploid plants of *Solanum melongena* L. In: Proceedings of the "Symposium on Plant Tissue Culture" Beijing, 25-30 May 1978. pp. 227-232

[16] Isouard G, Raquin C, Demarly Y. Obtention de plantes haploides et diploides par culture in vitro d'anthères d'aubergine (*Solanum melongena* L.). Comptes rendus de l'Académie des Sciences, Paris. 1979;**288**:987-989

[17] Dumas de Vaulx R, Chambonnet D. Culture *in vitro* d'anthères d'aubergine (*Solanum melongena* L.): stimulation de la production de plantes au moyen de traitements à + 35°C associés à de faibles teneurs en substances de croissance. Agronomie. 1982;**2**(10):983-988

[18] Chambonnet D. Culture D'antheres in vitro chez trois Solanaceaes maraicheres: le piment (*Capsicum annuum* L.), l'aubergine (*Solanum melongena* L.), la tomato (*Lycopersicon esculentum* Mill.) et obtention de plantes haploides [PhD thesis]. Academie de Montpellier; 1985. 90 p

[19] Karakullukçu Ş. Investigation of some factors inducing in vitro androgenesis and haploid plant development in different eggplant genotypes [PhD thesis]. Ankara, Turkey: Ankara University, Institute of Natural and Applied Sciences, Department of Horticulture; 1991. 131 p (in Turkish)

[20] Doksöz S. Development of dihaploid lines in some eggplant genotypes with *in vitro* androgenesis method [PhD thesis]. Tokat, Turkey: Gaziosmanpaşa University, Institute of Natural and Applied Sciences, Department of Horticulture; 2015. 49 p (in Turkish)

[21] Rotino GL. Anther culture in eggplant (*Solanum melongena* L.). In: Germanà MA, Lambardi M, editors. *In vitro* Embryogenesis in Higher Plants, Methods in Molecular Biology. Vol. 1359. New York: Springer Science+Business Media; 2016. pp. 453-466

[22] Dumas de Vaulx R, Chambonnet D, Pochard E. *In vitro* anther culture in red pepper (*Capsicum annuum* L.): Improvement of the rate of plant production in different genotypes by treatments at 35°C. Agronomie. 1981;**1**:859-864

[23] Hansson B. Temperature shock a method of increasing the frequency of embryoid formation in anther culture of swedc rape (*Brassica napus* L.). Sver. Utsaedsfoeren Tidskr. 1978;**88**(3):141-148

[24] Keller WA, Armstrong KG. Stimulation of embryogenesis and haploid production in *Brassica campestris* anther cultures by elevated temperature treatments. Theoretical and Applied Genetics. 1979;**55**:65-67

[25] Gu SR. Plantlets from isolated pollen culture of eggplant (*Solanum melongena* L.). Acta Botanica Sinica. 1979;**21**:30-36

[26] Miyoshi K. Efficient callus induction and subsequent plantlet formation from isolated microspore of eggplant (*Solanum melongena* L.). In: Vth International Congress of Plant Tissue and Cell Culture, France, 12-17 June 1994, Abstracts Book. 1994. pp. 3-25

[27] Corral-Martinez P, Seguì-Simarro JM. Efficient production of callus-derived doubled haploids through isolated microspore culture in eggplant (*Solanum melongena* L.). Euphytica. 2012;**187**:47-61

[28] Rotino G, Falavigna A, Fiume F, Nervo G, Restaino F. Possibility of eggplant (*Solanum melongena* L.) improvement through *in vitro* techniques. Genetica Agraria. 1987;**XLI**(3):314-315

[29] Fadel F, Wenzel G. Medium-genotype-interaction on androgenetic haploid production in wheat. Plant Breeding. 1990;**105**:278-282

[30] Dağüstü N. Factors affecting the anther culture of wheat (*Triticum aestivum* L.). Biotechnology & Biotechnological Equipment. 2002;**16**(1):30-34

[31] Alpsoy HC, Şeniz V. Researches on the in vitro androgenesis and

obtaining haploid plants in some eggplant genotypes. Acta Horticulturae. 2007;**729**:137-141

[32] Bal U, Ellialtioglu Ş, Abak K. Induction of symmetrical nucleus division and multi-nucleate structures in microspores of eggplant (*Solanum melongena* L.) cultured *in vitro*. Scientia Agricola. 2009;**66**:535-539

[33] Salas P, Prohens J, Seguí-Simarro JM. Evaluation of androgenic competence through anther culture in common eggplant and related species. Euphytica. 2011;**182**(2):261-274

[34] Özdemir B. The effects of ovary co-culture system to induce haploid embryo formation on eggplant (*Solanum melongena* L.) microspore culture Akdeniz University, Institute of Natural and Applied Sciences, Department of Horticulture [MSc thesis]. Antalya, Turkey; 2012. 63 p (in Turkish)

[35] Karakullukçu Ş, Abak K. Studies on anther culture in eggplant. I. Determination of suitable bud development stage. Doğa Turkish Agriculture & Forestry. 1993;**17**:801-810

[36] Karakullukçu Ş, Abak K. Studies on anther culture in eggplant. II. Effects of sugar and growth regulators. Doğa Turkish Agriculture & Forestry. 1993;**17**:811-820

[37] Ellialtioglu Ş, Tıpırdamaz R. The effect of the treatments for reducing of abscisic acid amount on androgenetic embryo formation in eggplant. Biyotechnology (Kükem) Journal. 1999;**24**(1):23-32 (in Turkish)

[38] Alpsoy HC. Researches on the in vitro androgenesis and obtaining haploid plants in some eggplant genotypes [PhD thesis]. Bursa, Turkey: Uludağ University, Institute of Natural and Applied Sciences, Department of Horticulture; 1999. 89 p (in Turkish)

[39] Ellialtioglu Ş. Investigations on some factors affecting haploid plant formation in anther culture of eggplant and tomatoes. TÜBİTAK Project No: TOGTAG-2607. Ankara, (Project Report unpublished); 2005. 134 p

[40] Ellialtioglu Ş, Başay S, Kuşvuran Ş. Comparison of in vitro and in vivo colchicine treatments for chromosome doubling in haploid eggplants obtained from anther culture. In: VI. National Vegetable Growing Symposium, 19-22 September 2006. Kahramanmaraş, Turkey. 2006. pp. 386-390

[41] Başay S, Ellialtioglu Ş, Şeniz V. Haploid embryo formation through anther culture in native and foreign eggplant varieties. In: Yaşar F, Çavuşoğlu Ş, Biçim M, editors. VIII. National Vegetable Growing Symposium, 23-26 June 2010, Van. 2010. 2010. pp. 588-590

[42] Başay S, Şeniz V, Tezcan H. Reactions of selected eggplant cultivars and lines to Verticillium wilt caused by *Verticillium dahliae* Kleb. African Journal of Biotechnology. 2011;**10**(18):3571-3573

[43] Ellialtioglu Ş, Başay S, Kuşvuran Ş. Investigations on the pollen dimorphism and its relationship with anther culture in eggplant. TABAD (Research Journal of Agricultural Sciences). 2012;**5**:149-152

[44] Başay S, Ellialtioglu ŞŞ. Effect of genotypical factors on the effectiveness of anther culture in eggplant (*Solanum melongena* L.). Turkish Journal Biology. 2013;**37**:499-505

[45] Geboloğlu N, Boncukcu SD, Durna P, Bayram M. Influence of sugar, honey and growth regulators on embryo induction of eggplant in anther culture. Akademic Journal of Agriculture, Ordu/Turkey. 2017;**6**:275-280

[46] Vural GE. Comparison of anter and shed-microspore cultures to

obtain haploid plant in eggplant (*Solanum melongena* L.) [master thesis]. Antalya, Turkey: Akdeniz University, Institute of Natural and Applied Sciences, Department of Agricultural Biotechnology; 2019 (in Turkish)

[47] Özdemir Çelik B. The effects of different application to induce haploid embryo and plant formation on eggplant (*Solanum melongena* L.) microspore culture [PhD thesis]. Antalya, Turkey: Akdeniz University, Institute of Natural and Applied Sciences, Department of Horticulture; 2018. 86 p (in Turkish)

[48] Karakullukçu Ş, Abak K. The response of some eggplant cultivars to anther culture. Annuals of Ankara University, Faculty of Agriculture. 1992;**42**(1-2-3-4):7-12

[49] Magioli C, Mansur E. Eggplant (*Solanum melongena* L.): Tissue culture, genetic transformation and use as an alternative model plant. Acta Botanica Brasilica. 2005;**19**(1):139-148

[50] Dunwell JM. A comparative study of environmental and developmental factors which influence embryo induction and growth in cultured anthers of *Nicotiana tabacum*. Environmental and Experimental Botany. 1976;**16**:109-118

[51] Mityko J, Andrasfalvy A, Csillery G, Fari M. Anther culture response in different genotypes and F1 hybrids of pepper (*Capsicum annuum* L.). Plant Breeding. 1995;**114**:78-80

[52] Kermicle JL. Androgenesis conditioned by a mutation in maize. Science. 1969;**166**:1422-1424

[53] Hagberg A, Hagberg G. High frequency of spontaneous haploids in the progeny of an induced mutation in barley. Hereditas. 1980;**93**:341-343

[54] Foroughi-Wehr B, Friedt W, Wenzel G. On the genetic improvement of androgenetic haploid formation in *Hordeum vulgare* L. Theoretical and Applied Genetics. 1982;**62**(3):233-239

[55] Tuberosa R, Sanguinetti MC, Conti S. Anther culture of eggplant (*Solanum melongena* L.) lines and hybrids. Genetica Agraria. 1987;**41**:267-274

[56] Yücel NK. Natural and *in vitro* hybridization studies between *Solanum melongena* L. and *Solanum torvum* Sw [PhD thesis]. Adana, Turkey: Cukurova University, Institute of Natural and Applied Sciences, Department of Horticulture; 2017. 155 p (in Turkish)

[57] Dunwell JM. Haploid cell cultures. In: Dixon RA, editor. Plant Cell Culture: A Practical Approach. Washington, DC: IRC Press, Oxford; 1985. pp. 21-36

[58] Çetinkaya Ş. Protected Cultivation of Vegetables (Örtüaltı Patlıcan Yetiştiriciliği). Vol. 104. Antalya, Turkey: BATEM; 2009. pp. 7-9

[59] Hatipoğlu R. Plant Biotechnology (Bitki Biyoteknolojisi). Adana: Cukurova University Faculty of Agriculture; 1999. 178 p

[60] Nitsch C. In: Thorpe TA, editor. Production of isogenic lines: Basic technical aspects of androgenesis, Plant Tissue Culture, Methods and Applications in Agriculture. New York: Academic Press; 1981. pp. 241-252

[61] Rotino G, Falavigna A, Restaino F. Production of anther-derived plantlets of eggplant. Capsicum Newsletter. 1987;**6**:89-90

[62] Ferrie AMR, Keller WA. Plant Tissue Culture Manuel: Production of Haploids in *Brassica* spp. via Microspor Culture. The Netherlands: Kluwer Academic Publishers; 1997. pp. 1-17

[63] Darlington CD, La Cour LF. Methoden der Chromosomen

Untersuchungten. Stuttgart, Deutschland: Frankische Verlagshandlung, W. Kellerund Ca; 1963

[64] Stösser R, Kaşka N, Anvarı SF, Eti S. Course Notes on 'Reproductive Biology in Horticulture'. Adana, Turkey; 18-22 March 1985. (Unpublished, in Turkish)

[65] Eti S. Über das Pollenschlauch-wachstum und die Entwicklung der Samenanlagen in Beziehung zum Frauchtansatz und zur Fruchtqualitaetbei der Mandarinensorte "Clemantine" (*Citrus reticulata* Blanco) [dissertation Universitaet Hohenheim]. 1987. 127 p

[66] Vergne P, Delvallee I, Dumas C. Rapid assessment of microspore and pollen development stage in wheat and maize using DAPI and membrane permeabilization. Stain Technology. 1987;**62**:299-304

[67] Ari E, Büyükalaca S. Determination of suitable flower bud morphology for androgenesis studies in *Anemone coronaria spp coccinea*. Mediterranean Agricultural Sciences. 2010;**23**(2):71-78

[68] Sunderland N, Dunwell M. Anther and pollen culture. In: Street HE, editor. Plant Tissue and Cell Culture. Blackwell Publ.; 1977. pp. 223-265

[69] Ari E. Anther culture studies on *Anemone coronaria* var. *coccinea* native to Turkey [PhD thesis]. Adana, Turkey: Cukurova University, Institute of Natural and Applied Sciences, Department of Horticulture; 2006. 169 p (in Turkish)

[70] Şeker E. Using cytological methods during anther culture [MSc thesis]. Ankara, Turkey: Ankara University, Institute of Natural and Applied Sciences, Department of Horticulture; 2015. 33 p (in Turkish)

[71] Özzambak E, Atasayar A. Anther culture of tomatoes and eggplants. Acta-Horticulturae. 1994;**336**:229-233

[72] Murashige T, Skoog F. A revised medium for rapid growth and bioassays with tobacco tissue cultures. Physiologia Plantarum. 1962;**15**:473-497

[73] Nitsch JP, Nitsch C. Haploid plants from pollen grains. Science. 1969;**163**:85-87

[74] Lichter R. Induction of haploid plants from isolated pollen of Brassica napus. Z. Pflanzenphysiol. 1982;**105**:427-434

[75] Ellialtioglu ŞŞ, Sönmez K, Evcen F, Gümrah E, Çetinkaya S. Studies on obtaining haploid plants from anther culture in different eggplant genotypes. In: 10. National Vegetable Growing Symposium, 2-6 September 2014, Tekirdağ/Turkey. 2014. (in Turkish)

[76] Ellialtioglu ŞŞ, Sönmez K, Evcen F. The effect of growth regulator and carbon source combinations on the formation of haploid embryos in eggplant anther cultures. In: Green Biotechnology Workshop, 21-23 September 2015; Darıca, Kocaeli, Turkey. 2015

[77] Ellialtioglu ŞŞ, Sönmez K, Geboloğlu N, Boyacı HF. Recent advances in anther culture of eggplants in the contexts of cytological observations and incubation conditions. In: The Eurasian Agriculture and Natural Sciences Congress, 20-23 September 2017, Bishkek, Kirghizistan. 2017

[78] Gönülşen N. Plant Tissue Cultures and Application Areas (Bitki Doku Kültürleri ve Uygulama Alanları). Menemen-İzmir, Turkey: Ministry of Agriculture, Forestry and Rural Affairs, Aegean Agricultural Research Institute; 1987. 140 p. Publ. No: 78

[79] Rivas-Sendra A, Campos-Vega M, Calabuig-Serna A, Seguí-Simarro JM. Development and characterization of an eggplant (*Solanum melongena*) doubled

haploid population and a doubled haploid line with high androgenic response. Euphytica. 2017;**213**(4):89

[80] Salas P, Rivas-Sendra A, Prohens J, Seguí-Simarro JM. Influence of the stage for anther excision and heterostyly in embryogenesis induction from eggplant anther cultures. Euphytica. 2012;**184**:235-250

[81] Rivas-Sendra A, Corral-Martínez P, Camacho-Fernández C, Seguí-Simarro JM. Improved regeneration of eggplant doubled haploids from microspore-derived calli through organogenesis. Plant Cell, Tissue and Organ Culture. 2015;**122**:759-765

[82] Seguí-Simarro JM. Androgenesis in Solanaceae. In: Germana M, Lambardi M, editors. In Vitro Embryogenesis in Higher Plants, Methods in Molecular Biology. Vol. 1359. New York: Humana Press; 2016. pp. 209-244

[83] Kyo M, Harada H. Control of the developmental pathway of tobacco pollen in vitro. Planta. 1986;**168**:427-432

Chapter 6

SSR Markers in the Genus *Pistacia*

Salih Kafkas

Abstract

Simple sequence repeats (SSRs) are one of the most powerful molecular marker systems due to abundance in the genomes, its codominant nature, and high repeatability. *P. vera* L. is cultivated species in the genus *Pistacia* due to commercial value of its edible nuts. Other species in the genus are in the wild and are important especially for rootstock sources as well as for ornamental and forest trees. There were a very limited number of SSR markers for *Pistacia* species until several years ago; however, next-generation sequencing (NGS) technology has allowed to develop plenty of SSRs since 2016 in the genus. There are currently about 1500 published SSR markers developed from cultivated *P. vera*. There are also several studies generating SSR loci from wild *Pistacia* species. In a conclusion, there are currently an adequate number of SSR markers for cultivated pistachio and that can be used in wild *Pistacia* species due to their high level of transferability rate between *Pistacia* species. These SSRs can be used for assaying diversity in natural populations, marker discovery, germplasm characterization, parental identification, genetic linkage mapping, and evolutionary studies in the genus *Pistacia*.

Keywords: *Pistacia*, SSR, microsatellite, pistachio, repeats

1. Introduction

Pistacia is a genus in the Anacardiaceae family which also contains cashew, mango, poison ivy, poison oak, pepper tree, and sumac plants [1]. The genus is estimated to be about 80 million years old [2]. It contains at least 11 species, and *P. vera* L. (pistachio) is the only cultivated one for its edible nuts [3]. In addition, its nuts are considerably larger than all the other species in the genus. The other species grow in the wild, and several of them have been used for many years as rootstock seed sources for *P. vera*. Furthermore, most of them have also been used as forest trees [4]. According to the Food and Agriculture Organization of the United Nations, the world production of pistachios in 2017 was 1,115,066 MT, ranking fifth in world tree nut production behind cashews (*Anacardium occidentale* L.), walnuts (*Juglans regia* L.), chestnuts (*Castanea* spp.), and almonds [*Prunus dulcis* (Mill.) D.A. Webb], and currently, Iran, the United States, and Turkey are the main pistachio producers in the world [5].

Pistacia species are dioecious and wind pollinated; however monoecious individuals within *P. atlantica* were also reported [6]. *P. vera* is believed to be the most ancestral species in the genus, and the other species are probably its derivatives [7]. There are two centers of diversity of cultivated pistachio: one comprises the Mediterranean region of Europe, Northern Africa, and the Middle East countries. The second comprises the Eastern part of Zagros Mountains from Crimea to the Caspian Sea. Pistachio cultivation extended westward from its center of origin to

Italy, Spain, and other Mediterranean regions of Southern Europe, North Africa, and the Middle East, as well as to China and more recently to the United States and Australia [8–10].

The *Pistacia* genus is distributed mainly across subtropical regions of the northern hemisphere and consists of both evergreen and deciduous species with shrub and/or tree-like growth habits [7]. Although *P. vera* is a commercially grown species in a number of semi-arid regions worldwide, the species remains quite underexploited when its wide native range and inherent genetic diversity are considered [8, 11, 12]. For instance, commercial pistachio cultivation is done in only a few countries in the world. Besides, pistachio production is done with a very limited number of cultivars in those countries, and most of them are seedling selections from the nature [1, 13]. This narrow genetic base in the production presents a threat in pistachio against new diseases and pests as well as changing ecological conditions. Therefore, the germplasm collections have great potentials to increase the genetic diversity and to develop pistachio cultivars for current production areas and/or to expand the regions where reliable commercial production is possible.

Dioecy and a long juvenile period are the primary difficulties encountered in breeding and genetic studies of *Pistacia*. The long juvenile period in combination with dioecious character causes large investments of time and land for characterization and evaluation of progenies in a breeding program. Furthermore, the genetic control of the most economically significant traits is not clearly understood, including disease and pest resistance, yield, nut quality characteristics, and alternate bearing in pistachio. Therefore, current technologies such as molecular markers are good facilities to overcome such difficulties in the breeding programs. Pistachio is a diploid plant which has a haploid chromosome number of $n = 15$. It is also highly heterozygous species due to dioecy [1, 14].

DNA markers have played a major role in breeding programs for several decades in plants. Several molecular markers such as randomly amplified polymorphic DNA (RAPD) [15], simple sequence repeat (SSR) [16, 17], sequence-related amplified polymorphism (SRAP) [18], amplified fragment length polymorphism (AFLP) [4], inter-simple sequence repeat (ISSR) [19], selectively amplified microsatellite polymorphic loci (SAMPL) [20], and single-nucleotide polymorphisms (SNPs) [21] have been used to assess the genetic diversity, fingerprinting, phylogenetic relationships, germplasm characterization, sex determination, genetic linkage mapping, and QTL analysis in cultivated and wild *Pistacia* species.

In pistachio, most of the cultivars in the production are earlier selections either by growers or breeders. Pistachio cultivars from controlled crosses are in a limited number, and selected genotypes in the current breeding programs are under evaluations in different countries. Using molecular tools in conventional breeding programs can be a good advance in pistachio. Initial molecular studies in pistachio were mainly on germplasm characterization by using different molecular marker techniques. Identification of molecular markers linked to sex determination was also studied that allows early selections of females in a cultivar breeding program [21–23]. Recently, the markers linked to sex have been used in a cultivar breeding program in Turkey [24].

SSRs are useful as molecular markers and very polymorphic due to the high mutation rate affecting the number of repeat units [25]. They are very useful for assaying diversity in natural populations or germplasm collections and for fingerprinting and parental identification. They are also very valuable markers especially for genetic linkage mapping and evolutionary studies [26] and have a high level of transferability between closely related species. The development of SSR markers from *P. vera* [14, 16, 17, 27–29] and from wild *Pistacia* species was performed in several studies [30–33].

The first complete genetic linkage map was constructed in pistachio by [34] who used SRAP, ISSR, and AFLP markers in an inter-specific F1 population derived from a cross between *P. vera* and *P. atlantica*. Recently, Khodaeiaminjan et al. [28] constructed the first complete SSR-based linkage map of pistachio using an intra-specific F1 population. More recently, Motalebipour et al. [35] constructed a genetic linkage map and performed the first QTL analysis in pistachio by using an inter-specific F1 population.

Next-generation sequencing (NGS) has provided a new perspective for research, owing to its high throughout and speed of data generation. It made easy to perform whole genome and transcriptome sequencing in a short time and low cost. Motalebipour et al. [14] performed genome survey study in pistachio and estimated genome size of pistachio as 600 Mb.

In this chapter, SSR abundance, distribution of their use as useful markers in the characterization of germplasm resources, taxonomy and phylogenetic analysis in the genus, as well as their transferability among *Pistacia* species are discussed.

2. SSR distribution in *P. vera*

Motalebipour et al. [14] performed a genome survey study in pistachio and estimated genome size of pistachio as 600 Mb. The authors used 26.77 Gb Illumina PE (pair-end) reads of Siirt cultivar to estimate genome size and to reveal SSR distribution in pistachio genome. Motalebipour et al. [14] detected a total of 59,280 di-, tri-, tetra-, penta-, and hexa-nucleotide SSR motifs, and the dinucleotide motifs were the most abundant type of repeats (63.2%) in pistachio (**Figure 1**), followed by tri- (18.0%), tetra- (12.8%), penta- (3.8%), and hexanucleotide motifs (2.2%).

Motalebipour et al. [14] searched repeat motifs in their genome survey study in pistachio, and AT/AT (23.0%) and TA/TA (21.6%) were the most abundant repeats, followed by AG/CT (7.6%) and GA/TC (6.2%), AAT/ATT (4.1%), CA/TG (2.6%), and TAA/ATT (2.6%) (**Figure 2**). The most abundant tetra- and penta-nucleotide

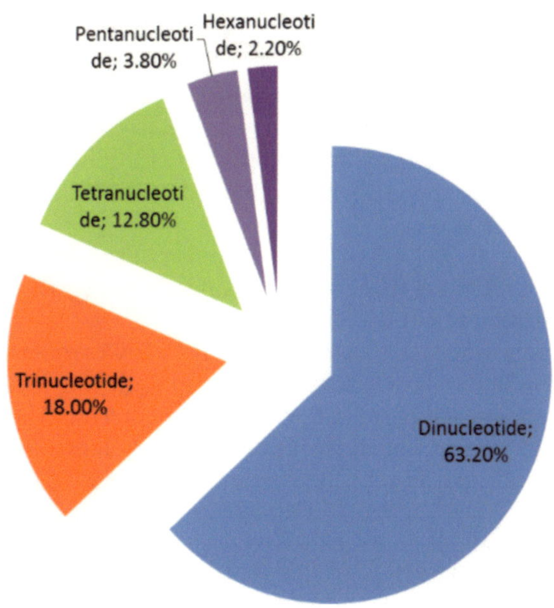

Figure 1.
Distribution of 59,280 SSRs in the pistachio genome based on repeat type. Obtained from Motalebipour et al. [14].

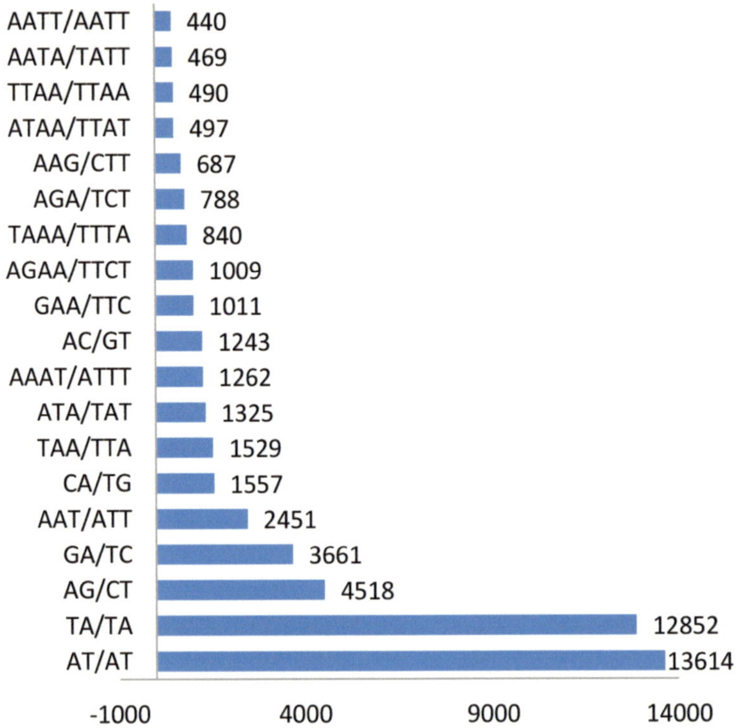

Figure 2.
Distribution of SSR motifs in pistachio at 40× coverage sequencing data. The X-axis represents motif types, and the Y-axis represents the count of motifs in the genome of pistachio. Obtained from Motalebipour et al. [14].

repeat motif types were AAAT/ATTT (2.1%) and AAAAT/ATTTT (0.44%), respectively (**Figure 2**). Among hexanucleotide motifs, AAAAAT/ATTTTT and GCCCAA/TTGGGC (0.07%) were the most abundant motifs. The authors calculated one SSR per 8.67 kb in the pistachio genome.

Khodaeiaminjan et al. [28] had about 10x Illumina data of three *P. vera* (Siirt, Ohadi, and Bağyolu) cultivars and one *P. atlantica* genotype (Pa-18) to find pairwise polymorphic SSR loci in silico. The authors detected a total of 3821, 3833, 2024, and 4597 SSRs in Siirt, Ohadi, and Bağyolu *P. vera* cultivars and in *P. atlantica* genotype, respectively. The dinucleotide motifs were the most abundant type of repeats (48.6%) in four genotypes followed by hexa- (21.3%) and trinucleotide motifs (19.4%) (**Figure 3**).

Khodaeiaminjan et al. [28] calculated the number of repeats in 4 genotypes, and 4 and 11 repeated motifs were the most abundant followed by 12, 8, 5, and 13 times repeated motifs (**Figure 4**).

Jazi et al. [29] performed a transcriptome study by RNA sequencing of a pooled sample representing 24 different tissues of 2 pistachio cultivars with contrasting salinity tolerance under control and salt treatment in pistachio. The authors searched SSR motifs in transcriptome sequences of pistachio, and 11,337 SSRs were defined as di- to hexanucleotide motifs in 11,130 contigs. Di- and trinucleotide repeats were the most abundant SSRs, accounting for 40–44% of total SSRs, followed by tetra- (9.5%), penta- (3.1%), and hexanucleotide repeats (2.2%). The pistachio transcriptome was rich in GA/TC (12.13%), AG/CT (11.02%), AT/AT (8.32%), TA/TA (8.04%), GAA/TTC (5%), and AGA/TCT (4.02%). To make these SSRs useful, a total of 7605 primer pairs were generated from the microsatellites with sufficient flanking sequences. However, none of them were tested in *Pistacia* species.

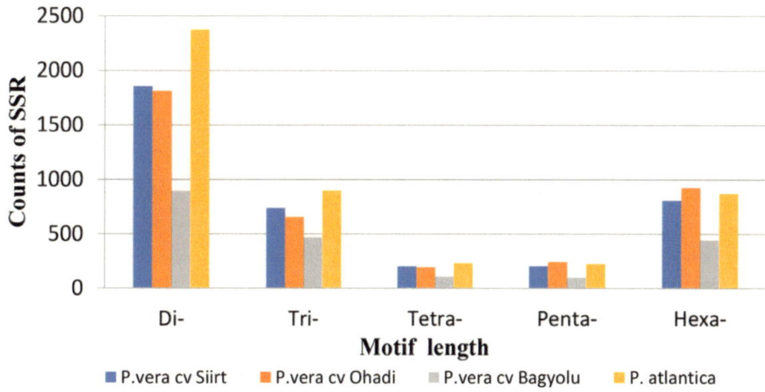

Figure 3.
The number of di-, tri-, tetra-, penta-, and hexanucleotide SSRs in Siirt, Ohadi, and Bağyolu P. vera *cultivars and in* P. atlantica. *Obtained from Khodaeiaminjan [36].*

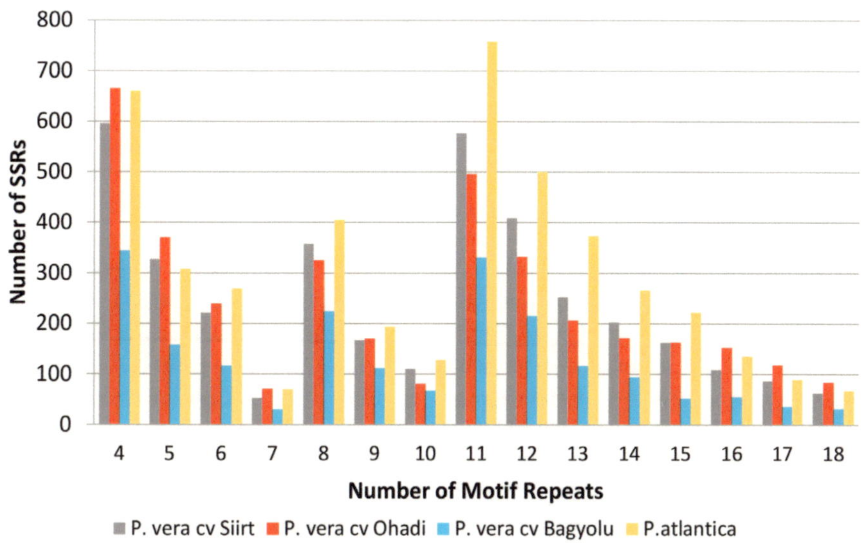

Figure 4.
The number of motif repeats in P. vera *Siirt, Ohadi, and Bağyolu cultivars and in* P. atlantica. *Obtained from Khodaeiaminjan [36].*

3. Development of SSRs from *P. vera*

In *P. vera*, the first SSR markers were developed by Ahmad et al. [37] using enrichment method from Kerman pistachio cultivar. The authors constructed dinucleotide (CT and CA) and trinucleotide (CTT)-enriched genomic DNA libraries. Eighty-nine clones randomly selected from 295 clones for "CT/GA"-enriched library and 57 clones (64%) contained the repeats. Sixty-two clones out of 250 clones, randomly selected from "CA/GT"-enriched library, and 37 (59%) clones contained the repeats. Of the total 151 clones, 67 (44%) clones had sufficient flanking sequence for primer design. Thirty-three clones were selected randomly for "CTT/GAA"-enriched library. Eight (24%) clones had sufficient flanking sequence for primer design. The authors tested these primer pairs in commercially important American, Iranian, Turkish, and Syrian pistachio cultivars, and 14 (56%) SSR loci successfully produced PCR products.

Kolahi-Zonoozi et al. [27] developed SSRs using the FIASCO protocol (Fast Isolation by AFLP of Sequences Containing Repeats) from genomic DNA of *P. vera* cv Akbari. A total of 234 clones were sequenced, and 125 (53.4%) contained SSR motifs. A total of 90 clones having repeats close to the edge of the insert was eliminated, and 35 clones remained and were used to design 42 primer pairs. The authors tested 42 SSR primer pairs in 45 Iranian pistachio cultivars for amplification and polymorphism. Sixteen primer pairs (38.1%) successfully produced scorable bands, and 12 pairs (28.6%) showed polymorphism in 45 pistachio cultivars. The most common repeat motifs in our study were dinucleotides. A total of 32 alleles was obtained with an average of 2.75. The *PIC* values ranged from 0.19 to 0.56 with an average of 0.33. The expected heterozygosity (*He*) varied from 0.081 to 0.518 with an average of 0.345, while the observed heterozygosity (*Ho*) ranged from 0.023 to 0.930 with a mean of 0.490.

Zaloğlu et al. [16] used enrichment method and developed SSR markers from *P. vera* cv Siirt. Genomic libraries enriched with the repeats of "CA," "GA," and "AAC" and "AAG" were constructed. A total of 94 clones (58.8%) contained repeats, and the CA-enriched library contained the highest number (32 clones; 86.5%), followed by the GA-enriched library (25 clones; 61%), whereas the AAC (16 clones; 45.7%)- and AAG (22 clones; 46.8%)-enriched libraries had the lowest number of clones containing repeats. From each library 57 clones were selected and a total of 228 clones was sequenced. A total of 84 primer pairs was designed, 59 generated (70.2%) bands, and 47 polymorphic in the characterization of 7 diverse pistachio cultivars. The number of alleles ranged from two to nine with an average of 3.6. A higher number and frequency of SSRs was obtained from dinucleotide-enriched libraries than trinucleotide libraries. The SSR loci from the CA enriched library (*Ho* = 0.49) were more homozygous than those from the other libraries (*Ho* = 0.59–0.61). The AAC-enriched library (*He* = 0.55) was the least informative, whereas the GA library (*He* = 0.71) was the most informative; the CA and AAG libraries had intermediate values. As a result, the GA-enriched library was the best, whereas the AAC library was the worst among the four libraries in terms of perfect repeats, number of alleles, polymorphism, and informativeness. The AAG-enriched library was also good because of its perfect repeats, observed heterozygosity, and numbers of amplified and polymorphic loci. The authors suggested to use the GA- and AAG-enriched libraries in further SSR marker development studies in pistachio. A higher frequency of SSRs was obtained from the dinucleotide-enriched (73.1%) libraries than the trinucleotide-enriched (46.3%) ones. The dinucleotide-enriched libraries had a higher number of alleles (*Na* = 3.8) and effective number of alleles (*Ne* = 3.0) than the trinucleotide-enriched libraries (3.3 and 2.5, respectively). The dinucleotide-enriched libraries (*He* = 0.67) were more informative than the trinucleotide ones (*He* = 0.59). The 59 SSR primer pairs were tested in 8 different *Pistacia* species, and 54 were transferable to at least 1 *Pistacia* species. The SSR loci in trinucleotide (72.7%)-enriched libraries had higher transferability than dinucleotide (54.7%) ones.

Topçu et al. [17] sequenced more clones from GA- and AAG-enriched libraries based on suggestion done by Zaloglu et al. [16]. A total of 192 clones was sequenced from each library, 135 primer pairs were designed, and 110 generated PCR products. Topçu et al. [17] tested 110 SSR loci in 12 diverse pistachio cultivars for amplification and polymorphism. A total of 46 loci from the GA library and 18 loci from the AAG library was polymorphic in *P. vera*. A total of 64 polymorphic loci produced 264 alleles with an average of 4.13 alleles per locus. The observed (*Ho*) and expected heterozygosity (*He*) values were 0.52 and 0.56, respectively, while average polymorphism information content (*PIC*) was 0.51. One hundred out of 110 SSR loci were transferable to at least one of the tested 10 *Pistacia* species. *P. eurycarpa* that is the closest species to *P. vera* had the highest number of transferable loci, whereas *Pistacia texana* and *P. lentiscus* that are the farthest species to *P. vera* had the lowest number of transferable loci.

Motalebipour et al. [14] obtained 40× sequencing data from *P. vera* cv. Siirt to develop SSR markers. A total of 59,280 SSR motifs was detected with a frequency of 8.67 kb in pistachio. The authors randomly selected 950 SSR loci and screened them in three *P. vera* cultivars (Siirt, Ohadi, and Bağyolu) in one *P. atlantica* genotype (Pa-18). A total of 610 loci (64.2%) generated amplification products, 197 (20.7%) loci were monomorphic, and the remaining 143 (15.1%) SSR loci failed to generate amplification products. Of the 610 that amplified, 204 produced polymorphic and easily scorable bands. Of these, 193 were perfect (94.6%), 8 (3.9%) were compound, and 3 (1.5%) were interrupted repeats. Dinucleotide motifs were the most abundant (63.2%), followed by tri- (18.0%), hexa- (12.8%), tetra- (3.8%), and pentanucleotide motifs (2.2%). Two SSR primer pairs amplified at two loci, and 206 SSR loci were obtained and used to study genetic diversity in *Pistacia*. The authors tested these 206 loci in 24 pistachio cultivars along with 20 wild accessions belonging to five *Pistacia* species (four accessions from each one). A total of 2036 alleles was obtained from 206 SSR loci ranging from 2 to 19 alleles with an average of 9.88 in testing 44 *Pistacia* accessions. Of the analyzed 206 SSR loci, 41 were polymorphic, and 136 had amplifications in all tested 6 *Pistacia* species. The observed heterozygosity (*Ho*) ranged from 0.0 to 0.82 with an average of *Ho* = 0.41. The average *He* value was 0.74, which ranged between 0.08 and 0.91. The *PIC* values ranged from 0.08 to 0.90, with an average of 0.71. All 206 SSR loci generated amplification products in 24 *P. vera* culticars, and a total of 897 alleles was produced with an average of 4.5 alleles per locus. Two hundred (97.1%) SSR loci were polymorphic in 24 pistachio cultivars. The observed heterozygosity (*Ho*) value ranged from 0.00 to 1.00 with an average of 0.46. The expected heterozygosity (*He*) values varied from 0.04 to 0.87 with an average of 0.55. The polymorphism information content values ranged between 0.04 and 0.85 with an average of 0.50.

Khodaeiaminjan et al. [28] used in silico approach to develop polymorphic SSR markers in pistachio. The authors compared 10× Illumina sequencing data of three *P. vera* (Siirt, Ohadi, and Bağyolu) cultivars and one *P. atlantica* genotype (Pa-18) to find pairwise polymorphic SSR loci in silico, and 750 loci were detected. The authors tested all 750 loci in 18 *P. vera* cultivars and 6 *P. atlantica* genotypes, and they obtained 625 polymorphic loci from 618 SSR primer pairs. A total of 3947 alleles was obtained from 625 loci with an average of 6.2 allele per locus. A total of 613 (98.1%) SSR loci in 18 *P. vera* cultivars and 544 SSR loci (87.0%) in six *P. atlantica* genotypes was polymorphic. The numbers of alleles were 2631 within *P. vera* and 2183 in *P. atlantica*. The lowest genetic diversity values were obtained from the Bagyolu-Pa-18 pairwise combination, while in silico SSR loci from Ohadi-Pa-18 pair produced high genetic diversity values. In the analysis of 24 *Pistacia* genotypes, a total of 613 (98.1%) loci was polymorphic, and the average number of alleles (*Na*) was 6.3 ranging from 2 to 20. The average expected heterozygosity (*He*) and observed heterozygosity (*Ho*) values were 0.67 and 0.53, respectively. The average polymorphism information content value was 0.63. The highest genetic diversity values were produced by SSR loci from the Ohadi-Pa18 and Ohadi-Siirt pairwise combinations, whereas the lowest values were in SSR loci from the Bagyolu-Pa18 pair.

4. SSRs in wild *Pistacia* species

4.1 *P. atlantica*

There is only one study using *P. atlantica* DNAs to develop SSRs by Khodaeiaminjan et al. [28]. The authors used 10× coverage sequencing data of monoecious *P. atlantica* genotype, namely, Pa-18, and generated 4597 SSR loci.

The authors tested 625 loci in 6 *P. atlantica* genotypes, and all had amplification, while 544 (87.0%) were polymorphic. The number of alleles (*Na*) ranged from 1.0 to 8.0 with a mean of 3.5. The mean of observed heterozygosity (*Ho*) was 0.51, while the average *PIC* and expected heterozygosity (*He*) values were 0.53 and 0.52, respectively.

Motalebipour et al. [14] developed 206 SSR loci from *P. vera* and tested them in 4 diverse *P. atlantica* genotypes. A total of 200 SSR loci generated amplification products with a high rate of transferability (97.1%). Thirty-nine (19.5%) of the amplified SSR loci were monomorphic, and the rest were polymorphic (80.5%). A total of 527 alleles was produced by 161 polymorphic SSR loci, ranging from 2 to 7 per locus with an average of 3.3 alleles per locus. The observed heterozygosity (*Ho*) ranged from 0.0 to 1.00 with an average of $Ho = 0.48$. The average *He* value was 0.56, which ranged from 0.22 to 0.84. The *PIC* values ranged from 0.19 to 0.82, with an average of 0.49.

Zaloğlu et al. [16] and Topçu et al. [17] developed 59 and 110 SSRs from *P. vera* and tested them for amplification in *P. atlantica*, and a total of 47 and 96 SSR loci was amplified, respectively. So, the authors published a total of 143 SSR loci for further studies in *P. atlantica* from both studies. The authors did not test their polymorphism levels in *P. atlantica*.

4.2 *P. khinjuk*

Arabnezhad et al. [31] were the first in developing SSRs from *P. khinjuk*, who constructed two enriched DNA libraries with dinucleotide (AG) and trinucleotide (ATG) microsatellite motifs. Thirty-six contained microsatellite motifs from 44 sequenced clones. Among them, a higher proportion of microsatellites (71%) were simple perfect, and the remaining SSRs identified as interrupted perfect (17%) and complex imperfect (11%) repeats. Nine of the sequences contained too short flanking DNA sequence to design primer pairs; thus only 27 primer pairs were designed and tested in six *Pistacia* species. Of 27 primer pairs, 25 pairs successfully amplified SSR loci in *P. khinjuk* with expected size. Five primer pairs were subsequently discarded due to low rate of amplification across six *Pistacia* species. The authors tested the remaining SSRs in a total of 18 *Pistacia* genotypes (13 *P. vera* cultivars and 1 genotype from each of *P. khinjuk, P. atlantica, P. mutica, P. integerrima*, and *P. palaestina*). The primer pairs produced 114 alleles in 18 *Pistacia* genotypes. In 13 *P. vera* cultivars, the average number of alleles per locus was 2.8, ranging from one to six. In all *Pistacia* accessions, the average values of *He* and *PIC* were 0.61 and 0.56, respectively, while values of these diversity parameters calculated 0.45 and 0.38 when only *P. vera* genotypes were considered.

Zaloğlu et al. [16] and Topçu et al. [17] developed 59 and 110 SSRs from *P. vera* using enrichment method, and a total of 39 and 96 SSR loci had amplification, respectively. It is still necessary to test 135 SSR loci for polymorphism in *P. khinjuk* for further studies.

4.3 *P. lentiscus*

The first SSR development study in *P. lentiscus* was published by Albaladejo et al. [30] who used di- (GA, GT, AT, GC), tri- (CAA, ATT, GCC), and tetranucleotide (GATA, CATA, ATAG) genomic-enriched libraries. The authors randomly selected 163 clones and 75 (46%) had microsatellite motifs. A total of 21 primer pairs was designed and tested in 16 individuals. Eight of 21 primer pairs displayed consistent and polymorphic patterns, whereas the others were discarded due to producing monomorphic and multibanding patterns and failing in amplification. Forty-two individuals from two populations sampled in Southern Spain were used for characterization of eight loci. A total of 59 alleles was detected, ranging from 3 to 13 per locus. The expected heterozygosity ranged from 0.139 to 0.895.

Motalebipour et al. [14] developed 206 SSR loci from *P. vera* and tested them in four *P. lentiscus* genotypes. A total of 151 SSR loci was amplified with a 73.3% transferability rate. Of the amplified SSR loci, 83 (55.0%) were polymorphic. A total of 217 alleles was obtained from 83 polymorphic SSR loci in *P. lentiscus,* ranging from 1 to 6, with an average of 2.6 alleles per locus. The observed heterozygosity (*Ho*) values ranged from 0.00 to 1.00 with an average of 0.50. The expected heterozygosity (*He*) values varied from 0.22 to 0.78 with an average of 0.49. The polymorphic information content (*PIC*) values ranged between 0.19 and 0.75 with an average of 0.41.

Zaloğlu et al. [16] and Topçu et al. [17] developed 59 and 110 SSRs from *P. vera* and tested them for amplification in *P. lentiscus*, and a total of 31 (52.5%) and 76 (69.1%) SSR loci was amplified, respectively. The authors did not test their polymorphism levels in *P. lentiscus*. Therefore, there are 107 SSR loci for *P. lentiscus* from both studies to use them in the future.

4.4 *P. chinensis*

Ding and Lu [33] developed 12 polymorphic microsatellite loci from *P. chinensis* using a microsatellite-enriched genomic library based on magnetic beads. These loci were characterized in 24 individuals from 3 populations located on Thousand Island Lake, Zhejiang Province, China. The number of alleles per locus varied from 3 to 16. The mean number of alleles per locus was between 3.3 and 4.0 at the population level. The observed and expected heterozygosity ranged from 0.12 to 0.87 and 0.23 to 0.89, respectively.

Motalebipour et al. [14] developed 206 SSR loci from *P. vera* and tested them in 4 diverse *P. chinensis* genotypes, and 177 SSR loci produced amplification products with 85.9% transferability rate. Of the amplified loci, 119 loci (67.2%) were polymorphic. A total of 365 alleles was amplified by 119 polymorphic SSR loci with an average of 3.1 alleles per locus. The average observed heterozygosity (*Ho*) was 0.48, and it ranged between 0.00 and 1.00. The expected heterozygosity (*He*) values varied from 0.22 to 0.84 with an average of 0.54. The polymorphism information content (*PIC*) values ranged between 0.19 and 0.82 with an average of 0.48 in *P. chinensis.*

Zaloğlu et al. [16] and Topçu et al. [17] developed SSRs from *P. vera* and tested them for amplification in *P. chinensis*. A total of 121 SSR loci had amplification PCR products in *P. chinensis,* with 71.6% transferability rate. However, the authors did not test polymorphism levels of the amplified SSR loci in *P. chinensis*. Chen et al. [32] developed 14 SSR loci from *P. weinmannifolia*, and 9 of them (64.3%) were amplified in *P. chinensis*.

4.5 *P. weinmannifolia*

Chen et al. [32] developed SSRs using the FIASCO protocol (Fast Isolation by AFLP of Sequences Containing Repeats) from genomic DNA of *P. weinmannifolia*. A total of 205 clones was sequenced, 147 contained SSR motifs, and 94 allowed primer design with sufficient flanking regions. The primer pairs were tested for polymorphism in 24 individuals from 2 populations, and 14 produced polymorphic microsatellite loci with an average of 4.1 alleles (ranging from 1 to 9) per locus in *P. weinmannifolia*. The expected (*H*e) and observed (*H*o) heterozygosities ranged from 0.000 to 0.906 and from 0.000 to 0.933, respectively. Ten of these loci contained dinucleotide repeat motifs, and four loci had complex repeat motifs. The authors tested 14 primer pairs in *P. chinensis* and *P. mexicana* for their transferability, and 9 (64.2%) loci in *P. chinensis* and 4 (28.6%) loci in *P. mexicana* were successfully transferable.

Zaloğlu et al. [16] developed 59 SSRs from *P. vera* and tested them for amplification in *P. weinmannifolia*, and a total of 23 SSR loci was amplified with a 39.0% transferability rate. However, the authors did not test their polymorphism levels within *P. weinmannifolia*.

4.6 *P. integerrima*

There is no study in the literature developing SSRs from *P. integerrima* tissues. Arabnezhad et al. [31] developed 27 SSR loci in *P. khinjuk*, and 18 (66.7%) were successfully amplified in *P. integerrima*. Two previous studies [16, 17] developed a total of 169 SSRs from *P. vera* and tested them for amplification in *P. integerrima*. A total of 147 SSR loci had amplification products with 87.0% transferability rate. There is a necessity to test the amplified loci for polymorphism within *P. integerrima*.

Motalebipour et al. [14] developed 206 SSR loci from *P. vera* and tested them in four diverse *P. integerrima* genotypes. A total of 193 SSR loci generated amplification products with a high rate of transferability (93.7%). Of the amplified SSR loci, 157 (81.3%) were polymorphic in *P. integerrima*. A total of 416 alleles was produced by 157 SSR loci with an average of 2.70 alleles per locus. The average observed (*Ho*) and expected (*He*) heterozygosities were 0.50 and 0.52, respectively. The expected heterozygosity (*He*) values varied from 0.22 to 0.78 with an average of 0.55. The polymorphism information content (*PIC*) values ranged between 0.19 and 0.75 with an average of 0.44.

4.7 *P. terebinthus*

There is no SSR development study also from *P. terebinthus* tissues in the literature. Motalebipour et al. [14] generated 206 SSR loci from *P. vera* and tested them in *P. terebinthus* genotypes, and 183 SSR loci produced amplification products with 88.8% transferability rate. Of amplified SSRs, 142 were polymorphic. A total of 416 alleles was produced by 142 polymorphic SSR loci. The number of alleles (*Na*) ranged from 1 to 7 with an average of 3.4 alleles per locus. The observed heterozygosity (*Ho*) ranged from 0.0 to 1.0 with an average of 0.47. The expected heterozygosity (*He*) values varied from 0.22 to 0.84 with an average of 0.56. The polymorphic information content (*PIC*) values changed between 0.19 and 0.82 with an average of 0.50.

Zaloğlu et al. [16] and Topçu et al. [17] developed 169 SSRs from *P. vera* and analyzed them in *P. terebinthus* for amplification. A total of 128 SSR loci had PCR products in *P. terebinthus*, with 71.6% transferability rate. However, the authors did not test polymorphism levels of the amplified SSR loci in *P. terebinthus*.

4.8 *P. texana, P. mexicana,* and *P. eurycarpa*

Chen et al. [32] developed 14 SSR loci from *P. weinmannifolia*, and 4 of them (28.6%) were amplified in *P. mexicana*. Zaloğlu et al. [16] and Topçu et al. [17] developed 59 and 110 SSRs from *P. vera* and tested them for amplification in *P. mexicana*, and a total of 33 and 77 SSR loci was amplified in *P. mexicana*, respectively. Topçu et al. [17] generated 110 SSRs from *P. vera* and tested them for amplification in *P. eurycarpa* and *P. texana*. A total of 100 and 76 SSR loci was amplified with a 90.9 and 69.1% transferability rates. However, Zaloğlu et al. [16]

and Topçu et al. [17] did not test polymorphism levels of the SSR loci in *P. mexicana, P. texana*, and *P. eurycarpa.*

5. Use of SSR markers in the genus *Pistacia*

There are only several studies characterizing *Pistacia* species and *P. vera* accessions using SSR markers. Pazouki et al. [38] used SSR markers and characterized 304 *Pistacia* accessions belonging to *P. khinjuk*, *P. vera*, and *P. atlantica* subsp. *kurdica*. The authors indicated lower level of polymorphism and variation within *P. atlantica* subsp. *kurdica* than *P. vera* and *P. khinjuk*. Motalebipour et al. [14] used 1505 alleles amplified by 136 SSR primer pairs for phylogenetic analysis of 6 *Pistacia* species. The closest species to *P. vera* was *P. atlantica*, and *P. integerrima*, *P. chinensis*, *P. terebinthus* followed it, while *P. lentiscus* was the most diverse species to the cultivated pistachio. The structure analysis confirmed the cluster analysis as well. Albaladejo et al. [30] used 8 SSR loci to characterize 42 *P. lentiscus* accessions belonging to 2 *P. lentiscus* populations. The number of alleles changed between 3 and 13 by obtaining a total of 59 alleles. The expected heterozygosity ranged from 0.139 to 0.895. Two *P. lentiscus* populations were separated clearly in cluster analysis. Chen et al. [32] characterized 24 *P. weinmannifolia* accessions using 14 SSR loci. The number of allele changed between one and nine with an average of 4.1. *Ho* values were between 0.000 and 0.933, while *He* values ranged from 0.000 to 0.906.

Ahmad et al. [37] used 14 SSR loci to characterize Iranian, Turkish, and Syrian pistachio cultivars. A total of 46 alleles was produced by 14 SSR loci ranging from 2 to 5 allele per loci. Cluster analysis placed most of the Iranian samples in one group, while the Syria samples were the most diverse and did not group in a single cluster. Kolahi-Zonoozi et al. [27] described 45 Iranian pistachio accessions by 12 SSR loci. The *PIC* values changed between 0.19 and 0.56 with an average of 0.33. The average *Ho* and *He* values were 0.490 and 0.345, respectively. Khodaeiaminjan et al. [28] characterized 18 pistachio cultivars from different origins by 2631 SSR alleles using 625 SSR loci. The constructed dendrogram separated pistachio cultivars mainly in two groups according to their geographical origin: one group contained the cultivars originated from Iran, and the second group included cultivars originated from Mediterranean countries such as Turkey, Syria, Greece, and Italy. Siirt cultivar (origin is Southeast part of Turkey) placed between two main groups. The results were in agreement with a hypothesis on diffusion of pistachio cultivars suggested by Kafkas et al. [19] who hypothesized that the Siirt cultivar is in a transition subcluster between Iranian and Mediterranean cultivars and pistachio cultivation diffused from its center of origin, the Iranian-Caspian region, via southeastern Turkey to Syria, the Mediterranean region of Europe, and North Africa. Motalebipour et al. [14] used 1505 alleles from 136 SSR primer pairs for genetic diversity analysis of 24 pistachio cultivars, and similar results were obtained with the study performed by Khodaeiaminjan et al. [28].

The SSR markers in the genus *Pistacia* were also used for genetic linkage map construction and QTL analysis. Khodaeiaminjan et al. [28] constructed the first SSR-based genetic linkage map of pistachio using an F1 segregating population derived from a cross between “Siirt” and “Bağyolu” cultivars. A total of 385 SSR markers was mapped along with 15 chromosomes, and the consensus map had 1511.3 cM length with an average of 25.6 SSR markers per LG, and the average distance between the markers was 3.9 cM with a 0.25 marker density. The first QTL study in pistachio was performed by Motalebipour et al. [35] who constructed a genetic linkage map of pistachio using an inter-specific F1 population and SSR markers. The authors mapped a total of 388 SSR markers along with 15 linkage groups. The length of consensus map was 1492 cM with an average marker distance of 3.7 cM. The QTL analysis was performed for 5 morphological traits such as leaf length, leaf width, number of leaflets, young shoot color, and leaf color, and 17 stable QTLs during 2 consecutive years were identified. The released SSR-based genetic linkage maps and reported QTLs can be useful genetic resources for future genetic studies in pistachio.

6. Conclusions

SSR is a very useful molecular marker system due to abundance in the genomes and its codominant inheritance as well as high repeatability. They have also a high level of transferability between closely related species as in the genus *Pistacia*. They have been used for assaying diversity in natural populations, marker discovery, germplasm characterization, parental identification, genetic linkage mapping, and evolutionary studies. There were a very limited number of SSR markers for *Pistacia* species until several years ago; however, next-generation sequencing (NGS) technology has allowed to develop plenty of SSRs since 2016 in *Pistacia*.

P. vera is the most important species in the genus *Pistacia* due to commercial value of its nuts. There are about 1500 published SSR markers, and 2/3 of them are polymorphic that were developed from *P. vera* tissues. There are also published polymorphic SSR markers for wild *Pistacia* species. They were developed mostly from cultivated pistachio due to their high transferability rate. The published SSRs were also used to construct SSR-based genetic linkage maps in pistachio.

In a conclusion, there are currently an adequate number of SSR markers for cultivated *P. vera* and for several wild *Pistacia* species such as *P. atlantica*. It is still necessary to develop polymorphic SSR loci for some other *Pistacia* species such as *P. integerrima* and *P. eurycarpa* which have been used as rootstock for cultivated pistachio.

Conflict of interest

The authors declare no conflict of interest.

Author details

Salih Kafkas
Department of Horticulture, Faculty of Agricuture, University of Çukurova, Adana, Turkey

*Address all correspondence to: skafkas@cu.edu.tr

References

[1] Kafkas S. In: Serdar U, Fullbright D, editors. Advances in Breeding of Pistachio, Achieving Sustainable Cultivation of Tree Nuts. Sawston, Cambridge, UK: Burleigh Dodds Science Publishing Limited; 2019

[2] Parfitt DE, Badenes ML. Phylogeny of the genus Pistacia as determined from analysis of the chloroplast genome. Proceedings of the National Academy of Sciences of the United States of America. 1997;**94**:7987-7992. DOI: 10.1073/pnas.94.15.7987

[3] Kafkas S. Phylogeny, evolution and biodiversity in the genus *Pistacia* (Anacardiaceae). In: Shama AK, Sharma A, editors. Plant Genome, Biodiversity and Evolution 1. Part C. Madison, USA: Science Publishers; 2006. pp. 525-557

[4] Kafkas S. Phylogenetic analysis of the genus *Pistacia* by AFLP markers. Plant Systematics and Evolution. 2006;**262**:113-124. DOI: 10.1007/s00606-006-0460-7

[5] Faostat. 2019. Available from http://www.fao.org/faostat/en/#data/QC [Accessed: 20 August 2019]

[6] Kafkas S, Kaşka N, Perl-Treves R. Unusual *Pistacia atlantica* Desf. (Anacardiaceae) Monoecious sex types in the Yunt Mountains of the Manisa Province of Turkey. Israel Journal of Plant Sciences. 2000;**48**:277-280. DOI: 10.1560/UFCU-7LF6-T0A3-UXWY

[7] Zohary M. A monographical study of the genus *Pistacia*. Palestine Journal of Botany, Jerusalem Series. 1952;**5**:187-228

[8] Maggs DH. Genetic resources in pistachio. Plant Genetic Resources Newsletter. 1973;**29**:7-15

[9] Hormaza JI, Dollo L, Polito VS. Determination of relatedness and geographic movements of *Pistacia vera* (pistachio; Anacardiaceae) germplasm by RAPD analysis. Economic Botany. 1994;**48**(4):349-358. DOI: 10.1007/BF02862231

[10] Hormaza JI, Pinney K, Polito VS. Genetic diversity of pistachio (*Pistacia vera* L., Anacardiaceae) germplasm based on randomly amplified polymorphic DNA (RAPD) markers. Economic Botany. 1998;**52**:78-87. DOI: 10.1007/BF02861298

[11] Zohary D. The genus *Pistacia* L. In: Padulosi S, Caruso T, Barone E, editors. Taxonomy, Distribution, Conservation and Uses of *Pistacia* Genetic Resources. Report of a Workshop, 29-30 June 1995, Palermo, Italy. Rome, Italy: International Plant Genetic Resources Institute; 1996. pp. 1-11

[12] Abdushukur AK, Chernova GM, Rakmonov AM, Nikolyi LV, Ablaeva E, Zaurov DE, et al. Genetic resources of *Pistacia vera* L. in Central Asia. Genetic Resources and Crop Evolution. 2009;**56**(3):429-443. DOI: 10.1007/s10722-009-9419-1

[13] Parfitt DE, Kafkas S, Batlle I, Vargas FJ, Kallsen CE. Pistachio, Badenes ML, Byrne DH2012. pp. 803-826

[14] Motalebipour E, Kafkas S, Khodaeiaminjan M, Coban N, Gozel H. Genome survey of pistachio (*Pistacia vera* L.) by next generation sequencing: Development of novel SSR markers and genetic diversity in *Pistacia* species. BMC Genomics. 2016;**17**:998

[15] KafkasS, Perl-TrevesR. Morphological and molecular phylogeny of *Pistacia* species in Turkey. Theoretical and Applied Genetics. 2001;**102**:908-915

[16] Zaloğlu S, Kafkas S, Dogan Y, Guney M. Development and characterization of SSR markers from pistachio (*Pistacia vera* L.) and their transferability to eight *Pistacia* species. Scientia Horticulturae. 2015;**189**:94-103. DOI: 10.1016/j.scienta.2015.04.006

[17] Topçu H, Coban N, Kafkas S. Novel microsatellite markers in *Pistacia vera* L. and their transferability across the genus *Pistacia*. Scientia Horticulturae. 2016;**198**:91-97. DOI: 10.1016/j.scienta.2015.11.012

[18] Ahmad R, Ferguson L, Southwick SM. Analyses of pistachio rootstocks by SSR and SRAP molecular markers. Journal of Horticultural Science and Biotechnology. 2005;**80**:382-386. DOI: 10.1080/14620316.2005.11511948

[19] Kafkas S, Ozkan H, Ak BE, Acar I, Atli HS, Koyuncu S. Detecting DNA polymorphism and genetic diversity in a wide pistachio germplasm: Comparison of AFLP, ISSR and RAPD markers. The Journal of American Society for Horticultural Science. 2006;**131**(4):522-529

[20] Karimi HR, Kafkas S. Genetic relationships among *Pistacia* species studied by SAMPL markers. Plant Systematics and Evolution. 2011;**297**:207-212. DOI: 10.1007/s00606-011-0508-1

[21] Kafkas S, Khodaeiaminjan M, Guney M, Kafkas E. Identification of sex-linked SNP markers using RAD sequencing suggests ZW/ZZ sex determination in *Pistacia vera* L. BMC Genomics. 2015;**16**:98. DOI: 10.1186/s12864-015-1326-6

[22] Hormaza JI, Dollo L, Polito VS. Identification of a RAPD marker linked to sex determination in *Pistacia vera* using bulked segregant analysis. Theoretical and Applied Genetics. 1994;**89**:9-13

[23] Khodaeiaminjan M, Kafkas NE, Guney M, Kafkas S. Development and genetic linkage mapping of novel sex-linked SNP markers in pistachio. Molecular Breeding. 2017;**37**:98. DOI: 10.1007/s11032-017-0705-x

[24] Kafkas S, Gozel H, Bozkurt H, Karci H, Topcu H, Paizila H, et al. Marker-assisted cultivar breeding in pistachio. In: 7th International Symposium on Almonds and Pistachios. 5-9 November 2017, Adelaide, Australia. 2017

[25] Levinson G, Gutman G. Slipped-strand mispairing: A major mechanism for DNA sequence evolution. Molecular Biology and Evolution. 1987;**4**:203-221. DOI: 10.1093/oxfordjournals.molbev.a040442

[26] Varshney RK, Graner A, Sorrells ME. Genic microsatellite markers in plants: Features and applications. Trends Biotechnology. 2005;**23**(1):48-55. DOI: 10.1016/j.tibtech.2004.11.005

[27] Kolahi-Zonoozi SH, Mardi M, Zeinalabedini M, Pirseyedi SM, Mahmoodi P, Tabatabaei I, et al. Development of 12 new SSR markers for genetic diversity and structure analysis in pistachio (*Pistacia vera* L.). Journal of Horticultural Science and Biotechnology. 2014;**89**:707-711. DOI: 10.1080/14620316.2014.11513141

[28] Khodaeiaminjan M, Kafkas S, Motalebipour E, Coban N. *In silico* polymorphic novel SSR marker development and the first SSR-based genetic linkage map in pistachio. Tree Genetics and Genomes. 2018. DOI: 10.1007/s11295-018-1259-8

[29] Jazi M, Seyedi S, Ebrahimie E, Ebrahimi M, Moro G, Botanga C. A genome-wide transcriptome map of pistachio (*Pistacia vera* L.) provides novel insights into salinity-related genes and marker discovery. BMC

Genomics. 2017;**18**(1):627. DOI: 10.1186/s12864-017-3989-7

[30] Albaladejo RG, Sebastiani F, Aparicio A, Buonamici A, Gonzalez-Martinez SC, Vendramins GG. Development and characterization of eight polymorphic microsatellite loci from *Pistacia lentiscus* L. (Anacardiaceae). Molecular Ecology Resources. 2008;**8**:904-906. DOI: 10.1111/j.1755-0998.2008.02110.x

[31] Arabnezhad H, Bahar M, Pour AT. Evaluation of genetic relationship among Iranian pistachios using microsatellite markers developed from *Pistacia khinjuk* stocks. Scientia Horticulture. 2011;**128**:249-254. DOI: 10.1016/j.scienta.2011.01.028

[32] Chen S, Wu X, Ji Y, Yang J. Isolation and characterization of microsatellite loci in *Pistacia weinmannifolia* (Anacardiaceae). International Journal of Molecular Sciences. 2011;**12**(11):7818-7823. DOI: 10.3390/ijms12117818

[33] Ding Z, Lu J. Isolation and characterization of polymorphic microsatellite loci for *Pistacia chinensis* Bunge (Anacardiaceae). Silvae Genetica. 2019;**68**:26-28. DOI: 10.2478/sg-2019-0005

[34] Turkeli Y, Kafkas S. First genetic linkage map in pistachio constructed using an interspecific cross between *Pistacia vera* L. and monoecious *Pistacia atlantica* Desf. Scientia Horticulturae. 2013;**151**:30-37. DOI: 10.1016/j.scientia.2012.11.024

[35] Motalebipour E, Kafkas S, Khodaeiaminjan M, Gozel H. SSR-based genetic linkage map and QTL analysis for leaf and shoot traits in pistachio using an interspecific F1 population between *Pistacia vera* L. and monoecious *Pistacia atlantica* Desf. Molecular Breeding. 2018;**38**:134. DOI: 10.1007/s11032-018-0893-z

[36] Khodaeiaminjan M. Antepfıstığında yeni nesil sekanslama teknolojisi kullanılarak SSR markörlerin geliştirilmesi ve referans genetic haritanın oluşturulması [thesis]. Adana: University of Cukurova; 2017

[37] Ahmad R, Ferguson L, Southwick SM. Identification of pistachio (*Pistacia vera* L.) nuts with microsatellite markers. The Journal of American Society for Horticultural Science. 2003;**128**:898-903. DOI: 10.21273/JASHS.128.6.0898

[38] Pazouki L, Salehi Shanjani P, Hagidimitriou M, Pirseyedi SM, Naghavi MR, Avanzato D, et al. Genetic diversity and relationship among *Pistacia* species and cultivars. Conservation Genetics. **11**:311-318. DOI: 10.1007/s10592-009-9812-5

Section 2

Soil Management

Chapter 7

Soil Management in Sustainable Agriculture

Koç Mehmet Tuğrul

Abstract

People need food to live, and this is largely due to natural resources. However, time is also required to meet these limited resources and increased consumption demands, and for a renewal cycle. This cycle can be traditional, industrial and commercial, as well as be sustainable. So, what is sustainable agriculture? Sustainable agriculture is the way to increase productivity in agriculture and to increase the level of economic prosperity by protecting all living things on earth, living spaces and natural resources. It is clear that the continuity of all the living things is possible with the food provided by natural resources. At this point, sustainable production, consumption and preservation of the natural balance are of great importance. Today, the world population is rapidly increasing and resources are consumed at the same rate; creating awareness about sustainable food, transferring this consciousness to future generations in a more permanent way, increasing the number of conscious producers and consumers, strengthening awareness for sustainable foods, respecting the natural balance, and gaining a sense of responsibility, saving consumption habits should be our main target.

Keywords: sustainable agriculture, new approaches, natural balance, soil management, cultivation

1. Introduction

Today, the greatest success in agriculture will be to achieve the desired increase in production by reducing the negative environmental conditions. This can only be achieved by implementing sustainable methods and sustainable solutions in agriculture. The fact that the agricultural activities and practices are compatible with the environment and being permanent is great importance in terms of contributing to the sustainability of the ecology. There are many definitions and explanations about sustainable agriculture [1]. Sustainable Agriculture includes all of the systems and practices that will improve the protection of the environment and natural agricultural resources necessary to ensure the production of adequate and high quality foodstuffs at affordable costs which the rapidly growing world population needs. To be fully self-sufficient for sustainable agriculture is not a requirement. Long-term stability and efficiency is required. For this purpose, the minimum and most economical and fastest way of implementation of each application in agriculture is one of the priorities that should be focused on the protection of agricultural areas and natural resources.

If an awareness of sustainable practices is to be created, it is necessary to rethink in detail what the concept of agriculture means first. The questions such as: What is agriculture and how should a production be made to meet increasing agricultural demands? Is agriculture only an activity on the field, or is it possible to produce more qualified by applying new production techniques? What methods should be applied to obtain sufficient product without damaging nature? are required to find answers firstly [2].

All of the work done on soil in order to grow the necessary and useful plants and animals for the survival of the people and to obtain the products is called agriculture. In order to meet the growing agricultural needs in a healthy manner, water resources should be protected and soil should be developed and original seeds should be stored and reproduced for the future. At the same time, an increase in soil fertility, protection of water, protection of valuable seeds and biodiversity need to be taken into account.

In general, many methods are applied under the definitions of traditional, organic, industrial, ecological, smart and integrated, and each method differs from the others [3]. Sustainable agriculture mainly focuses on increasing the productivity of the soil and reducing the harmful effects of agricultural practices on climate, soil, water, environment and human health. Reduces the use of non-renewable sources and inputs from petroleum-based products and uses renewable resources to generate production. In general, it focuses on the needs, knowledge, skills and socio-cultural values of the local people.

2. Principles of sustainable agriculture

Some general principles of sustainable agriculture can be listed as follows:

- Soil must be protected and developed: Soil is absolutely necessary for good and healthy products. Soil should be enriched with natural fertilizers such as organic and green manure and compost. Natural fertilizers are healthier for soil, plants, water, air and people than chemical fertilizers.

- Water and water resources should be protected: As in life, it needs absolute water in agriculture. In arid regions, the best way to protect water is to grow plants that are suitable for the ecology of the region or that only need water during the rainy season. Green manure and mulch are useful in keeping water in soil. The contour barriers protect the water by preventing the water flow. Another method for preserving water is to apply drip irrigation instead of traditional irrigation methods and to make irrigation time planning.

- To control pests and diseases naturally: Instead of chemical control, natural or integrated protection management should be applied to balance nature, products, pests, diseases, weeds and soil. In this regard, techniques such as choosing durable varieties, keeping proper distance between plants in planting, determining the timing of agricultural practices correctly, using natural predators and crop rotation are important for the success of the method.

- Cultivate different agricultural products: This is called product rotation. According to the characteristics of the products, for 3–6 years rotation or cultivating multiple crops are the methods of preventing diseases and pests. Thus, nutrients are kept in the soil and diversity in agriculture is ensured and healthy food is provided.

- Start with small changes first: Most agricultural techniques have been developed over a long period of time. However, new methods may not always be successful. New ideas should first be tried in small areas, and should be applied when it becomes clear and successful.

3. Why sustainable agriculture is important?

The world population is growing at a great pace. There are countries with a population expressed in billions of Asian countries, and in Europe and the Americas it is estimated that the population will soon find billions. This will certainly create a serious need for food in the future. One of the main objectives of industrial agriculture is to ensure that everyone has access to basic needs in the present and future years.

Industrial agriculture, on the one hand, uses more chemical input to meet the increasing demand, on the other hand, agricultural and soil resources are polluted by chemical residues and production potential is reduced. In fact, this is a contradiction. At this point, the sustainable farming method protects both the soil and the environment and ensures the production and the long-term agricultural production. In summary, the benefits of sustainable agriculture are as follows:

- With sustainable agriculture method, it is possible to produce more than one product in small areas and high efficiency.

- An enterprise with sustainability will have a positive impact on the ecosystem. Efficient soils will have a habitat for animals, but will also contribute to agricultural production.

- The fertilization of the soil will ensure long-term use and increase of productivity.

- In addition to the benefits to agriculture, contributes to the creation of new areas of employment.

4. Sustainable agriculture practices

As a result of long years of practices and scientific studies, several common sustainable agricultural practices have been put forward in **Figure 1**.

4.1 Precision chemical application

It can be defined as a set of methods that include mechanical and biological controls to reduce the use of pesticides and control pest populations [5]. In this method, Variable Rate Application Technologies (VRA) is applied and unlike traditional agriculture, instead of homogeneous input, it is the application of measurement of productivity differences in the field and appropriate input according to the spatial needs resulting from these differences.

4.2 Conservation reserve program (CRP)

CRP is a land conservation program administered by the Farm Service Agency (FSA). In return for a contract with farmers involved in the program, it is an incentive for farmers to make agricultural production that is sensitive to the environment and improves the environmental health and quality [6]. Contracts for land enrolled in CRP are 10–15 years in length. The long-term aim of the program is to restore valuable land cover to help improve water quality, prevent soil erosion and reduce loss of wildlife areas.

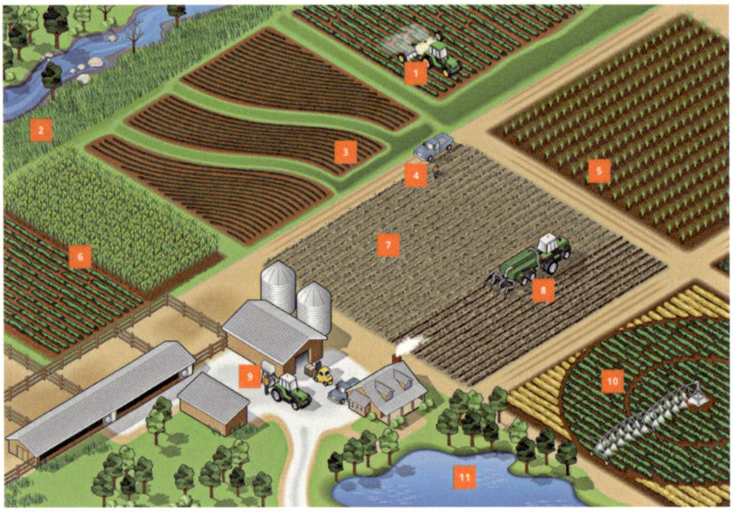

Figure 1.
Common sustainable agricultural practices [4].

4.3 Terraces

In particular, in order to make agriculture in high slope areas, it is the name given to the arrangement of the land in the form of steps and supported by walls. Thus agricultural applications are possible in these areas.

4.4 Scouting

Scouting is the most fundamental act of traveling in crop fields and make observations. The farmer is required to identify how different areas of development change in his land. If there are problems during the growing season, these problems affect the yield at the time of harvest, so the farmer tries to reduce them. If the problems are not noticed or resolved during the growing season, they may limit the yield, thus reducing the revenue generated. Traditional methods include walking in the field and observing plants manually, while methods such as global positioning systems (GPS) and drones (UAVs) help to make a more accurate decision by making fast and reliable measurements with the help of special equipment and precision sensors.

4.5 Cover crops

Cover plants (alfalfa, vetch, etc.) can be cultivated during off-season periods when the soil is bare and can be grown between the main plant rows. These products prevent soil erosion, renew soil nutrients, keep weeds under control, and protect soil health by reducing the need for herbicides [5].

4.6 Crop rotation/diversity

It is the process of producing various products in the field one after the other from year to year respectively. Thus, different parts of the soil are utilized with different products, and pests and diseases that are specific to each product are prevented from spreading.

4.7 No-till/conservation tillage

Intensive or traditional agriculture causes physical and chemical degradation of soil, loss of organic matter, reduced biological activity in the soil and consequently a decrease in crop production. On the contrary, the method of sustainable agriculture envisages a sustainable and profitable farming system based on three basic rules, including soil-free agriculture, continuous soil surface covered with plant or plant debris, and crop rotation [7, 8].

4.8 Precision nutrient management

Fertilization, which constitutes 10–15% of the costs of agricultural inputs, is critical for increasing product productivity by up to 50%. The application time and method are of great importance in the fertilization process which is applied to soil in order to meet the basic nutrients (nitrogen, phosphorus, potassium etc.) which are not enough in agricultural soils. Data's such as climate and weather conditions, soil characteristics and product types are important in determining the appropriate fertilization time.

4.9 Reducing fuel use

Mechanization tools that reduce labor requirements in agriculture generally use fossil fuels. Nowadays, the use of fossil fuel energies directly or indirectly in agriculture has not been economically profitable for producers. In developing countries, large amounts of fossil fuels are used in agricultural production, in particular fertilizer production and machinery use. It is not possible to carry out modern agricultural production processes without using fuel. However, the use of combined agricultural tools and machinery in one pass and the use of renewable energy sources instead of fossil fuels will reduce both the cost of fuel in agriculture and reduce the carbon emissions and make the agriculture sensitive to the environment.

4.10 Irrigation

Effective irrigation is possible by determining the optimum water amount using different parameters such as soil humidity, effective precipitation rate and evapotranspiration and by determining the correct irrigation time with climate, weather forecasts and real-time weather data. In this way, effective and economical irrigation will be provided by protecting the limited water resources and the environmental and agricultural negative effects of leaching, salinity and fungal diseases caused by excess water will be prevented.

4.11 Water storage ponds

Agricultural ponds are important water sources for irrigated areas. These structures collect water from small sources and allow for efficient storage and use of large flow rates when needed and help to regulate water flow.

5. Sustainable soil management

Sustainable land management includes many components. The multiplicity of components and the different prescriptions are due to the delicate but complex structure of the method and its applications. According to the FESLM: An

International Framework for Evaluating Sustainable Land Management definitions by FAO, sustainable land management combines socio-economic principles with environmentally sensitive technologies, policies and activities [9]. In order for sustainable land management to be feasible, five objectives have been identified as Efficiency, Security, Protection, Vitality and Acceptability, and the implementation and findings of the SLM regulation have been identified as the main pillars to be tested and monitored. Each target has its own characteristics and can be explained as follows:

- Efficiency: The return obtained from SLM is more than just evaluating with financial gains, it is evaluated to include the benefits that will be obtained from the protective, health and esthetic purposes of land use.

- Security: The management models that support the balance between land use and the existing environmental conditions reduce the production risks, whereas only those approaches that emphasize commercial anxiety increase this risk.

- Protection: Soil and water resources should be taken under strict protection for future generations. Locally, there may be additional protection priorities, such as the protection of genetic diversity or the need to protect specific plant or animal species.

- Vitality: If the applied land uses do not match the local conditions, the use cannot survive.

- Acceptability: If the social effects of land use methods are negative, it is inevitable to fail over time. The part directly affected by social and economic impact is not always clear.

Considering this framework, it should be produced safely in the field, established a production model that will protect the natural resources, the model should be economically feasible and socially acceptable. However, it should also be accepted that the system cannot be sustainable with the practices where the agricultural structure is not properly managed and the land is constantly destroyed. This method requires, in principle, to protect and improve soil fertility, to prevent and correct soil degradation and to prevent environmental damage.

5.1 Maintaining and improving soil productivity

5.1.1 Managing soil nutrients

In agriculture, healthy nutrition of the plants and increasing the use of fertilizers depends on the application of nutrients at the time of need, with sufficient and correct methods. Correct plant nutrition management is in interaction with many factors. For example, increasing fertilizer usage efficiency depends on reducing the losses of plant nutrients from soil due to leaching, denitrification, evaporation, surface flow. In fertilizer applications not suitable for the technique, the nitrogen is leaching from the soil or away from the gaseous state and the nutrients such as phosphorus and potassium are transformed into non-volatile forms. As a matter of fact, while 50% of the nitrogen applied to the soil is lost in various ways, 90% of the phosphorus cannot be taken by plants [10, 11]. Studies have shown that fertilizer nitrogen use efficiency is very low for wheat, paddy and corn, and nitrogen utilization rate is between 29 and 42% [12]. High nitrogen losses lead to significant environmental problems such as groundwater pollution, lake and river water eutrophication.

On the other hand, soil quality, soil organic matter and nutrient availability also show significant differences between methods such as minimum soil tillage, conventional tillage, conservational tillage and no-till agricultural systems. Totally used soil quality indicators in minimum data sets include total organic carbon, volume weight, aggregate resistance, usable moisture content, pH and EC [13]. Soil water retention capacity, soil water movement in soil, soil compaction and soil temperature also show significant changes depending on the agricultural system. Therefore, soil management has a special place in terms of fertilizer usage efficiency. In this respect, soil management includes more factors such as chemical fertilizers and the use of organic fertilizers (applied fertilizer type, dose, fertilizer application time, method) and irrigation. Fertilizer application methods are extremely important in terms of fertilizer economy. With the method to be applied, the efficiency of the fertilizers is increased and the larger areas can be fertilized with less fertilizer. In the case of slow and controlled conversion of fertilizers into a useful form, the loss of nutrients, especially nitrogen, is prevented and the plant will be used for a longer period of time and increased usage efficiency.

Soil analysis and soil sampling technique are very important in terms of fertilizer usage efficiency. In fact, it is a known fact that the physical and chemical properties of soils are highly variable in agricultural areas. Regionally, even on field level, soil properties show significant differences depending on distance. In fertilization without considering this feature of the land, some parts of the land will be applied more than the need and in some places less fertilizer will be applied. In this case, fertilizer will be deposited or washed in the soil in areas where fertilizer is given, and in areas where less fertilizer is needed, the yield will be low. Increased fertilizer use efficiency and the decrease in nutrient loss are proportional to each other [14]. Therefore, precision farming practices are one of the most important components of sustainable soil fertility and plant nutrition management.

The aim of this course is to evaluate the soil conditions, product characteristics and the variable productivity related to agricultural conditions within the boundaries of agricultural land in variable rate fertilization technologies in precision farming and to determine the time and amount of fertilization. Variable Rate Fertilization Maps used in this direction indicate the variable applications to different geographical coordinates based on the analysis of these conditions. This technology is also effective in deciding the nature of the fertilizer to be used. Different areas within the field can be evaluated separately and variable nutritional needs can be calculated. With the use of advanced technologies, the topographic structure of the field (different slope levels and depressions), the soil color which varies according to the organic matter content and the temporal yield variability in the field are taken into consideration. With the inclusion of land sampling data, all information is classified and analyzed in different databases; accordingly, the need for qualitative and quantitative fertilizers of the field subfields is determined.

5.1.2 Managing soil physical conditions

The soils under natural vegetation normally support the population of organisms and soil animals in an active biological activity. They live in plant roots and trash, digging and loosening the soil and use it as a nest. The vegetation is normally compressed by exposure to the effects of rain and soil processing and the effects of humans, animals and machinery. A certain proportion of compaction makes it suitable for the growth of plant roots in the soil and increases the ability of plants to retain the water they need to survive. Exposing the soil to compressing and then drying may cause the surfaces to crust. This reduces the water penetration rate and may cause water to flow from the surface and soil erosion.

Larger land resources were needed to supply food to the growing population, and soils were put under intensive use for overproduction. On the other hand, as a result of the pressure of increasing population, the deterioration in the fertile soil resources and the result of the structuralization show the effects of the loss of the area. As a result of the increase in the need for land resources, many countries around the world need to map their land in detail and use the land according to their capabilities. When the sustainability of natural resources is mentioned, first of all, soil erosion and its negative effects on the environment are one of the first issues that come to mind. Under normal conditions, climate, soil, topography and vegetation are the main elements that complement each other. Soil erosion is the result of this interaction. It is clear that the risk of erosion in agricultural areas is high, and if the conservation measures required by sustainable agricultural techniques are not taken, it will be possible to reach irreversible levels. Moreover, our resources, which are already limited by accelerated soil erosion, may be under great threat in the future.

Managing the physical properties of the soil includes the protection of the soil structure necessary for agricultural production, as well as the application of agricultural techniques and processing techniques to increase the long-term efficiency of the soil. Under these conditions, environment-friendly, healthy, economic and quality production conditions will be provided. Soil cultivation is also important for weed control, and this is usually one of the most important reasons for cultivating the soil. However, the introduction of herbicides has resulted in zero or minimum soil tillage techniques that eliminate the need to soil cultivation. Zero and minimum soil tillage techniques protect the soil from the direct impact of rain and wind by leaving crop remains on the surface. Surface residues prevent soil aggregates from being dispersed, transported by water or wind, the infiltration capacity of the soil is preserved, consequently there is no flow on the soil surface and erosion problem decreases. Generally, 56% water and 28% wind erosion are effective in soil degradation types. Among these reasons, agriculture has an important place with 28% (**Figure 2**) [15].

Intensive and timeless machine operations cause compression on the soil surface, especially in deeper layers and deterioration of the soil structure. Soil compaction is a state of degradation of soil aggregates and reduced pores between aggregates. Reduction of pore density reduces soil aeration, water drainage and water penetration into the deep layers, causing surface flow in rainy conditions. Soil compaction also complicates germination of the seed, limits the growth of plant roots, affects the biodiversity of the soil and causes the surface soil crusting.

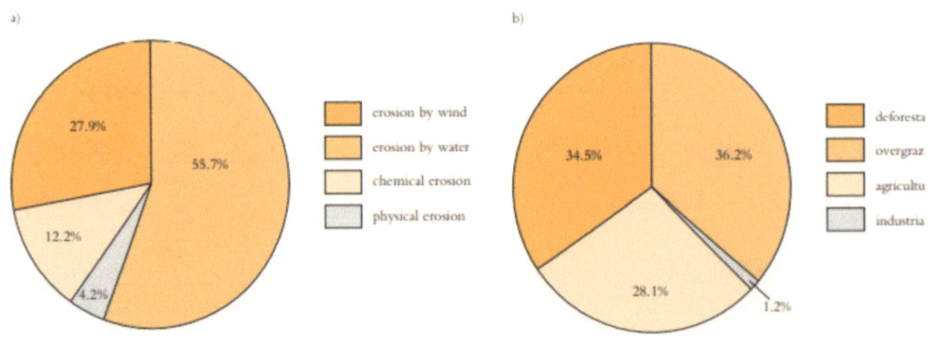

Figure 2.
Types and causes of soil degradation [15].

Some of the issues to be taken into consideration for the protection of soil physical structure can be listed as follows [15];

- Reduce the number and frequency of vehicle traffic, avoid unnecessary operations.
- Select suitable machines for the soil properties and the work to be carried out, check the tire pressure to reduce the pressure on the surface and reduce it if necessary.
- Agricultural practices that will increase soil organic matter and encourage soil structure, such as soil aeration, water leakage, heat transfer and root growth should be favored.
- In grazing systems, grazing intensity and timing should be planned well.

5.1.3 Water management

The most important means of ensuring healthy growth of the plant in sustainable agriculture is the sufficient amount of moisture in the root area of the soil during the plant's growing season. The first source of this moisture is the natural rainfall. In cases where sufficient water cannot be met by rainfall, the water needed should be given by irrigation water. Inadequate or too much soil moisture in the plant root area usually results in a decrease in yield.

The sustainability of water resources is a social, physical, economic and ecological concept. Sustainable water management encompasses the water needs of future generations, drinking and using, irrigation, industrial and recreational water conservation and ecosystem conservation services. In order to ensure sustainability, the following points should be taken into account:

- Irrigation system should be continuously controlled, pumps should be operated at optimum performance, water amount should be measured and water distribution evenness should be ensured.
- The irrigation time and amount should be planned by determining the plant water requirement and the most effective use of water should be ensured.
- Irrigation should be avoided in the middle of day and windy weather, irrigation should be done at night, and if possible drip irrigation method should be used.
- The system should be operated at optimum pressure, pipelines should be checked and leaks should be prevented.
- In any case, water, water sources and drainage channels should be avoided contamination.
- To reduce waterborne erosion; it should be ensured that the water is infiltrated to the soil with the principle of agriculture and irrigation method which is perpendicular to the direction of inclination.
- Production planning should be made considering the water quantity and the distance of water resources.

- Discharge of untreated farm wastes and wastewater into natural surface waters should be avoided. Measures should be taken to reduce the negative effects of irrigation on the ecosystem.

In addition, drone and sensor technologies can be used to collect necessary data for the development of an effective irrigation methodology. According to this:

- Determination of soil water potential with soil moisture sensors,

- Thermal images obtained from drones concerning soil and crop moisture content,

- Nitrogen deficiency can be measured by multispectral camera,

- A variable-rate irrigation program is created in line with weather data and weather forecasts.

- Variable rate applications can be done in optimum timing in fields varying in the field in terms of water requirement.

5.1.4 Pests and diseases management

Integrated Pest Management (IPM), as one of the effective methods used in modern agriculture, takes into account all plant protection methods available in the application. IPM implies the integration of appropriate measures that minimize the risks for human health and the environment by preventing the development of pest populations and by ensuring the use of plant protection products and other forms of intervention at economic and ecologically justified and reduced levels.

A well-designed integrated pest management program (IPM) includes three main steps for maximum effectiveness and minimum environmental impact in pest, weed and disease control [16]:

Find: Producers should first identify pests, diseases or weeds. Then, physical, chemical, biological and regulatory compliance options should be decided.

Watch: Reproduction rates are noted after the identification of harmful species. The determination of the effects of the protection methods and the limit threshold where the plant protection products will be used should be determined.

Select: When the density of harmful species reaches the threshold, many protection options are activated. With other protection methods, the use of pesticides that cause the least damage to the environment is the most effective protection method with harmful species. In addition, early harvesting or other physical protection methods can help minimize crop damage. When deciding on the protection method, the existence of useful species should be taken into consideration, and harmful species can be fought with the species which are the enemy of the pests without any application.

IPM for the prevention or suppression of harmful organisms as well as chemical control; crop rotation, use of appropriate breeding techniques (planting dates and densities, protected cultivation, pruning and direct sowing), use of tolerant varieties and certified seed and planting materials, use of balanced fertilization, liming and irrigation/drainage applications and prevention of spread of harmful organisms by hygiene measures (regular cleaning of machinery and equipment), which can be considered as a number of methods are important for sustainable agriculture [16]. Energy-based innovative cultural techniques: leguminous rotations, use of organic wastes as well as farm based by-products, integrated pest management

(IPM), pest and disease prediction, biological and cultural pest control, mulching and mechanical weed control, conservational tillage or no-till, mixed sowing and trap crops should be applied within the system [17].

5.1.5 Cover crop and rotation

Covered plants provide important contributions to agricultural production at the point of protection of soil, temperature, humidity or light at the desired level, pest and weed control. The reduction of soil cultivation in sustainable agriculture has brought with it the weed problem. Many plants such as clover, vetch, trefoil, oats, rye, sorghum vary widely according to usage and production purpose. For example, cereals are preferred for weed control, and legumes cultivation is preferred for providing nitrogen to the crop plant. The most important point in the cultivation of cover plants is to know the balance between the cost and the benefits of the system.

The system should both reduce the input cost and increase the product efficiency. Apart from its main purpose, cover crops have many other contributions to agriculture and production. The use of these plants allows increasing the amount of organic matter in the soil by protecting plant biomass and vegetative waste in the field. In this way, soil weathering improves, root growth of plants is encouraged and surface water flow decreases and aggregate formation increases. In addition, an increase in the population of living things such as microorganisms and worms, which contribute to the improvement of the nutritional cycle and soil structure, is achieved.

On the other hand, it is possible to reduce soil tillage, increase soil organic matter, benefit from different depths of nutrients, protect soil moisture, increase soil water holding capacity and weed control. For this reason, cover crop and crop rotation in sustainable agriculture is one of the important applications to reduce production inputs and to make economic agriculture.

6. Conclusions

The management of agricultural areas by traditional methods, the evaluation and processing of soil characteristics using traditional habits alone are not sufficient for the past, present situation and future productivity of the soil. Therefore, an evaluation of the tillage systems where soil tillage is appropriate to the management objectives and the effects on soil functions can be determined precisely. In determining the soil tillage system, the most suitable tillage system should be selected by evaluating the soil structure and quality, not only for the purpose of loosening and aerating the soil and destroying weeds. In order to compare soil management and processing systems, different indicators can be used in soil quality assessments according to soil conditions.

Today, the most important issue of researchers is the question of whether or not food can be produced enough to feed so many people in parallel with the rapidly increasing population. As a matter of fact, while focusing on this issue, it should not only be focused on the subject to feed of human, and should not be overlooked for the healthy and sustainable feeding. In particular, the non-cultivation agricultural system and protective agriculture in general are facing an ecologically and economically large potential for cultivated areas, whose productivity is decreasing day by day and becoming more open to erosion every day. On the other hand, the relationship between fertilizer, pesticide, tillage and crop rotation issues in sustainable systems and their effects on product yield and income should be well established.

According to most of the researches, agricultural production programs will begin to decrease as a result of rapid soil deterioration with the applied agricultural production programs, carbon balance will deteriorate and it will be difficult to obtain a healthy, sufficient and qualitative product in the not too distant future. Therefore, it is now necessary to increase the agricultural production in a way to protect nature and it is inevitable that sustainable agricultural techniques will be applied to reduce soil erosion, salinization, pollution of water resources and other damages. When planning production growth in agriculture, we are faced with the need and the necessity to develop new methods that guarantee natural resources instead of intensive input techniques, which cause irreversible microorganism losses in agricultural areas.

By applying the yield mapping system in agricultural production, it is necessary to determine the changes in the product characteristics in the land and thus the effective and economic planning of the amount of agricultural inputs to be used. In this direction, precision farming and variable rate applications are the most suitable methods to achieve maximum output by using the optimum and limited input. In contrast to traditional agricultural activities, this practice does not apply the amount of input to be applied to the field equal to each point, and applies variable rate according to the input maps created in line with the yield map. This application determines the need of appropriate input considering the specific conditions and the requirements of the land and weather conditions. Data maps are generated with the help of geospatial data, geographic information system (GIS) technologies and software which are acquired by various sensors on the harvesting machines. Drone and satellite technologies facilitate the creation of visuals that provide important information about land, soil and product structure. In this context, high resolution terrain and plant structure visuals, high resolution relief, slope and product maps can be obtained and thus it is possible to create drainage maps, to evaluate the effect of the slope factor in land efficiency and to obtain various data and base map that can be used in farm management.

Nowadays, with the introduction of Industry 4.0 technology, it is possible to reduce the costs of using natural resources at the required level by ensuring the communication of objects in agriculture. Similarly, all the factors necessary for production with smart systems in the farm are analyzed and presented to the manufacturer simultaneously. With the machines that are in contact with each other and working synchronously, a quick decision can be taken, resource wastage is prevented and quality products are produced. With systems equipped with digital sensors, it is aimed to maximize productivity by providing detailed and real-time information such as the type and amount of fertilizer to be given to the regions, weather conditions, plant mineral need, irrigation time, soil condition, estimated harvest time. Workload and cost are reduced with machines that work together and work synchronously. The producer is given the opportunity to manage and observe the whole farm from a tablet or telephone and by reducing the labor force, efficient, fun, high quality and natural production facilities are created.

DOI: http://dx.doi.org/10.5772/intechopen.88319

Author details

Koç Mehmet Tuğrul
Sugar Institute, Ankara, Turkey

*Address all correspondence to: kmtugrul@yahoo.com

References

[1] Pretty J. Agricultural sustainability: Concepts, principles and evidence. Philosophical Transactions or the Royal Society B. 2008;**363**:447-465. DOI: 10.1098/rstb.2007.2163

[2] Aydemir G, Oğuz O. Sustainable Agriculture (in Turkish). WWF-Türkiye. 2018. Available from: wwf.org.tr.

[3] Krall S. What is sustainable agriculture? Deutsche Gesellschaft für Internationale Zusammenarbeit (GIZ) GmbH. 2015. Available from: www.giz.de/nachhaltige-landwirtschaft [Accessed: 05 June 2019]

[4] Ussec. A Framework for Sustainable Farming. Available from: https://ussec.org/resources/soybean-growth-cycle/ [Accessed: 05 June 2019]

[5] Concerned Scientists. What is Sustainable Agriculture? 2019. Available from: https://www.ucsusa.org/food-agriculture/advance-sustainable-agriculture/what-is-sustainable-agriculture [Accessed: 05 June 2019]

[6] USDA. Conservation Reserve Program. United States Department of Agriculture Farm Service Agency. 2019. Available from: https://www.fsa.usda.gov/programs-and-services/conservation-programs/conservation-reserve-program/ [Accessed: 05 June 2019]

[7] Giller KE, Andersson JA, Corbeels M, Kirkegaard J, Mortensen D, Erenstein O, et al. Beyond conservation agriculture. Frontiers in Plant Science. 2015;**6**:870. DOI: 10.3389/fpls.2015.00870

[8] Pittelkow CM, Liang X, Linquist BA, van Groenigen KJ, Lee J, Lundy ME, et al. Productivity limits and potentials of the principles of conservation agriculture. Nature. 2015;**517**:365-368. DOI: 10. 1038/nature13809

[9] FAO. FESLM: An International Framework for Evaluating Sustainable Land Management. World Soil Resources Report 73. 1993. Available from: http://www.fao.org/3/T1079E/t1079e04.htm#nature%20of%20sustainable%20land%20management%20(slm) [Accessed: 05 June 2019]

[10] SmithaDR, OwensbPR, LeytemcAB, Warnemuendea EA. Nutrient losses from manure and fertilizer applications as impacted by time to first runoff event. Environmental Pollution. 2007;**147**(1):131-137

[11] Zhao H, Li X, Jiang Y. Response of nitrogen losses to excessive nitrogen fertilizer application in intensive greenhouse vegetable production. Sustainability. 2019;**11**:1513. DOI: 10.3390/su11061513

[12] Ma J, Li XL, Xu H, Han Y, Cai ZC, Yagi K. Effects of nitrogen fertilizer and wheat straw application on CH_4 and N_2O emissions from a paddy rice field. Australian Journal of Soil Research. 2007;**45**:359-367

[13] Bayram M, Günal H, Özgöz E. Importance of soil quality in determination of sustainable soil tillage methods. Yuzuncu Yil University Journal of Agricultural Sciences. 2015;**25**(3):337-346

[14] Liu G, Yang X. Spatial Variability Analysis of Soil Properties within a Field. In: Computer and Computing Technologies in Agriculture. Boston, MA: Springer; Vol. II. 2007. pp. 1341-1344. DOI: 10.1007/978-0-387-77253-0_75

[15] FAO. Cherish the Earth. Soil Management for Sustainable Agriculture and Environmental Protection in the Tropics. Available from: http://wgbis.ces.iisc.ernet.in/energy/HC270799/LM/SUSLUP/

FAO/CHRSHNGL.PDF [Accessed: 05 June 2019]

[16] NPIC. Integrated Pest Management (IPM). National Pesticide Information Center. Available from: http://npic.orst.edu/pest/ipm.html [Accessed: 05 June 2019]

[17] Edwards CA. The importance of integration in sustainable agricultural systems. Agriculture, Ecosystems and Environment. 1989;**27**(1-4):25-35

Chapter 8

Managing Soil Nitrogen under Rain-Fed Lowland Rice Production Systems in the Forest Agroecological Zones in Ghana

Mohammed Moro Buri and Roland Nuhu Issaka

Abstract

Rice is the second most important cereal in Ghana after maize. However, current production levels are about 47% of the country's requirements resulting in huge annual imports of the crop. One major constraint to production has been low soil nutrients and poor nitrogen management. Nitrogen is not only a major nutrient but also most often the most limiting nutrient element in lowland ecologies. With the introduction of improved soil and water management ("sawah" system) for lowland rice production, a study was conducted to determine the optimum nitrogen rates required. A randomized complete block design arranged in a split plot consisting of five levels of nitrogen as main treatments and three improved rice varieties as sub-treatments was adopted. Results showed that the total number of tillers per m^2 increased significantly with increasing levels of N as was total dry matter production. However, total number of panicles did not show the same relationship. Total biomass yield increased significantly and linearly with increasing levels of N. Paddy yield significantly increased from 1.7 t ha^{-1} (control) to a maximum of 9.4 t ha^{-1} (90 kg N ha^{-1}) before declining to 5.8 t ha^{-1} (150 kg N ha^{-1}) in the order $0 < 30 < 60 < 150 < 120 = 90$ kg N ha^{-1}, respectively. This result significantly and positively reflected on grain harvest index (GHI) in the order $0.27 < 0.38 < 0.46 < 0.47 < 0.57 < 0.68$ for 0, 30, 60, 150, 120 and 90 kg N ha^{-1}, respectively. Nitrogen at 90–120 kg ha^{-1} was therefore recommended. These rice varieties in addition to other improved ones will also perform well in other environments with similar biophysical characteristics across the country.

Keywords: grain yields, mineral fertilization, "sawah" technology, soil nutrients

1. Introduction

Poor and declining soil fertility remains the most important biophysical (abiotic) stress that accounts for the decline in agricultural productivity particularly in rice-growing environment in sub-Saharan Africa and in Ghana in particular [1–13]. Another notable and critical factor contributing to low agricultural productivity especially rice in Ghana is the low use of fertilizers [10, 14]. In highly weathered soils with low clay content and low activity clay minerals [13] as those of West Africa including Ghana, technology development for increased and sustainable

IntechOpen

nutrient management under improved soil and water management is very paramount. In Ghana where over 80% of rice farmers are poorly resourced, rice production levels will continue to be low unless technology development for increased and efficient use of inputs such as fertilizer is critically and urgently promoted.

Rice consumption has been on the increase in Ghana over the past few decades. According to the Ministry of Food and Agriculture, Ghana [15] rice has become the second most important staple food after maize, and its consumption keeps increasing. This has led to large annual imports of the crop as production constantly falls short of demand. On the average, annual rice import for Ghana is about 400,000 tons. The self-sufficiency ratio of rice in Ghana declined from 38% in 1999 to 24% in 2006. Rice yields in Ghana average about 2 tons per hector due to inherent poor soils and improper soil management practices [5, 6, 8]. With a potential available lowland area of over 700,000 ha, rice is cultivated across all the agroecological zones of Ghana. However, there are significant differences in the production potential (area available and suitability) among these ecosystems due to differences in soil, climate, and economic conditions.

The high rain forest and semi-deciduous rain forest agroecological zones have a comparative advantage due to their good rainfall and better water availability throughout a greater part of the year. While the impact of fertilizer use for crop production is considered large in regions of extremely low soil fertility (particularly N and P), the application of chemical fertilizer to crops in Ghana is one of the lowest in West Africa [3, 16]. Rice is grown within these zones with very little or sometimes no application of mineral fertilizers. There are also no structures put in place to manage water. Efficiency of applied fertilizer is therefore very low due to poor water control. With the recent introduction of improved soil and water management, rice framers' yield of at least 4.0 t ha^{-1} is ensured [4, 7]. However, for higher yields and improved/sustained productivity, mineral fertilization is necessary. Previous studies have shown that rice responds to mineral fertilization in these lowlands [5–7]. Hence, for site-specific management and the bulk of rice growers being resource-poor small-scale farmers, it is necessary to develop technologies (optimum levels of critical nutrients such as nitrogen) that are easily transferrable and adaptable.

With the increasing use of lowlands in the forest agroecological zones for rice production, this chapter looks at the relevance and how effective nitrogen can be managed for increased rice grain yields and developing a sustainable production system.

2. Forest agro-ecology and rice-growing soils

In Ghana, there are two main forest agroecological zones, namely, the high rain forest and semi-deciduous rain forest (**Figure 1**). These agroecological zones lie between latitude 5°N and 7°N and longitude 0°W and 3°W. The agro-ecology covers a total land area of about 3.45 million hectares representing 18.9% of total land area of country. The agroecological zones cover the whole of the Western, Western North, parts of the Ashanti, Central and Eastern regions. Lowlands (inland valleys, floodplains) are spread across the area where rice cultivation is gradually intensifying due to water availability.

Rice is grown mainly in the valley bottoms, valley fringes, floodplains, colluvial foot slopes, and generally hydromorphic sites. Water is readily available throughout greater portions of the year. Rice is mainly grown under partially irrigated and rain-fed conditions. Within very limited areas, however, rice is also grown under irrigation. The main soils of rice-growing sites include valley bottoms (Gleysols), foot slopes, and valley fringes (Humic Ferralsol and Gleyic Lixisols).

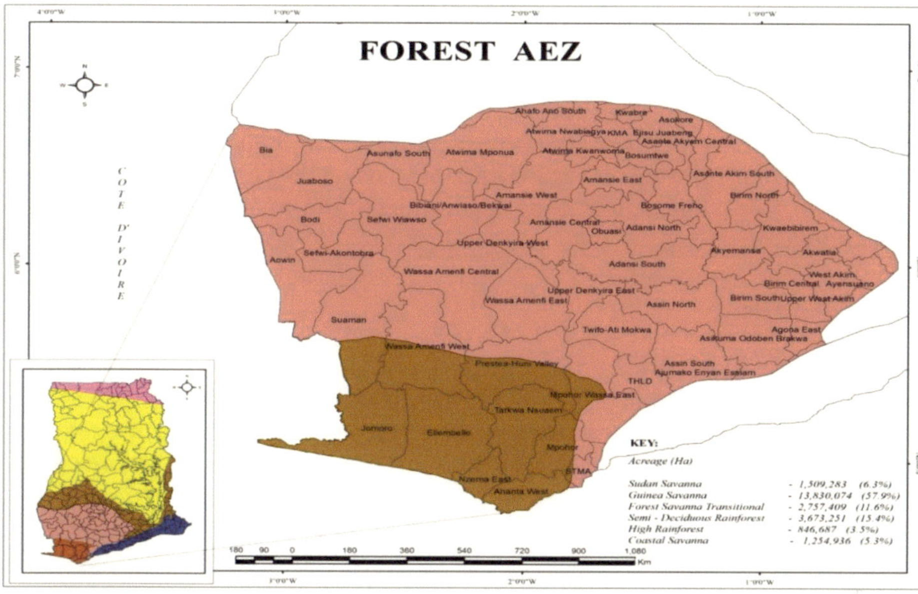

Figure 1.
Location of the two forest agroecological zones of Ghana.

3. Materials and methods

3.1 Experimental setup

The site was initially slashed and vegetative cover removed. The area was then ploughed using a mini-tractor (power tiller). The plowed site was demarcated into four main blocks through the construction of bunds. Using a split plot design with nitrogen rate (level) as main treatment and rice variety as sub-treatment, each block was divided into six main plots using minor bunds (100 cm wide and 50 cm high) representing six nitrogen rates (0, 30, 60, 90, 120, 150) kg N per ha. Each main plot was again subdivided to three plots, each measuring 2 m × 2 m for the three rice varieties (Sikamo, Jasmine 85, Marshall). The characteristics of these rice varieties are indicated in **Table 1**. Each subplot was then puddled and manually leveled. A composite soil sample (0–30 cm) was initially collected from the site for laboratory analysis before land preparation. Three-week-old rice seedlings were transplanted to their respective plots using the specified varieties at a spacing of 20 cm × 20 cm and at two seedlings per hill. A uniform level of 60 kg P ha^{-1} as

Rice variety	Days to maturity	Av. yield (t ha^{-1})	Yield potential (t ha^{-1})	Comments
Sikamo	130–135	6.5	8.5	Nonaromatic
Jasmine 85	120–130	6.5	8.5	Aromatic
Marshall	120–130	6.0	8.0	Nonaromatic
AGRA	125–130	6.0	8.0	Aromatic
Amankwatia	120–125	6.0	8.0	Aromatic
CRI-Dartey	120–125	6.5	9.0	Aromatic

Table 1.
Characteristics of the varieties used and other existing varieties.

P_2O_5 using triple superphosphate as phosphorus source, 60 kg K ha^{-1} as K_2O using Muriate of Potash as potassium source, and 50% N using urea as nitrogen source was applied to each subplot immediately after transplanting as basal fertilizer. All fertilizer was uniformly broadcasted on the field manually. The remaining 50% N was applied as split, at 25 days after transplanting (maximum tiller formation) and 55 days after transplanting (at panicle initiation) using the same broadcast method. Weed control was manual, mainly by handpicking. Crop growth was then monitored until harvest.

3.2 Soil analysis

Soil sample was air dried at room temperature. Dried sample was then ground and passed through a 2 mm sieve. Soil pH was measured in a soil/water ratio of 1:2.5 [17]. Total carbon was measured by the method of [18] and total nitrogen by the micro Kjeldahl method [17]. Available P was measured by the method of [19]. Exchangeable cations (Ca, Mg, K, Na) were extracted using a 1.0 M ammonium acetate solution and determined by atomic absorption spectrometry [20]. Exchangeable acidity was determined by titration and eCEC calculated as sum of exchangeable cations and acidity.

3.3 Growth characteristics

Number of tillers was counted after maximum tiller formation stage and mean number of tillers determined, while plant height was measured at harvest.

3.4 Yield characteristics

At maturity, an area of $1m^2$ excluding border rows was measured out in each subplot and harvested. Grain and stover yield were measured and yield per hectare estimated. Panicles were also collected from non-border rows and mean individual weight per panicle determined. The weight of 1000 grains was measured using an electronic balance. Grain harvest index (HI) was calculated as ratio of grain yield to total yield (grain + stover).

3.5 Statistical analysis

The statistical software STATISTIX 8 was used to analyze the data, and LSD (0.05) was used as the mean separator.

4. Characteristics of growing environment

The agroecological zones have a bimodal rainfall pattern (**Figure 2**) and therefore have two main cropping seasons (major and minor). The major season has its peak rainfall in June to July, while that of the minor is in September to October. The two agroecological zones have a comparative advantage over other agroecological zones due to their good rainfall and higher water availability throughout a greater part of the growing season. The physicochemical properties of the soils of these zones are shown in **Table 2**. The soils are typically low in inherent fertility and poor in plant nutrients particularly total nitrogen (N) and available phosphorus (P). Soil texture ranges from pure sandy soils through loam to clay soils. Under such low levels of fertility, improved/efficient nutrient management is critical if higher rice yields are to be obtained and when increased and sustained total productivity are to be achieved.

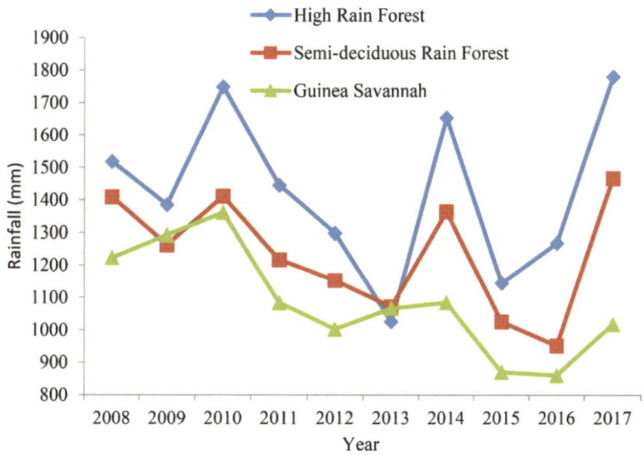

Figure 2.
Rainfall amounts and distribution in the three main agroecological zones of Ghana.

Parameter	Units	High rain forest	Semi-deciduous rain forest	Savannah
pH (water)	—	5.7	5.7	4.6
Total carbon	$g\ kg^{-1}$	10.3	12	6.1
Total nitrogen	$g\ kg^{-1}$	0.90	1.10	0.65
Av. phosphorus	$mg\ kg^{-1}$	1.4	4.9	1.5
Ex. potassium	cmol (+) kg^{-1}	0.22	0.42	0.22
Ex. calcium	cmol (+) kg^{-1}	2.25	7.50	2.10
Ex. magnesium	cmol (+) kg^{-1}	1.12	4.10	1.00
Ex. sodium	cmol (+) kg^{-1}	0.26	0.32	0.12
Ex. acidity	cmol (+) kg^{-1}	0.65	0.31	1.00
Effective CEC	cmol (+) kg^{-1}	4.49	12.65	4.44
Clay	$g\ kg^{-1}$	110	127	66
Silt	$g\ kg^{-1}$	240	620	607
Sand	$g\ kg^{-1}$	650	300	327

Table 2.
Mean physicochemical properties of soils of the main agroecological zones.

5. Responses to nitrogen fertilizer application

5.1 Effect of nitrogen on growth parameters

5.1.1 Number of tillers m^2

The number of effective tillers produced is a good indicator as it is a major determinant of yield. Tiller number increased with increasing N levels, but the increased was more pronounced from 0 kg N to 30 kg N than from 30 to 60, 60 to 90, 90 to 120, and 120 to 150 kg N ha^{-1} (**Table 3**). Generally, total number of tillers per m^2 significantly increased by 53, 70, 72, 77, and 103% over the control for 30, 60, 90, 120, and 150 kgN ha^{-1}, respectively. There was also a corresponding increase in total dry matter production with increasing levels of N. These observations are

Nitrogen rate (kg ha^{-1})	Rice variety			
	Sikamo	Jasmine 85	Marshall	Mean
0	218	198	232	216
30	303	358	333	331
60	368	377	358	368
90	385	370	368	374
120	383	410	355	383
150	440	407	475	440
Mean	350	353	354	

LSD (0.05) Fertilizer = 86; LSD (0.05) Variety = 36; LSD (0.05) Fertilizer × Variety = 112.

Table 3.
Effect of the interaction of nitrogen levels and rice varieties on the number of tillers m^{-2}.

similar to other findings. In 2006 [21], working on the effect of N and P fertilizers reported application of N up to 120 kg ha^{-1} increased the number of panicles per m^2 significantly apparently by increasing the number of productive tillers. However, the authors also reported that there was a reduction in the number of panicles per m^2 at the highest N application, attributing this observation to excessive vegetative growth of the rice crop. However, paddy yield did not show a similar trend with increasing levels of N. At higher levels of N (> 90 kg ha^{-1}), more tillers tended to be unproductive resulting in lower paddy yield. There were also no significant differences in the number of effective tillers produced in the variety × N rate interaction in line with an observation earlier made by [22] who noted that interactions between N and variety were not significant for all measured traits for four lowland NERICA varieties in Nigeria and those of [23], who worked on the effect of minerals N and P on the yield and yield components of flooded lowland rice in Ethiopia.

5.1.2 Plant height

Plant height was significantly affected by N application (**Table 4**). Plant height was similar for 0 and 30 kg N ha^{-1} levels but significantly shorter than for 60, 90, 120, and 150 kg N ha^{-1}. The initial nutrient levels were probably good enough to produce plants of similar height to 30 kg N ha^{-1}. Nitrogen is a major contributor

Nitrogen rate		Rice variety		
(kg ha^{-1})	Sikamo	Jasmine 85	Marshall	Mean
0	84	85	72	80
30	101	95	94	97
60	119	105	99	108
90	122	117	105	115
120	128	118	110	119
150	124	115	112	117
Mean	108	105	99	

LSD (0.05) Fertilizer = 14; LSD (0.05) Variety = 5; LSD (0.05) Fertilizer × Variety = 16.

Table 4.
Effect of the interaction of nitrogen levels and rice varieties on plant height (cm).

to crop growth, size, and total dry matter production. The increase in height with increasing levels of N could not be explained better. While [23] in a similar study in Bida, Nigeria, observed that there were significant increases in plant height with increasing levels of N when compared with the Control, Metwally [24] also reported that plant height was significantly affected by nitrogen rate of 110 and 165 kg N ha^{-1} over the control. Ref. [24] further indicated that the interaction between mineral N rates and organic materials had a significant effect on plant height. [25], however, reported that there were no significant differences in N rates × variety interaction, while significant N effects were only found in plant height. In this study, comparing the three varieties, Sikamo and Jasmine 85 had similar plant heights which were significantly taller than Marshall. Two varieties (Sikamo and Jasmine 85) interacted with 60 kg N ha^{-1} level and above to give significantly taller plants. Similar taller plants were also observed when Marshall interacted with 60 kg N ha^{-1} level and above. Generally when N was not applied, plants were significantly shorter.

5.2 Effect of nitrogen on yield parameters

5.2.1 Total biomass

The total biomass (straw + grain) increased with increasing levels of N (**Table 5**). Total biomass increased from 9.9 t ha^{-1} at 0 kg N ha^{-1} to a maximum of 18.5 t ha^{-1} at 150kg N ha^{-1}. At N rates of 30, 60, and 90 kg N ha$^{-1,}$ biomass yields were significantly higher than 0 kg N ha^{-1}. Higher N rates of 120 and 150 kg N ha^{-1} further significantly produced higher biomass yields. Total biomass increased by 4.0, 5.4, 6.1, 8.4, and 8.6 t ha^{-1} over the control for 30, 60, 90, 120, and 150 kg N ha^{-1}, respectively. Between varieties, total biomass production for Sikamo was similar to Jasmine 85 but significantly higher than Marshall. The effect of both N and variety interaction showed that Sikamo at 120 and 150 kg N ha^{-1} gave significantly higher biomass than Sikamo or Jasmine 85 fertilized at 0 or 30 kg N ha^{-1}. Marshall fertilized from 0 to 90 kg N ha^{-1} produced lower total biomass than both Sikamo and Jasmine 85. Generally Sikamo and Jasmine 85 were taller than Marshall (**Table 3**), and higher N rates had more tillers than the control (**Table 4**). This largely explains the observed differences in biomass production.

Reference [23] while looking at the effect of water management and N rates in a similar study reported that there were significant differences in straw and grain yield in other treatments compared with the control. According to the authors, yield and N use efficiency generally increased with increasing levels of N but declined at

Nitrogen rate	Rice variety			
(kg ha^{-1})	Sikamo	Jasmine 85	Marshall	Mean
0	10.27	9.60	9.90	9.92
30	14.30	14.47	13.00	13.92
60	16.73	16.70	13.17	15.53
90	16.57	15.97	15.47	16.00
120	20.20	18.03	16.67	18.30
150	19.67	17.77	18.00	18.50
Mean	16.29	15.42	14.37	

LSD (0.05) Fertilizer = 2.612; LSD (0.05) Variety = 1.148; LSD (0.05) Fertilizer × Variety = 3.475.

Table 5.
Effect of different levels of nitrogen on total biomass (t ha^{-1}) for the three varieties.

80 kg N ha^{-1}. Ref. [24] while investigating the effect of mineral N fertilizer on rice reported that increasing N fertilizer levels resulted in a corresponding increase in straw yields, stating that the highest straw yields were obtained with the highest N rates of 165 and 110 kg N ha^{-1}. Ref. [24] attributed these observations mainly due to the fact that N fertilizer increased dry matter, leaf area index, and number of tillers. In this study, while total biomass increased with increasing levels of N up to 150 kg N ha^{-1}, grain yield declined after 90 kg N ha^{-1}. After 90 kg N ha^{-1}, further N addition seemed to contribute more to vegetative growth (greater straw production) at the expense of reproductive growth (grain production).

5.2.2 Mean panicle weight

The mean weight of individual panicles was determined for each level of N applied (**Figure 3**). Panicle weight was significantly affected by N application. Lowest individual panicle weights (< 3.0 g $panicle^{-1}$) were obtained under the control where N was not applied. Individual panicle weight increased significantly (> 4.0 g $panicle^{-1}$) with 30kgN ha^{-1} additions, rising to above 5.0 g per $panicle^{-1}$ at 90 and 120 kg N ha^{-1}. Significantly lower panicle weights were recorded at 150 kg N ha^{-1} than 90 and 120 kg N ha^{-1}. These results are in conformity with other findings. Ref. [23] reported that plant height, grain yield, panicle weight, 1000 grain weight, and grain harvest index (GHI) were significantly influenced by N and genotype treatments. In the same vain, [24] also reported that mineral N and organic material application to rice significantly affected the number of grains per panicle. Treatments that received mineral N fertilizer in addition to organic materials had significantly higher panicle weights over those that did not, and it increased with increasing levels of fertilizer and organic materials. In this study, the significantly higher panicle weights of 90 and 120 kg N ha^{-1} significantly contributed to higher grain yields recorded for those treatments, particularly at 90 kg N ha^{-1} (**Figure 3**).

5.2.3 Grain yield

Grain yield produced for the different levels of N applied is presented in **Figure 4**. Grain yield ranged from 1.7 t ha^{-1}(lowest) to 9.4 t ha^{-1} (highest) across N levels and varieties. Grain yield was significantly higher for Sikamo and Jasmine 85 fertilized at 90 kg N ha^{-1} than all the other N x variety interactions except Marshall × 90 kg N ha^{-1} and both Sikamo and Jasmine fertilized at 120 kg N ha^{-1}. Grain yield for all the varieties was almost similar at both 60 and 150 kg N ha^{-1}. Generally grain

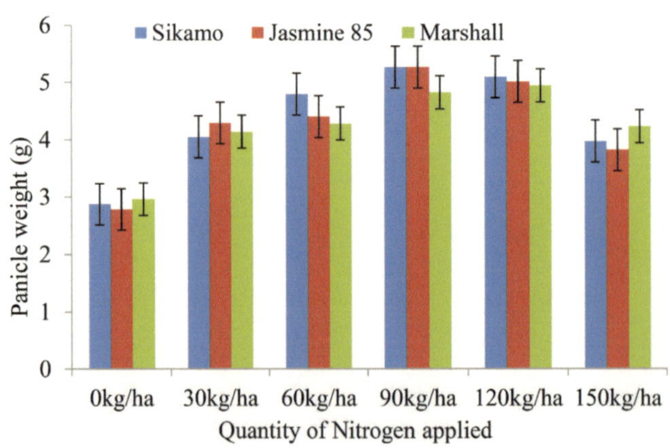

Figure 3.
Effect of varying levels of nitrogen on individual panicle weight (g) of rice.

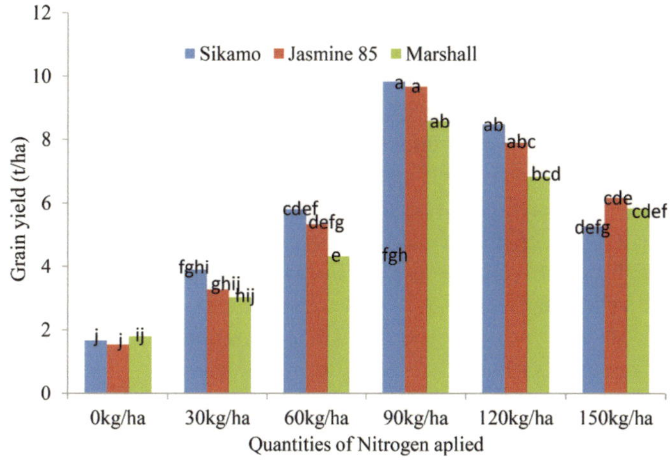

Figure 4.
Effect of varying levels of nitrogen on rice paddy yield.

yield increased with increasing levels of N from 1.7 t ha^{-1} (0kgN ha^{-1}) to a maximum of 9.4 t ha^{-1} (90 kg N ha^{-1}) and thereafter declined, indicating that higher levels of N suppressed yield. This is in accordance with the earlier findings of [25] who reported that excessive nitrogen application to rice in China caused environmental pollution, increased cost of farming, reduced grain yield, and contributed to global warming. Furthermore, [24] indicated that, filled grain percentage was affected by nitrogen fertilizer and organic materials, adding that plants that did not receive nitrogen produced the lowest number of filled grains while those that received 165 kg N ha^{-1} produced the highest filled grains, followed by those that received 110 kg N ha^{-1}. Ref. [21] working on the effect of N and P fertilizers on yield, and yield components of rice also reported that N had a marked effect on grain yield and that grain yield increased from 3240 to 3962 kg ha^{-1} with an increase in the levels of N from the control (no N) to 60 kg N ha^{-1} and decreased further with increase in applied N fertilizer. Ref. [21] further reported that the magnitude of increase in grain yield over the control due to application of 30 and 60 kg of N ha^{-1} was 13.5% and 22.3%, respectively. Grain yield was generally very high compared to the mean grain yield of 2.0 t ha^{-1} reported by the Ministry of Food and Agriculture, Ghana [15]. Such high levels of grain yield for the rain-fed lowlands could be attributed to the use of good varieties, fertilizer additions, and improved soil and water management under the "sawah" system (bunded and leveled fields). Ref. [8] reported that lowland rice significantly responded to N, P, and K additions in selected sites in southern Ghana. Ref. [26] also observed that while bunding significantly increased yield across sites in La Cote d'Ivoire by almost 40% and controlled weeds, mineral fertilizer N application significantly increased yield by 18% with N use efficiency being 12 kg compared to 4 kg of rice grain per kg of N applied in open field. Ref. [26] further indicated that across environments, about 60% of observed variability in rice grain yield was explained by water control and agronomic management (N application, weed control). With improved soil and water management under the "sawah" system, N use efficiency is increased, and higher grain yields are obtained when compared to open fields with poor soil management and no water control [4]. Under this study, N utilization was improved due to improved water management. Hence moderate levels of N recorded higher grain yields. Evaluating the response of four rain-fed NERICA varieties to N fertilization, [22] also reported that even though the interactions between N and variety were not significant for all measured traits, yield response to N was linear and significantly increased with increasing levels of N up to 100 kg N ha^{-1}.

With results showing a linear trend and yield increase of 3 tons ha^{-1} (100 kg N ha^{-1}) over the control, the authors recommended further studies to establish optimum levels for the rain-fed lowlands of the northern Guinea savanna zone of Nigeria. In a similar study, [27] reported that N fertilization significantly increased dry matter and grain yield with maximum yield (6.4 t ha^{-1}) obtained at 120 kg N ha^{-1} during year 1 and maximum yield (6.3 t ha^{-1}) obtained at 90 kg N ha^{-1} in year 2. Ref. [27] further observed that other yield components such as panicle length and panicle number per m^2 were significantly affected by N fertilization with panicle number per m^2 showing the highest correlation ($r = 0.70$ and 0.78) for 2 years. In this study, however, mean maximum yields were obtained at 90 kg N ha^{-1} for all three varieties over the period confirming the findings of [9] who recommended 90 kg N ha^{-1} as the optimum rate.

5.2.4 Weight of 1000 grains

The effect of varying levels of N on 1000 grains is presented in **Table 6**. Lowest 1000 grain weight recorded was 22.04 g, while the highest was 26.91 g, both under Jasmine 85. The application of N significantly affected the weight of 1000 grains over the control. However, there were no significant differences in 1000 grain weight between 30, 60, 120, and 150 kg N ha^{-1} application. Jasmine 85 interacted with 60 kg N ha^{-1} to produce the highest 1000 grain weight, followed closely by Sikamo at 90 kg N ha^{-1}.

5.2.5 Correlation between grain yield and yield components.

Table 7 shows the relationships between grain yield and yield components. All the yield components strongly correlated with grain yield with plant height, biomass, and panicle weight giving the highest correlations. This signifies that changes in these components will affect grain yield, as was observed.

5.2.6 Grain harvest index (GHI)

The grain harvest index (GHI) calculated for the different levels of N applied is shown in **Figure 5**. This is a measure of the ratio of economic yield (grain) to total yield (grain + straw). The higher the value, the better or higher the returns/gain from any fertilizer additions. GHI was significantly affected by N application for all the three varieties. The lowest GHI (0.27) was recorded for the control (no N applied), while the highest GHI (0.68) was recorded at 90 kg N ha^{-1}. Harvest index was in

Nitrogen		**Rice variety**		
Rate (kg ha^{-1})	**Sikamo**	**Jasmine 85**	**Marshall**	**Mean**
0	22.32	22.04	22.20	22.19
30	24.17	25.46	26.65	25.42
60	26.46	26.91	26.66	26.68
90	26.88	26.74	26.66	26.76
120	26.51	26.42	26.69	26.54
150	25.44	25.58	25.86	26.17
Mean	16.29	15.42	14.37	

LSD (0.05) Fertilizer = 1.35; LSD (0.05) Variety = 0.66; LSD (0.05) Fertilizer x Variety = 1.89.

Table 6.
Effect of different levels of nitrogen on 1000 grain weight (g) for the three varieties.

Growth parameter	Grain yield
Plant height	0.7474***
Biomass	0.7533***
Tillers m^{-2}	0.5881***
Panicle weight	0.7567***
1000 seed weight	0.5718***

****indicates significance at 1% probability level.*

Table 7.
Correlation between grain yield and yield components.

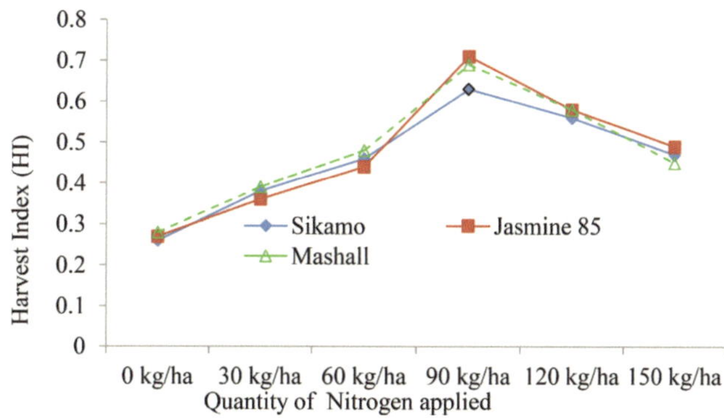

Figure 5.
Effect of varying levels of nitrogen on grain harvest index (GHI).

the order 0 < 30 < 60 < 150 < 120 < 90 kg N ha^{-1}. GHI showed a similar trend for the three varieties and was significantly and positively correlated with grain yield. A similar observation was also reported by [23]. The above observations clearly show that higher doses of nitrogen for rice production in these lowlands do not only result in significant yield reductions but also lead to higher cost of production for the mostly resource-poor farmers as cost of mineral fertilizer is high.

6. Conclusion

Results show that fertilizer use significantly affects rice yield. However, higher rates of N tended to suppress grain yield but promote straw production. The optimum rate was observed to be 90 kg N ha^{-1}, but this could be increased to 120 kg N ha^{-1} depending on soil type, rainfall regime, and affordability of individual farmer. In the lowlands therefore appropriate crop, soil and water management practices can result in high rice grain yield of over 9000 kg ha^{-1}. The introduction of such improved technologies can help to significantly improve yields over the current national mean of 2000 kg ha^{-1} and contribute to enhancing food availability and security in the country. The three rice varieties are highly productive under these nitrogen rates.

6.1 Recommendations

It is therefore recommended that, to sustain rice production and for increased yields, N application is best within 90–120 kg ha^{-1} based on location, specific rainfall amounts, and soil types. The three tested rice varieties (Sikamo, Jasmine 85, Marshall) are all suitable for cultivation within the high rain forest and semi-deciduous rain forest agroecological zones of the country. Furthermore these rice varieties, in addition to other improved varieties like AGRA, Amankwatia, and CRI-Dartey, are suitable and recommended for lowlands in the other agroecological zones with similar biophysical and physicochemical characteristics. Land preparation methods and water management remain key and very critical factors, and the adoption of the "sawah" technology (bunding, puddling, and leveling) with easy-to-adapt water control structures is most suitable for these areas.

Acknowledgements

The authors are very grateful to JIRCAS who partially provided funding for this study to be conducted. We are also grateful to all those who contributed in diverse ways to make this study possible particularly the technical and support staff at the Soil Chemistry, Fertility and Plant Nutrition Division of the CSIR-Soil Research Institute.

Author details

Mohammed Moro Buri[1*] and Roland Nuhu Issaka[2]

1 CSIR – Soil Research Institute (SRI), Kumasi, Ghana

2 Africa Rice Centre, Central Agricultural Research Institute (CARI), Suakoko, Bong County, Liberia

*Address all correspondence to: moro_buri@yahoo.com

References

[1] Sanchez PA, Buresh RJ, Leakey RB. Trees, soils and security. Philosophical Transactions of the Royal Society, Series B. 1997;**352**:352-370

[2] Issaka RN, Ishida M, Kubota D, Wakatsuki T. Geographical distribution of soil fertility parameters of West Africa Inland Valleys. Geoderma. 1997;**75**:99-116

[3] Issaka RN, Masunaga T, Kosaki T, Wakatsuki T. Soils of inland valleys of West Africa: General fertility parameters. Soil Science and Plant Nutrition. 1996;**42**:71-80

[4] Issaka RN, Buri MM, Wakatsuki T. Effect of soil and water management practices on the growth and yield of rice in the forest agro-ecology of Ghana. Journal of Food, Agriculture and Environment. 2008;**7**(1):214-218

[5] Senayah JK, Issaka RN, Dedzoe CD. Characteristics of major lowland rice-growing. Agricultural Food Science Journal of Ghana. 2008;**6**:445-458

[6] Buri MM, Ishida F, Kubota D, Masunaga T, Wakatsuki T. Soils of flood plains of West Africa: General fertility status. Soil Science and Plant Nutrition. 1999;**45**:37-50

[7] Buri MM, Masunaga T, Wakatsuki T. Sulfur and zinc levels as limiting factors to rice production in West Africa lowlands. Geoderma. 2000;**94**(2000):23-42

[8] Buri MM, Issaka RN, Wakatsuki T, Otoo E. Soil organic amendments and mineral fertilizers: Options for sustainable lowland rice production in the forest agro-ecology of Ghana. Agricultural and Food Science Journal of Ghana. 2004;**3**:237-248

[9] Buri MM, Issaka RN, Wakatsuki T. Determining optimum rates of mineral fertilizers for economic rice grain yields under the "Sawah" system in Ghana. West Africa Journal of Applied Ecology. 2008;**12**:19-31

[10] Buri MM, Issaka RN, Fujii H, Wakatsuki T. Comparison of soil nutrient status of some rice growing environments in the major agroecological zones of Ghana. Journal of Food, Agriculture and Environment. 2010;**8**(1):384-388

[11] Buri MM, Issaka RN. Study on the Utilization of Fertilizers in the Ashanti region of Ghana. CSIR Technical Report No. CSIR – SRI/CR/BMM/2010/01; 2010

[12] Buri MM, Issaka RN, Wakatsuki T, Kawano N. Improving the productivity of lowland soils for rice cultivation in Ghana: The role of the 'Sawah' system. Journal of Soil Science and Environmental Management. 2011;**8**:300-310

[13] Abe S, Buri MM, Issaka RN, Kiepe P, Wakatsuki T. Soil fertility potential for rice production in West African lowlands. JARQ. 2010;**44**:343-355

[14] IFDC. Ghana Fertilizer Assessment. Muscle Shoals, Alabama, USA: International Fertilizer Development Center; 2012. p. 40

[15] MoFA. National Rice Development Strategy of the Ministry of Food and Agriculture. Ghana; 2010

[16] FAO. Food and Agricultural Organization of the United Nations Current World Fertilizer Trends and Outlook to 2015; 2011

[17] IITA. Selected Methods for Soil and Plant Analysis. Manual Series No. 1. Ibadan, Nigeria: IITA; 1979

[18] Bremner JM. Total nitrogen. In: Black CA, editor. Methods of Soil Analysis, Part 2 Chemical and

Microbiological Properties. 1965. pp. 1149-1178

[19] Bray RH, Kurtz LT. Determination of total, organic and available forms of phosphorus in soils. Soil Science. 1945;**59**:39-46

[20] Thomas GW. Exchangeable cations. In: Page AL, editor. Methods of Soil Analysis. Madison: American Society of Agronomy; 1982. pp. 159-165

[21] Gebrekidan H, Syoum M. Effects of mineral N and P fertilizers on yield and yield components of flooded lowland rice on vertisols of Fogera plain, Ethiopia. Journal of Agriculture and Rural development in the Tropics and Subtropics. 2006;**107**(2):161-176

[22] Kamara AY, Ekelene F, Omoigu KO, Chikoye D. Influence of nitrogen fertilization on yield and yield components of rain-fed lowland NERICA rice in the northern Guinea Savanna of Nigeria. African Journal of Agricultural Research. 2011;**6**(13):3092-3097

[23] Fageria NK, Santos AB. Yield and yield components of lowland rice genotypes as influenced by nitrogen fertilization. Communications in Soil Science and Plant Analysis. 2015;**32**:1405-1429. DOI: 10.1080/00103624.2015.1043443

[24] Metwally TF. Impact of organic materials combined with mineral nitrogen on rice growth, yield, gain quality and soil organic matter. International Journal of ChemTech Research. 2015;**8**(4):1533-1542

[25] Shaobing P, Buresh RJ, Huang J, Zhong X, Zou Y, Yang J, et al. Improving nitrogen fertilization in rice by site specific N management. A review. Agronomy for Sustainable Development. 2010;**30**(2010):649-656

[26] Becker M, Johnson DE. Improved water control and crop management effects on lowland rice productivity in West Africa. Nutrient Cycling in Agro-ecosystems. 2001;**59**:119-127

[27] Fageria NK, Baligar VC. Lowland rice response to nitrogen fertilization. Communications in Soil Science and Plant Analysis. 2001;**32**(9 & 10): 1405-1429

Chapter 9

Plant Nutrition and Sustainable Crop Production in Nigeria

Romanus Osabohien and Toun Ogunbiyi

Abstract

The aim of this study is to examine the determining factors of plant nutrition and sustainable crop production in Nigeria. The study applied an in-depth review of literature and observed that different biotic and abiotic factors interact together to determine the outcome of plant nutrition and sustainable crop production in Nigeria. These factors include; types of fertilizers applied, atmospheric emissions, level of technological development, infrastructural facilities, climatic conditions, irrigation method, and level of skilled labour force. The study recommended that there should be increased and equal access to credit facilities, social protection incentives, and more innovation and technological involvement in the agricultural sector in order to increase productivity and efficiency.

Keywords: credit access, crop production, plant nutrition, productivity

1. Introduction

Farmers experience various problems in the quest to produce enough crops in order to meet the demand of the ever-teaming population and still keep constant and without comprising the standard of the available resources for generations to come. Mechanisms are needed to enhance soil and plants nutrients in order to increase crop yields, and plant nutrients are one of the requirements to enhance crop production [1, 2].

Plants' nutritional quality directly affects human nutrition in terms of productivity. It is therefore no gainsaying that the quality of food consumed in a country determines the quality of its populace. In many developing and developed countries, deficiency of micronutrients in pastures and crops has a negative effect on the health of both plants and animals [3]. In addition, the adequate provision of sunlight, air and water is a major prerequisite for optimum plant yield and improved crop management. In a bid to achieve these, various countries have devised means to reduce the negative effects of both abiotic and biotic factors in plants. After dedicating sufficient time and capital for farming, the goal of many Nigerian farmers is to produce sustainable crop yield [1].

However, certain factors come into play which might not augur well for agricultural yield in general and crop production in particular. Considering the rate of food insecurity in most developing countries, resulting from unfavourable weather condition owing to global climate change, the improved sustainable management of plant nutrition has been considered a precondition to reduce the challenge of prevailing hunger in the affected countries, Nigeria not excluded [1].

IntechOpen

Owing to inadequate mechanisation and the small-scale nature of agricultural production, Nigeria has not been able to achieve self-sufficiency in food production.

According to Obasi et al. [4], which noted that the sub-Saharan Africa region is among the countries that have continued to experience significant food shortages, more than 40% of the region population is estimated to be suffering from hunger and poor nutrition. Just like many other developing countries, the Food and Agricultural Organisation identified that widespread poverty, poor economic conditions, institutional failure and constraints in logistics, among many other challenges, significantly affect crop production in Nigeria.

In a bid to tackle these challenges, the Nigerian government over the years has intensified efforts towards improving both plant nutrient and crop production mainly through better land use, human resource development in the agricultural sector, research in diversification of types of crops and seeds, fight against pests and diseases and increased use of fertilisers. However, despite the resources devoted to crop production in Nigeria, the productive efficiency of farmers for most crops still fall below 60% [4]. Globally, both socioeconomic and ecological factors interact to determine plant nutrients and sustainable crop production. Efforts to intensify agricultural production in Nigeria has been a continuous process which is taking place through several pathways; therefore, we examined how some of these factors affects plant nutrition and crop production in the case of Nigeria.

2. Literature review

Crop production has continued to play a major role in sustaining economic growth in Nigeria. However, its sustainability has been threatened with major challenges overtime. These challenges range from deficiency in plant nutrient as a result of unfavourable biotic and abiotic factors which includes unfavourable climatic conditions, low level of technological development in the agricultural sector, misapplication of fertilisers, infrastructural decay and so on. Various policies have been recommended overtime to address the issue of low crop production in Nigeria. However, it is salient to know how some of these factors have affected plant nutrition and crop production in Nigeria; examined below are some of the factors as identified in the literature.

In an attempt to correct the deficiencies of nutritional elements in crops, a wide range of Nigerian farmers often apply organic and inorganic fertilisers as both play a prominent role in improving soil fertility. However, fertiliser application is a necessary condition for crop yield but not a sufficient condition for an improved crop yield. According to Awodun et al. [5] cited in Ayeni et al. [6], both organic manure and fertilisers play different roles in improving soil fertility, but they both cannot supply all the nutrients in plants that can solely feed a teeming Nigerian population. Nottidge [7] further identified that fertiliser application leads to nutrient imbalance and low infiltration rate, all of which hinders the uptake of nutrients by plants. Also, Ayeni et al. [6] identified that the constant use of inorganic fertilisers can increase the level of soil acidity thereby leading to soil damage.

It has been globally recognised that the most serious threat to agricultural productivity is environmental issue [8]. For countries with higher temperature, the consequences of climate change tend to be more severe. This is most especially true for many developing countries with little adaptive capacity [9]. In recent times, atmospheric emission has been on the increase due to the improper use of agro chemicals, low level of land and environmental management and inadequate manure management. According to Yobannes [9], one of the most important emissions that affects crop productivity and plant nutrient is nitrous oxide, which is determined by fertiliser application, irrigation methods and animal feeds.

Ufiobor [10] further identified that one of the major factors that determines crop yield in Nigeria is the climatic condition. From 1970 to 2018, temperature has increased from an average of 1.4–1.9°C [11] cited in Ufiobor [10]. The northwest, northeast and southwest of the country are especially being affected by extreme harsh weather conditions. The consequence of this is that higher temperature will decrease soil moisture which will have an attendant effect on plant nutrients and crop production.

In the developed economies, most especially Europe and North America, sustainable crop production has been increasing rapidly due to the developed nature of their farming system which has been made possible as a result of innovation and technological enhancement [10]. Farmers in these countries have accepted the evolving change and are now actively engaged in research and training for a sustainable cropping system [10]. In these countries, the government has also implemented programmes to support rigorous scientific investigation that will improve plant nutrition to produce healthy food for its populace. However, Nigeria has not yet witnessed the kind of agricultural evolution that has taken place in developed countries. One major constraint to agricultural development in Nigeria has been the slow response to technological adoption which in turn leads to low productivity and poor farming system, which affects plant nutrients. Nigeria majorly depends on traditional farming system which has an effect on the use of farmlands as the farming system is mainly carried out without the use of machines.

Just like other developing countries, the role of labour force in determining the level of output in all the sectors cannot be undermined. The agricultural system in Nigeria is highly labour intensive as labour force is a crucial part of its production system. According to Ufiobor [10], labour force accounts for over 90% of its total farm operations. Ufiobor [10] further envisaged that this could be as a result of the fact that many of its educated youth have shown little interest in the agricultural sector over the years, thus causing a shortage of skilled labour force in the agricultural sector that can also affect the nutritional value of plants and total crop production itself.

According to the Nations Encyclopedia [12], major crops cultivated in Nigeria include sesame, beans, nuts, cashew, beans, groundnut, cassava, cocoa, gum Arabic, millet, melon, rice, palm kernels, rubber sorghum, banana, plantain, beans and yams. However, the most widely produced crops are cassava and yams in the south and millet and sorghum in the north. Nigerian farmers also grow many fruits and vegetables. In recent years, the use of fertiliser in many countries has been increasing overtime. However, the use of organic wastes for pasture has been more feasible in the developed countries especially China, than in all other countries including Nigeria [13]. This is an indication of the fact that the Nigerian government has not really encouraged the use and development of organic fertiliser in Nigeria which might be responsible for the low level of manure generated for the purpose of farming.

The International Food Policy Research Institute [14] identified that there are signs of an increase use of fertilisers in countries where fertiliser subsidies are being granted to farmers by the government. Prominent among these countries are Malawi, Nigeria and Zambia. The use of fertiliser by Nigerian farmers is however quite common especially among the shareholder farmers. In some cases, these farmers also use some inorganic fertiliser which covers 70% of plots of lands [14]. Since the 1970s, efforts by the Nigerian government to stimulate the demand for fertiliser have been on the increase. This aim has been achieved by growing commercial fertiliser sector through price reduction, extension services to boost soil

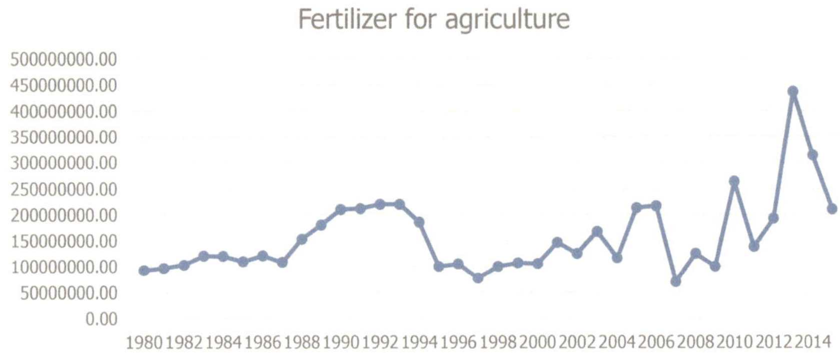

Figure 1.
Agricultural fertiliser use. Source: Authors' computation using excel [13].

fertility, increased use of technology and increased access to credit facilities by farmers [15] (**Figure 1**).

In the year 1980, fertiliser production (kilogrammes per hectare of arable land) in Nigeria was 9,220,000. Over the past 36 years, its highest value was 436,957,273 in the year 2013, while its lowest value was 70,115,000 in the year 2007. The upward and downward movement of this trend is an indication that the level of fertiliser production in Nigeria has not witnessed a stable movement.

According to the International Food Policy Research Institute [14], the types of fertiliser commonly produced and used in Nigeria include urea, nitrogen-phosphorus-potassium (NPK) and superphosphate (SSP). The most commonly used NPK blends are 15-15-15, 20-10-10, 12-12-17 + 2 MgO and 25-10-10. NPK fertilisers are further formulated to be site and crop specific. In a bid to further boost the effective procurement and distribution of fertiliser, the Nigerian government at various times has introduced several measures for its production, procurement and distribution.

In Nigeria, emphasis on increased agricultural productivity of farmers from the perspective of soil conditioning has been on chemical fertiliser, while there has been less emphasis on the impact of the bio-organic input [16]. Even as the quest to ensure the eradication of hunger and poverty has been on the increase, the Nigerian government has taken measures to ensure national self-sufficiency through local fertiliser production, supplemented by importation to ensure adequate and timely fertiliser supply to all farmers. The government also offers a subsidy on the market price of fertiliser so as to make fertiliser affordable to smallholder farmers. Given that the agenda of most successive government is to boost local food production and ensure national self-sufficiency, various efforts has to be intensified to synergise the use of both organic and inorganic medium of improving soil fertility for plant nutrition.

3. Methodology

The method used in this study is the survey of literature and stylised facts approach. Relevant data was collected from Food and Agricultural Organization data (FA) data base and the National Bureau of Statistics (NBS) of the Nigerian statistical bulletin for the indicators of crop production and other major agriculture commodities in Nigeria. Tables were used to describe the yields and prices of various agriculture commodities and determinants. **Table 1** presents crop

Year	2000	2001	2002	2003	2004	2005	2006	2007	2008	2009	2010	2011	2012	2013	2014	2015	2016	2017
Crop production	80.5	79.4	83.3	87.8	93.7	99.5	106.9	97.5	104.3	90.1	104.5	96.3	107.5	104.3	117.8	120.1	118.9	—
Total agriculture employment	57.3	57.6	60.7	58.2	51.9	51.2	49.6	48.6	44.0	40.4	30.6	33.1	35.9	38.3	36.8	36.4	36.3	36.5
Male agriculture employment	51.1	51.7	55.8	52.7	45.3	44.4	42.7	41.7	35.8	30.9	19.4	22.0	25.3	28.2	26.7	26.3	26.1	26.2
Female agricultural employment	61.9	61.9	64.2	62.3	57.0	56.3	54.9	54.0	50.5	48.0	39.5	42.2	44.7	46.7	45.2	44.8	44.9	45.3
Agriculture land area	78.3	76.3	77.5	78.8	78.8	79.8	80.5	80.9	79.8	75.8	76.9	78.0	79.1	77.7	77.7	77.7		

Source: Authors' compilation.

Table 1.
Crop production and its determinants.

	Jan	Feb	Mar	Apr	May	Jun	Jul	Aug	Sep	Oct	Nov	Dec
Item labels	42736.0	42767.0	42795.0	42826.0	42856.0	42887.0	42917.0	42948.0	42979.0	43009.0	43040.0	43070.0
Agric eggs (medium size price of one)	47.4	42.9	43.9	46.2	45.7	45.3	44.3	42.9	45.7	42.1	40.8	41.3
Beans: brown, sold loose	353.6	337.1	353.3	357.2	365.9	374.3	382.3	370.3	404.8	382.6	369.8	362.0
Beans: white black eye, sold loose	305.5	309.9	318.5	324.0	332.3	339.6	344.0	335.7	358.1	342.1	337.1	314.9
Beef bone in	1001.2	995.6	1010.3	1035.4	1123.7	1129.0	1128.9	1151.4	1078.5	1081.3	1065.6	1067.7
Beef, boneless	1249.5	1270.7	1281.7	1323.1	1378.9	1393.4	1376.9	1276.9	1324.9	1312.7	1286.9	1236.4
Bread sliced 500 g	302.9	299.7	297.5	296.6	307.7	320.6	314.5	304.2	310.6	305.9	299.0	290.1
Bread unsliced 500 g	270.3	264.9	262.3	277.7	282.3	286.8	285.6	258.7	286.3	280.3	274.3	268.2
Broken rice (Ofada)	377.4	392.0	421.5	425.5	460.1	472.3	473.4	415.3	431.6	416.2	417.2	319.9
Catfish (obokun), fresh	899.6	885.1	884.1	886.7	900.7	916.5	910.6	921.2	894.5	906.9	902.0	919.8
Catfish, dried	2204.1	2150.2	2189.6	2189.1	2213.7	2255.2	2215.3	2176.8	2159.0	2083.7	2214.3	2146.6
Catfish, smoked	817.1	825.3	834.3	837.1	845.9	848.4	853.2	852.9	839.5	845.1	847.7	1007.4
Chicken feet	765.3	785.4	907.5	768.1	819.6	832.8	832.7	834.7	817.2	800.1	790.9	1156.0
Chicken wings	919.1	963.5	980.9	886.6	925.6	952.2	958.0	946.9	979.3	960.7	953.6	950.6
Dried fish sardine	959.1	935.5	958.7	972.7	991.6	989.5	968.8	972.6	947.6	967.0	968.8	1077.8
Evaporated tinned milk carnation, 170 g	136.3	140.5	143.6	151.5	157.9	162.8	162.7	158.8	174.4	170.6	169.4	159.7
Evaporated tinned milk (peak), 170 g	157.7	166.1	177.1	197.2	198.4	194.7	195.9	190.5	206.0	198.7	196.0	191.4
Frozen chicken	1419.8	1429.0	1555.5	1606.6	1606.0	1645.3	1623.8	1529.3	1570.3	1580.1	1571.5	1708.6
Gari white, sold loose	219.6	260.9	273.7	288.5	293.0	315.6	317.1	310.1	302.0	268.1	251.9	199.7
Gari yellow, sold loose	255.8	250.5	302.0	320.9	326.8	354.6	350.5	345.8	335.9	305.0	289.2	219.8
Groundnut oil: 1 bottle, specify bottle	477.9	482.9	494.1	494.2	500.2	503.3	505.1	508.3	478.0	504.0	503.8	660.4
Iced sardine	1880.0	1915.7	1902.2	1928.0	1919.6	1916.2	1915.2	1904.7	1878.5	1914.5	1903.7	1545.3
Irish potato	300.9	307.3	311.8	318.9	315.9	319.3	311.4	310.5	307.5	291.1	290.6	314.2
Mackerel: frozen	759.0	764.6	774.9	778.2	785.4	794.6	795.8	797.8	785.6	828.9	844.6	1015.6

	Jan	Feb	Mar	Apr	May	Jun	Jul	Aug	Sep	Oct	Nov	Dec
Maize grain: white, sold loose	167.1	172.9	174.0	182.9	188.9	190.3	191.3	192.4	168.1	191.2	185.1	154.4
Maize grain: yellow, sold loose	168.8	174.7	178.8	185.6	190.6	191.4	193.0	193.4	168.2	191.1	189.9	161.9
Mudfish (aro), fresh	994.4	998.7	1079.2	1008.9	1047.3	1071.1	1080.3	998.6	1069.9	1074.7	1081.6	1060.0
Mudfish, dried	1812.0	1955.1	2084.8	2319.9	2388.1	2416.3	2395.8	2161.7	2204.0	2190.2	2144.2	1621.4
Onion bulb	258.9	241.4	246.9	205.6	203.6	214.8	213.1	236.7	238.3	217.6	228.5	311.1
Palm oil: 1 bottle, specify bottle	420.6	434.8	442.0	452.4	458.2	471.3	478.8	492.9	439.9	473.6	475.5	551.8
Plantain (ripe)	234.3	234.3	236.7	240.7	241.5	249.7	251.3	254.0	231.7	247.4	244.9	240.8
Plantain (unripe)	212.5	214.4	215.8	221.8	224.7	228.5	223.5	226.3	216.0	229.9	233.3	253.1
Rice: agric, sold loose	324.0	355.4	360.9	324.8	347.7	352.1	354.1	349.6	351.1	325.9	315.1	317.5
Rice: local, sold loose	286.2	306.3	308.9	299.3	323.8	325.5	323.3	320.2	316.4	292.9	278.9	278.9
Rice: medium, grained	312.1	352.7	378.0	332.8	350.4	348.2	344.4	313.5	339.5	320.6	309.9	304.9
Rice: imported, high-quality, sold loose	402.0	410.6	418.7	388.5	410.5	415.8	409.2	384.3	398.0	368.9	360.8	371.2
Sweet potato	129.4	127.2	132.9	132.4	130.3	135.8	138.0	138.6	139.0	115.7	111.4	120.2
Tilapia fish (epiya), fresh	792.2	795.3	798.1	800.8	805.2	817.6	823.3	820.3	795.6	822.4	813.8	1158.8
Titus, frozen	884.8	942.2	964.6	935.0	969.6	1012.2	1014.5	998.0	974.0	1050.4	1125.9	1109.2
Tomato	247.5	236.6	268.6	285.7	339.7	375.0	394.1	431.3	322.4	291.5	286.4	276.2
Vegetable oil: 1 bottle, specify bottle	495.3	507.4	513.4	525.7	524.8	546.3	552.9	559.0	505.5	547.1	540.2	565.0
Wheat flour: prepacked (golden penny 2 kg)	626.5	621.4	623.1	627.3	646.2	632.1	630.5	647.3	627.0	641.4	649.2	639.2
Yam tuber	210.6	215.6	255.9	250.3	279.2	292.1	294.1	343.4	259.5	223.6	212.3	211.6

Source: Authors.

Table 2.
Prices of items of various agricultural/food commodities in 2017 (January to December).

Item labels	Jan	Feb	Mar	Apr	May	Jun	Jul	Aug	Sep	Oct	Nov	Dec
Agric eggs (medium size price of one)	38.8	41.3	41.2	42.2	41.8	41.4	41.2	41.8	41.2	41.7	42.4	42.6
Beans: brown, sold loose	385.5	387.5	395.3	395.2	407.7	418.8	411.0	410.2	416.1	411.0	396.7	386.8
Beans: white black eye, sold loose	344.5	348.6	357.0	355.3	367.2	369.4	361.3	361.7	362.7	354.2	346.4	344.7
Beef bone in	1045.6	1017.3	1024.6	980.3	1011.1	1034.6	1028.1	1061.8	1027.7	1003.3	998.7	997.7
Beef, boneless	1298.7	1274.7	1262.5	1257.7	1275.8	1271.5	1262.7	1281.0	1281.6	1260.6	1251.4	1270.8
Bread: sliced, 500 g	313.4	314.2	305.9	306.5	306.0	308.0	305.3	308.0	303.8	299.2	305.5	296.9
Bread: unsliced, 500 g	282.1	284.4	278.6	278.6	278.1	275.6	272.7	278.3	278.6	272.9	277.9	271.1
Broken rice (Ofada)	413.2	405.7	399.5	400.9	405.0	404.4	403.6	397.3	384.5	370.2	385.6	381.6
Catfish (obokun), fresh	1014.6	1037.3	1008.3	993.5	1013.1	1062.4	1064.7	1088.9	1089.3	1080.4	1112.1	1060.8
Catfish, dried	1852.9	1857.2	1793.6	1753.5	1779.7	1770.5	1792.1	1807.1	1766.5	2362.3	1815.0	1808.6
Catfish, smoked	1540.8	1896.9	1524.7	1514.5	1520.9	1527.8	1504.4	1512.4	1499.6	1482.3	1490.8	1519.8
Chicken feet	850.0	778.6	773.0	758.9	782.5	784.8	930.5	803.0	808.1	775.7	985.3	737.8
Chicken wings	950.8	955.0	935.8	921.5	925.5	960.6	963.0	992.5	1012.5	982.5	981.5	944.0
Dried fish sardine	1455.5	1405.7	1410.1	1371.8	1400.0	1378.6	1395.4	1382.9	1374.3	1365.7	1331.6	1343.0
Evaporated tinned milk carnation, 170 g	172.2	166.5	158.5	160.2	165.3	166.2	165.7	163.0	166.3	163.3	162.4	163.3
Evaporated tinned milk (peak), 170 g	196.7	191.1	187.6	188.1	192.9	189.9	187.8	184.2	189.3	187.2	187.6	190.0
Frozen chicken	1547.5	1548.9	1544.9	1537.9	1536.5	1537.2	1541.4	1533.2	1534.2	1558.2	1551.2	1625.1
Gari: white, sold loose	229.3	210.2	209.1	208.2	216.0	216.7	200.7	198.8	195.9	188.4	183.6	166.6
Gari: yellow, sold loose	271.3	250.1	244.0	246.5	256.5	259.8	248.7	246.8	253.7	236.3	222.4	196.0
Groundnut oil: 1 bottle, specify bottle	528.6	571.1	585.2	584.5	601.5	612.4	603.6	603.4	608.4	592.5	603.9	585.5
Iced sardine	861.1	885.1	899.7	907.9	941.5	923.5	940.5	937.2	931.6	916.8	921.0	906.4
Irish potato	232.9	250.9	261.8	282.1	300.6	297.1	296.6	309.5	291.9	300.1	330.1	298.6

DOI: http://dx.doi.org/10.5772/intechopen.88975

Item labels	Jan	Feb	Mar	Apr	May	Jun	Jul	Aug	Sep	Oct	Nov	Dec
Mackerel, frozen	915.3	875.3	915.9	896.4	927.2	908.9	922.9	926.2	944.5	921.0	937.8	934.2
Maize grain: white, sold loose	190.5	185.2	231.6	173.5	180.5	183.7	179.4	180.1	179.9	171.1	165.0	161.5
Maize grain: yellow, sold loose	180.0	199.8	193.8	191.1	195.7	197.3	190.4	191.1	189.8	178.4	168.9	160.4
Mudfish (aro), fresh	1099.8	1098.6	1069.9	1039.0	1062.7	1070.6	1065.6	1104.8	1072.1	1052.1	1067.1	1032.2
Mudfish, dried	2134.7	2066.0	1945.3	1912.7	1928.2	1941.9	1907.2	1897.1	1908.0	1874.4	1894.3	1861.7
Onion bulb	248.7	252.0	235.3	234.4	234.0	240.0	246.9	252.1	251.3	232.8	287.7	259.4
Palm oil: 1 bottle, specify bottle	480.9	510.1	492.1	486.8	494.9	500.0	496.4	501.1	501.0	495.1	496.2	474.5
Plantain (ripe)	273.7	257.6	259.8	262.3	273.5	276.6	270.8	275.8	272.6	245.6	254.1	227.3
Plantain (unripe)	248.7	232.2	242.5	235.7	247.1	248.5	248.6	247.1	244.3	226.6	225.5	215.7
Rice: agric, sold loose	322.8	322.8	326.6	322.5	327.8	331.1	325.1	327.0	328.4	323.3	329.1	321.6
Rice, local, sold loose	274.7	276.3	283.6	281.5	286.2	280.8	276.3	280.3	277.8	277.9	278.5	280.8
Rice: medium grained	308.5	314.8	317.2	323.6	325.4	323.1	319.4	322.5	319.5	314.9	318.9	318.0
Rice: imported, high-quality sold loose	360.8	365.2	363.3	369.4	374.6	373.5	370.8	375.0	371.3	373.0	376.6	370.6
Sweet potato	113.4	126.4	130.3	137.4	150.5	148.7	163.3	167.1	167.3	154.7	147.7	140.7
Tilapia fish (epiya), fresh	864.0	885.7	890.1	889.5	924.1	939.4	935.2	947.8	979.2	923.1	934.1	887.5
Titus, frozen	894.7	901.5	905.8	898.6	921.0	950.3	924.8	946.5	941.5	923.9	1105.2	926.4
Tomato	272.0	267.1	267.1	289.5	307.1	317.7	336.3	336.7	328.3	306.6	294.4	271.5
Vegetable oil: 1 bottle, specify bottle	540.3	549.0	533.5	548.0	553.2	552.1	547.9	540.4	546.8	536.1	536.5	507.8
Wheat flour, prepacked (golden penny 2 kg)	655.1	646.0	657.6	649.9	659.5	660.1	657.6	661.1	651.9	656.8	659.6	662.9
Yam tuber	226.5	230.9	254.5	279.6	291.3	285.8	280.8	293.0	280.3	252.2	239.2	212.7

Source: Authors.

Table 3.
Prices of items of various agricultural/food commodities in 2018.

Item	Element	Unit	1980	1985	1986	1990	1991	1995	2000	2005	2010	2015
Bast fibres, other	Area harvested	ha		1000	1000	1107	1000	1000	1000	1000	1000	1008
Bast fibres, other	Yield	hg/ha		6000	7000	7859	8800	9712	10,992	12,193	13,310	14,388
Bast fibres, other	Production	tonnes		600	700	870	880	971	1099	1219	1331	1451
Carrots and turnips	Area harvested	ha	20,000	20,000	20,000	22,000	22,303	24,285	26,492	27,750	25,300	25,704
Carrots and turnips	Yield	hg/ha	75,000	75,000	75,000	78,636	78,464	81,531	85,097	85,586	86,921	88,010
Carrots and turnips	Production	tonnes	150,000	150,000	150,000	173,000	175,000	198,000	225,440	237,500	219,911	226,222
Cashew nuts, with shell	Area harvested	ha	40,000	40,000	40,000	50,000	75,000	155,000	259,000	309,000	382,509	131,529
Cashew nuts, with shell	Yield	hg/ha	6250	6250	6250	6000	6000	6129	17,992	19,223	20,698	7386
Cashew nuts, with shell	Production	tonnes	25,000	25,000	25,000	30,000	45,000	95,000	466,000	594,000	791,726	97,149
Cassava	Area harvested	ha	1,200,000	1,075,000	1,095,000	1,634,130	2,551,000	2,944,000	3,300,000	3,782,000	3,481,900	6,216,434
Cassava	Yield	hg/ha	95,833	112,465	113,132	116,533	101,936	106,671	97,000	109,902	122,155	92,727
Cassava	Production	tonnes	11,500,000	12,090,000	12,388,000	19,043,008	26,004,000	31,404,000	32,010,000	41,565,000	42,533,180	57,643,271
Chillies and peppers, dry	Area harvested	ha	25,700	27,200	27,500	28,700	29,000	30,200	30,410	30,750	34,000	38,077
Chillies and peppers, dry	Yield	hg/ha	15,370	15,441	15,455	15,505	15,517	15,563	15,624	15,610	16,765	17,077
Chillies and peppers, dry	Production	tonnes	39,500	42,000	42,500	44,500	45,000	47,000	47,512	48,000	57,000	65,022
Chillies and peppers, green	Area harvested	ha	69,000	80,000	80,000	89,048	75,000	72,766	88,706	91,500	94,808	96,625
Chillies and peppers, green	Yield	hg/ha	91,304	87,500	90,000	84,224	86,667	84,105	80,677	78,798	77,381	76,941
Chillies and peppers, green	Production	tonnes	630,000	700,000	720,000	750,000	650,000	612,000	715,657	721,000	733,631	743,442

Item	Element	Unit	1980	1985	1986	1990	1991	1995	2000	2005	2010	2015
Cocoa, beans	Area harvested	ha	700,000	700,000	700,000	715,000	726,000	788,000	966,000	1,198,902	1,272,430	1,057,174
Cocoa, beans	Yield	hg/ha	2186	2286	2114	3413	3691	2576	3499	3678	3137	2857
Cocoa, beans	Production	tonnes	153,000	160,000	148,000	244,000	268,000	203,000	338,000	441,000	399,200	302,066
Coconuts	Area harvested	ha	32,000	34,500	34,500	35,500	37,000	28,500	36,000	39,000	39,000	38,701
Coconuts	Yield	hg/ha	28,125	29,565	30,145	33,239	34,865	52,281	44,444	53,590	67,645	69,744
Coconuts	Production	tonnes	90,000	102,000	104,000	118,000	129,000	149,000	160,000	209,000	263,815	269,920
Coffee, green	Area harvested	ha	7000	12,000	2400	3434	3500	3122	3190	3670	1990	1534
Coffee, green	Yield	hg/ha	5000	5000	5000	8824	9143	9896	12,006	13,597	12,063	12,899
Coffee, green	Production	tonnes	3500	6000	1200	3030	3200	3090	3830	4990	2400	1979
Cotton lint	Production	tonnes	29,324	10,524	36,290	95,000	103,000	95,000	147,000	190,000	220,000	
Cottonseed	Production	tonnes	55,075	24,000	63,000	180,000	195,000	153,000	247,000	323,000	370,000	
Cow peas, dry	Area harvested	ha	1,463,000	1,405,000	1,405,000	1,805,000	1,885,740	3,585,000	3,583,000	4,140,000	2,859,760	3,635,700
Cow peas, dry	Yield	hg/ha	3486	4349	4569	7490	7180	4884	6001	6800	11,778	6343
Cow peas, dry	Production	tonnes	510,000	611,000	642,000	1,352,000	1,354,000	1,751,000	2,150,000	2,815,000	3,368,250	2,306,200
Fibre crops nes	Production	tonnes				11	12					
Fonio	Area harvested	ha	30,000	38,000	44,000	65,000	72,000	108,000	133,000	198,000	151,766	187,560
Fonio	Yield	hg/ha	6000	6579	6136	6000	5972	5370	5714	4798	5211	4405
Fonio	Production	tonnes	18,000	25,000	27,000	39,000	43,000	58,000	76,000	95,000	79,087	82,617
Fruit, citrus nes	Area harvested	ha	550,000	570,000	570,000	580,000	630,000	643,589	727,596	731,000	790,000	821,533

Item	Element	Unit	1980	1985	1986	1990	1991	1995	2000	2005	2010	2015
Fruit, citrus nes	Yield	hg/ha	32,727	35,088	35,088	35,914	39,683	42,263	44,668	45,179	48,101	48,757
Fruit, citrus nes	Production	tonnes	1,800,000	2,000,000	2,000,000	2,083,000	2,500,000	2,720,000	3,250,000	3,302,611	3,800,000	4,005,520
Fruit, fresh nes	Area harvested	ha	145,000	184,500	196,000	197,349	208,520	238,082	284,711	218,500	177,000	180,210
Fruit, fresh nes	Yield	hg/ha	64,138	65,041	66,327	65,873	63,337	62,902	63,575	64,119	67,797	68,225
Fruit, fresh nes	Production	tonnes	930,000	1,200,000	1,300,000	1,300,000	1,320,713	1,497,578	1,810,060	1,401,000	1,200,000	1,229,484
Garlic	Production	tonnes									587	800
Ginger	Area harvested	ha	400	16,000	30,000	84,000	100,000	148,000	158,000	181,000	52,330	64,356
Ginger	Yield	hg/ha	5000	5000	5000	5000	5000	5338	6203	6906	31,000	44,198
Ginger	Production	tonnes	200	8000	15,000	42,000	50,000	79,000	98,000	125,000	162,223	284,440
Groundnuts, with shell	Area harvested	ha	563,000	594,000	793,000	707,000	112,7000	1,767,000	1,934,000	2,187,000	2,789,180	2,801,756
Groundnuts, with shell	Yield	hg/ha	8366	10,455	11,299	16,492	12,076	8936	15,000	15,903	13,621	12,376
Groundnuts, with shell	Production	tonnes	471,000	621,000	896,000	1,166,000	1,361,000	1,579,000	2,901,000	3,478,000	3,799,240	3,467,446
Karite nuts (shea nuts)	Area harvested	ha	100,000	94,000	87,000	184,000	204,000	235,000	232,000	257,239	342,750	409,963
Karite nuts (shea nuts)	Yield	hg/ha	11,000	10,638	11,839	15,707	15,980	16,340	15,905	15,940	9500	8786
Karite nuts (shea nuts)	Production	tonnes	110,000	100,000	103,000	289,000	326,000	384,000	369,000	410,029	325,610	360,177
Kola nuts	Area harvested	ha	140,000	175,000	175,000	125,000	130,000	105,000	91,000	94,250	270,143	244,705
Kola nuts	Yield	hg/ha	9643	9829	9943	10,800	12,692	9048	9011	9045	5366	6608
Kola nuts	Production	tonnes	135,000	172,000	174,000	135,000	165,000	95,000	82,000	85,250	144,950	161,711

Item	Element	Unit	1980	1985	1986	1990	1991	1995	2000	2005	2010	2015
Maize	Area harvested	ha	465,000	1,556,000	2,800,000	5,104,000	5,142,000	5,472,000	3,159,000	3,589,000	4,149,310	6,771,189
Maize	Yield	hg/ha	13,161	11,735	12,679	11,301	11,299	12,666	13,001	16,598	18,502	15,599
Maize	Production	tonnes	612,000	1,826,000	3,550,000	5,768,000	5,810,000	6,931,000	4,107,000	5,957,000	7,676,850	10,562,050
Maize, green	Area harvested	ha	46,000	156,000	172,000	150,000	155,000	167,706	162,619	161,500	183,916	200,356
Maize, green	Yield	hg/ha	35,000	29,231	25,233	30,533	30,645	33,704	34,662	35,697	36,774	37,699
Maize, green	Production	tonnes	161,000	456,000	434,000	458,000	475,000	565,240	563,667	576,500	676,338	755,319
Mangoes, mangosteens, guavas	Area harvested	ha	80,000	80,000	80,000	85,000	88,000	106,000	125,000	125,500	130,000	131,132
Mangoes, mangosteens, guavas	Yield	hg/ha	50,000	50,000	50,000	59,294	59,091	59,528	58,400	58,247	65,385	68,239
Mangoes, mangosteens, guavas	Production	tonnes	400,000	400,000	400,000	504,000	520,000	631,000	730,000	731,000	850,000	894,833
Melon seed	Area harvested	ha	76,000	183,000	150,000	230,000	231,000	285,000	575,000	694,000	469,690	967,937
Melon seed	Yield	hg/ha	12,368	8033	10,200	9043	9481	10,070	6000	6499	10,802	5758
Melon seed	Production	tonnes	94,000	147,000	153,000	208,000	219,000	287,000	345,000	451,000	507,340	557,328
Millet	Area harvested	ha	2,824,000	2,346,000	3,917,000	4,778,000	4,560,000	5,107,000	5,814,000	4,685,000	4,364,140	1,591,803
Millet	Yield	hg/ha	8336	15,277	10,495	10,749	9011	10,893	10,501	15,300	11,848	9331
Millet	Production	tonnes	2,354,000	3,584,000	4,111,000	5,136,000	4,109,000	5,563,000	6,105,000	7,168,000	5,170,430	1,485,387
Nuts, nes	Area harvested	ha		100	300	2500	2500	5421	2965	2550	2800	2799

Item	Element	Unit	1980	1985	1986	1990	1991	1995	2000	2005	2010	2015
Nuts, nes	Yield	hg/ha		20,000	20,000	24,000	20,000	6118	17,416	20,196	25,000	25,915
Nuts, nes	Production	tonnes		200	600	6000	5000	3316	5163	5150	7000	7253
Oil palm fruit	Area harvested	ha	2,300,000	2,200,000	2,220,000	2,300,000	2,450,000	2,938,000	3,080,000	3,350,000	3,200,000	3,076,881
Oil palm fruit	Yield	hg/ha	25,000	25,909	26,577	26,957	26,531	26,549	26,688	25,373	25,000	25,683
Oil palm fruit	Production	tonnes	5,750,000	5,700,000	5,900,000	6,200,000	6,500,000	7,800,000	8,220,000	8,500,000	8,000,000	7,902,277
Oil, palm	Production	tonnes	650,000	615,000	650,000	730,000	760,000	860,000	899,000	1,170,000	970,820	
Oilseeds nes	Production	tonnes				548	759	820	596	700	814	600
Okra	Area harvested	ha	200,000	230,000	250,000	260,000	231,278	259,393	292,135	350,000	397,290	1,859,900
Okra	Yield	hg/ha	21,000	20,652	19,600	20,231	22,916	24,287	25,719	27,143	27,275	11,118
Okra	Production	tonnes	420,000	475,000	490,000	526,000	530,000	630,000	751,342	950,000	1,083,620	2,067,900
Onions, dry	Area harvested	ha	30,000	30,000	35,000	36,667	63,403	87,996	115,501	264,174	179,984	434,500
Onions, dry	Yield	hg/ha	133,333	133,333	142,857	137,649	60,904	56,654	51,342	44,763	74,797	22,967
Onions, dry	Production	tonnes	400,000	400,000	500,000	504,719	386,152	498,539	593,008	1,182,520	1,346,218	997,900
Onions, shallots, green	Area harvested	ha	10,000	10,000	10,000	10,476	6046	10,128	10,794	11,250	13,232	14,366
Onions, shallots, green	Yield	hg/ha	100,000	100,000	105,000	119,320	223,291	215,057	206,277	196,000	175,089	163,848
Onions, shallots, green	Production	tonnes	100,000	100,000	105,000	125,000	135,000	217,815	222,656	220,500	231,684	235,383
Palm kernels	Area harvested	ha									450,000	
Palm kernels	Production	tonnes	279,000	360,000	355,000	356,000	369,000	543,000	577,000	465,000	233,000	
Papayas	Area harvested	ha	55,000	55,000	55,000	65,000	66,000	80,000	89,315	91,500	92,865	93,445

Item	Element	Unit	1980	1985	1986	1990	1991	1995	2000	2005	2010	2015
Papayas	Yield	hg/ha	72,727	72,727	72,727	79,538	80,303	81,000	83,371	82,568	80,763	93,680
Papayas	Production	tonnes	400,000	400,000	400,000	517,000	530,000	648,000	744,626	755,500	750,000	875,401
Pineapples	Area harvested	ha	95,000	95,000	95,000	100,000	105,505	105,802	117,005	116,500	180,000	184,551
Pineapples	Yield	hg/ha	63,158	73,684	73,684	76,300	75,825	75,613	75,733	76,395	82,631	81,270
Pineapples	Production	tonnes	600,000	700,000	700,000	763,000	800,000	800,000	886,110	890,000	1,487,350	1,499,840
Plantains and others	Area harvested	ha	180,000	185,500	187,200	162,000	178,000	250,000	386,000	447,000	449,220	486,048
Plantains and others	Yield	hg/ha	57,889	60,000	60,203	75,000	75,225	65,280	51,010	57,964	59,559	62,977
Plantains and others	Production	tonnes	1,042,000	1,113,000	1,127,000	1,215,000	1,339,000	1,632,000	1,969,000	2,591,000	2,675,530	3,060,962
Potatoes	Area harvested	ha	5500	7000	7600	7700	9400	13,600	212,000	260,000	265,992	328,009
Potatoes	Yield	hg/ha	72,727	61,429	60,526	70,130	70,213	69,853	28,255	29,846	38,584	36,727
Potatoes	Production	tonnes	40,000	43,000	46,000	54,000	66,000	95,000	599,000	776,000	102,6311	120,4676
Pulses, nes	Area harvested	ha	115,000	120,000	120,000	131,832	74,309	87,216	102,492	119,432	130,000	140,095
Pulses, nes	Yield	hg/ha	4609	4167	4167	4162	3870	3966	4087	4104	4308	4609
Pulses, nes	Production	tonnes	53,000	50,000	50,000	54,870	28,756	34,592	41,887	49,018	56,000	64,570
Rice, paddy	Area harvested	ha	550,000	670,000	700,000	1,208,000	1,652,000	1,796,000	2,199,000	2,494,000	2,432,630	3,121,562
Rice, paddy	Yield	hg/ha	19,818	21,343	20,233	20,695	19,528	16,258	14,998	14,302	18,386	20,042
Rice, paddy	Production	tonnes	1,090,000	1,430,000	1,416,322	2,500,000	3,226,000	2,920,000	3,298,000	3,567,000	4,472,520	6,256,228
Rubber, natural	Area harvested	ha	73,000	73,000	73,000	22,5000	268,000	297,000	330,000	339,500	360,541	365,622
Rubber, natural	Yield	hg/ha	6164	8219	8219	6533	5784	4209	3242	3998	4019	4228

Item	Element	Unit	1980	1985	1986	1990	1991	1995	2000	2005	2010	2015
Rubber, natural	Production	tonnes	45,000	60,000	60,000	147,000	155,000	125,000	107,000	135,716	144,912	154,571
Seed cotton	Area harvested	ha	476,000	220,000	285,000	575,000	643,000	431,000	538,000	659,000	398,570	401,441
Seed cotton	Yield	hg/ha	1801	1591	3509	4800	4806	5824	7416	7906	15,115	6913
Seed cotton	Production	tonnes	85,733	35,000	100,000	276,000	309,000	251,000	399,000	521,000	602,440	277,523
Sesame seed	Area harvested	ha	135,000	116,700	104,700	110,000	115,000	133,000	151,000	196,000	324,570	329,460
Sesame seed	Yield	hg/ha	3037	2999	3343	4000	4000	4511	4768	5102	4603	5218
Sesame seed	Production	tonnes	41,000	35,000	35,000	44,000	46,000	60,000	72,000	100,000	149,410	171,900
Sorghum	Area harvested	ha	3,286,000	4,862,000	5,147,000	4,185,000	5,538,000	6,095,000	6,885,000	7,284,000	4,960,130	5,899,134
Sorghum	Yield	hg/ha	11,229	10,101	10,540	10,000	9691	11,480	11,200	12,600	14,397	11,875
Sorghum	Production	tonnes	3,690,000	4,911,000	5,425,000	4,185,000	5,367,000	6,997,000	7,711,000	9,178,000	7,140,970	7,005,025
Soybeans	Area harvested	ha	270,000	205,000	210,000	729,000	468,000	617,000	517,000	601,000	281,890	609,333
Soybeans	Yield	hg/ha	2778	2927	3238	2990	3098	4652	8298	9401	12,951	9658
Soybeans	Production	tonnes	75,000	60,000	68,000	218,000	145,000	287,000	429,000	565,000	365,080	588,523
Spices, nes	Area harvested	ha	2900	3000	3000	3600	1428	1971	2634	3250	3600	4087
Spices, nes	Yield	hg/ha	13,793	14,667	13,333	13,333	12,962	13,112	13,299	13,846	15,278	15,746
Spices, nes	Production	tonnes	4000	4400	4000	4800	1851	2585	3503	4500	5500	6436
Sugar cane	Area harvested	ha	22,000	21,400	22,400	22,400	22,200	19,270	24,000	44,000	45,680	88,135
Sugar cane	Yield	hg/ha	395,455	402,804	400,446	410,714	400,000	305,656	289,583	207,727	186,055	164,516

Item	Element	Unit	1980	1985	1986	1990	1991	1995	2000	2005	2010	2015
Sugar cane	Production	tonnes	870,000	862,000	897,000	920,000	888,000	589,000	695,000	914,000	849,898	1,449,963
Sweet potatoes	Area harvested	ha	12,000	12,000	13,000	28,000	31,000	299,000	823,000	989,000	1,298,486	1,499,015
Sweet potatoes	Yield	hg/ha	83,333	66,667	63,846	51,071	59,355	39,064	29,988	32,406	26,701	25,652
Sweet potatoes	Production	tonnes	100,000	80,000	83,000	143,000	184,000	1,168,000	2,468,000	3,205,000	3,467,073	3,845,248
Taro (cocoyam)	Area harvested	ha	65,000	70,000	91,000	141,000	166,000	229,000	587,000	667,000	520,130	826,800
Taro (cocoyam)	Yield	hg/ha	32,000	33,143	40,989	51,844	49,940	51,616	66,201	75,982	56,853	39,631
Taro (cocoyam)	Production	tonnes	208,000	232,000	373,000	731,000	829,000	1,182,000	3,886,000	5,068,000	2,957,090	3,276,700
Tobacco, unmanufactured	Area harvested	ha	20,680	19,500	20,000	22,000	18,522	17,463	37,000	25,000	14,789	9500
Tobacco, unmanufactured	Yield	hg/ha	6286	5385	4500	4091	4859	5268	5946	6000	6131	5263
Tobacco, unmanufactured	Production	tonnes	13,000	10,500	9000	9000	9000	9200	22,000	15,000	9066	5000
Tomatoes	Area harvested	ha	32,500	35,000	35,500	37,500	38,000	55,000	210,000	250,000	272,950	557,500
Tomatoes	Yield	hg/ha	100,000	100,000	100,000	100,000	100,000	103,455	60,038	81,714	65,945	75,862
Tomatoes	Production	tonnes	325,000	350,000	355,000	375,000	380,000	569,000	1,260,794	2,042,861	1,799,960	4,229,330
Vegetables, fresh nes	Area harvested	ha	210,000	260,000	265,000	350,000	380,000	440,000	620,754	725,000	724,335	753,081
Vegetables, fresh nes	Yield	hg/ha	46,286	48,231	48,792	50,314	53,289	59,273	63,552	71,517	82,818	91,171
Vegetables, fresh nes	Production	tonnes	972,000	1,254,000	1,293,000	1,761,000	2,025,000	2,608,000	3,945,000	5,185,000	5,998,811	6,865,947
Wheat	Area harvested	ha	10,000	55,000	67,000	60,000	50,000	20,000	52,000	60,000	74,399	60,000
Wheat	Yield	hg/ha	24,000	20,545	19,701	8333	12,000	21,800	14,038	11,000	14,844	10,000

Item	Element	Unit	1980	1985	1986	1990	1991	1995	2000	2005	2010	2015
Wheat	Production	tonnes	24,000	11,3000	13,2000	50,000	60,000	43,600	73,000	66,000	110,441	60,000
Yams	Area harvested	ha	498,000	840,000	924,000	1,276,000	1,639,000	2,118,000	2,647,000	2,957,000	2,868,990	5,389,870
Yams	Yield	hg/ha	105,382	56,405	56,374	106,771	103,453	107,734	98,984	114,981	13,0109	84,748
Yams	Production	tonnes	5,248,000	4,738,000	5,209,000	13,624,000	16,956,000	22,818,000	26,201,000	34,000,000	37,328,180	45,677,939
Cereals (rice milled eqv)	Area harvested	ha	7,165,000	9,527,000	12,675,000	15,400,000	17,014,000	18,598,000	18,242,000	18,310,000	16,132,376	17,631,248
Cereals (rice milled eqv)	Yield	hg/ha	10,363	11,979	11,195	10,939	10,310	11,582	11,113	13,568	14,357	13,254
Cereals (rice milled eqv)	Production	tonnes	7,425,030	11,412,810	14,189,687	16,845,500	17,540,742	21,540,240	20,271,766	24,843,189	23,160,948	23,367,984
Cereals, total	Area harvested	ha	7,165,000	9,527,000	12,675,000	15,400,000	17,014,000	18,598,000	18,242,000	18,310,000	16,132,376	17,631,248
Cereals, total	Yield	hg/ha	10,870	12,479	11,567	11,479	10,941	12,105	11,715	14,217	15,280	14,435
Cereals, total	Production	tonnes	7,788,000	11,889,000	14,661,322	17,678,000	18,615,000	22,512,600	21,370,000	26,031,000	24,650,297	25,451,307
Citrus fruit, total	Area harvested	ha	550,000	570,000	570,000	580,000	630,000	643,589	727,596	731,000	790,000	821,533
Citrus fruit, total	Yield	hg/ha	32,727	35,088	35,088	35,914	39,683	42,263	44,668	45,179	48,101	48,757
Citrus fruit, total	Production	tonnes	1,800,000	2,000,000	2,000,000	2,083,000	2,500,000	2,720,000	3,250,000	3,302,611	3,800,000	4,005,520
Coarse grain, total	Area harvested	ha	6,605,000	8,802,000	11,908,000	14,132,000	15,312,000	16,782,000	15,991,000	15,756,000	13,625,346	14,449,686
Coarse grain, total	Yield	hg/ha	10,104	11,754	11,012	10,705	10,011	11,649	11,256	142,16	14,728	13,243
Coarse grain, total	Production	tonnes	6,674,000	10,346,000	13,113,000	15,128,000	15,329,000	19,549,000	17,999,000	22,398,000	20,067,337	19,135,079
Fibre crops primary	Area harvested	ha	476,000	221,000	286,000	576,107	644,000	432,000	539,000	660,000	399,570	
Fibre crops primary	Yield	hg/ha	616	503	1293	1664	1613	2222	2748	2897	5539	

Item	Element	Unit	1980	1985	1986	1990	1991	1995	2000	2005	2010	2015
Fibre crops primary	Production	tonnes	29,324	11,124	36,990	95,881	103,892	95,971	148,099	191,219	221,331	
Fruit primary	Area harvested	ha	1,105,000	1,170,000	1,183,200	1,189,349	1,276,025	1,423,473	1,729,627	1,730,000	1,819,085	1,896,920
Fruit primary	Yield	hg/ha	46,805	49,684	50,093	53,660	54,934	55,699	54,288	55,902	59,166	60,973
Fruit primary	Production	tonnes	5,172,000	5,813,000	5,927,000	6,382,000	7,009,713	7,928,578	9,389,796	9,671,111	10,762,880	11,566,040
Oil crops, cake equivalent	Area harvested	ha	3,852,000	3,553,200	3,797,200	4,686,500	5,071,000	6,199,500	6,831,000	7,726,000	7,736,900	
Oil crops, cake equivalent	Yield	hg/ha	1287	1680	1920	2239	2147	2262	3110	3253	3159	
Oil crops, cake equivalent	Production	tonnes	495,908	597,030	728,900	1,049,229	1,088,695	1,402,352	2,124,177	2,513,260	2,444,181	
Oil crops, oil equivalent	Area harvested	ha	3,952,000	3,647,200	3,884,200	4,870,500	5,275,000	6,434,500	7,063,000	7,983,239	8,052,650	
Oil crops, oil equivalent	Yield	hg/ha	2599	2958	3101	3040	2994	2933	3397	3619	3322	
Oil crops, oil equivalent	Production	tonnes	1,026,982	1,078,950	1,204,640	1,480,674	1,579,318	1,887,136	2,399,149	2,888,867	2,675,077	
Pulses, total	Area harvested	ha	1,578,000	1,525,000	1,525,000	1,936,832	1,960,049	3,672,216	3,685,492	4,259,432	2,989,760	3,775,795
Pulses, total	Yield	hg/ha	3568	4334	4538	7264	7055	4862	5947	6724	11,453	6279
Pulses, total	Production	tonnes	563,000	661,000	692,000	1,406,870	1,382,756	1,785,592	2,191,887	2,864,018	3,424,250	2,370,770
Roots and tubers, total	Area harvested	ha	1,780,500	2,004,000	2,130,600	3,086,830	4,396,400	5,603,600	7,569,000	8,655,000	8,435,498	14,260,128
Roots and tubers, total	Yield	hg/ha	96,018	85,744	84,948	108,833	100,171	101,126	86,093	97,763	10,3505	78,294
Roots and tubers, total	Production	tonnes	17,096,000	17,183,000	18,099,000	33,595,008	44,039,000	56,667,000	65,164,000	84,614,000	87,311,834	1.12E+08
Tree nuts, total	Area harvested	ha	40,000	40,100	40,300	52,500	77,500	16,0421	26,1965	31,1550	38,5309	13,4328
Tree nuts, total	Yield	hg/ha	6250	6284	6352	6857	6452	6129	17,986	19,231	20,730	7772

Item	Element	Unit	1980	1985	1986	1990	1991	1995	2000	2005	2010	2015
Tree nuts, total	Production	tonnes	25,000	25,200	25,600	36,000	50,000	98,316	471,163	599,150	798,726	104,402
Vegetables primary	Area harvested	ha	617,500	821,000	867,500	955,691	971,030	1,117,275	1,527,001	1,881,174	1,891,815	3,942,033
Vegetables primary	Yield	hg/ha	51,142	47,320	46,651	48,894	48,980	52,794	54,208	59,090	63,911	40,898
Vegetables primary	Production	tonnes	3,158,000	3,885,000	4,047,000	4,672,719	4,756,152	5,898,594	8,277,564	11,115,881	12,090,760	16,122,242

Source: Authors.

Table 4.
Crop production in Nigeria (1980–2015): harvest area, yield and production.

production index, employment in agriculture (male, female and total employment in the agricultural sector) and agricultural land.

From **Table 1**, crop production in Nigeria shows an increase and decrease trend; it was observed among those that are employed in the agriculture; the number of women in agriculture is more than the number of men in agriculture. The price of various agricultural items across Nigeria in 2017 is presented in **Table 2**.

In **Table 3**, prices of agriculture commodities resulting from production are presented. Such commodities include eggs; beans: brown; beef; rice (Ofada); catfish (obokun), fresh; catfish, dried; catfish, smoked; chicken feet; chicken wings; dried fish sardine; evaporated tinned milk carnation 170 g; and evaporated tinned milk (peak) 170 g. Frozen chicken; gari, white, sold loose; gari, yellow; groundnut oil; iced sardine; Irish potato; mackerel; maize grain; mudfish (aro), fresh; mudfish, dried; onion bulb; palm oil; plantain (ripe); plantain (unripe); sweet potato; tilapia fish (epiya) fresh; titus (frozen); tomato; vegetable oil; wheat flour, prepacked (golden penny 2 kg); and yam tuber, among other commodities not included. The prices of those commodities vary from January to December in 2017. This is also similar in 2018 as presented in **Table 3**.

Table 4 presents the area of crops harvested (ha), yield of crop production (hg/ha) and output level of various crops (tonnes) from 1980 to 2015.

Various crops presented in **Table 4** include bast fibres, carrots and turnips, cashew nuts (with shell), cassava, chillies and peppers (dry), chillies and peppers (green), cocoa, beans, coconuts, coffee (green), cotton (lint), cottonseed, cow peas (dry), fibre crops (nes), fruit, citrus (nes), garlic, groundnuts, karite nuts (shea nuts), kola nuts, maize, maize (green), mangoes, mangosteens, guavas, melon seed, millet, nuts (nes), nuts (nes), oil palm fruit, oilseeds (nes), okra, onions (dry), shallots (green), palm kernels, papayas, potatoes, pulses (nes), rice (paddy), rubber (natural), seed cotton, sesame seed, sorghum, soybeans, spices (nes), sugar cane, sweet potatoes, taro (cocoyam), tobacco, unmanufactured, tomatoes, vegetables (fresh nes), wheat, yams, cereals (rice milled eqv), cereals (total), citrus fruit (total), coarse grain (total), fibre crops primary, fruit primary, oil crops, cake equivalent, pulses (total), roots and tubers (total), tree nuts (total), vegetables primary, etc.

Therefore, to ensure sustainable crop production, the agricultural sector needs to be invested on through various means like credit facilities and incentives such as social protection for the mitigation of risk and shocks [1]. Also, the nutritional level of plants should be improved through fertiliser application among other means to enhance crop yields [17].

4. Conclusion

The study aims at examining factors that improve agricultural production, especially crop yields that can be made possible by plant nutrients. Increase in crop production (food and cash crops) will lead to food security in the long run. The study employed a review of literature and stylised fact approach using tables to know the level of crop production in Nigeria. From the stylised facts and the reviewed literature, authors noticed that there are fluctuations of prices of food items in Nigeria.

With respect to the factors contributing to crop and agricultural production, employment in agriculture was observed to be a major factor. Also, the proportion of women in agriculture is higher than the proportion of men in agriculture; this invariably implies that women actually contribute more to production level. In this regard, to further enhance productivity, there should be equal access to production resources such land, credit facilities, access to social protection incentives to mitigate risks and shocks and more innovation and technological advancement in the agricultural sector thereby improving the sustainability of crop production.

Author details

Romanus Osabohien* and Toun Ogunbiyi
Department of Economics and Development Studies, Covenant University, Ota, Nigeria

*Address all correspondence to: romanus.osabohien@covenantuniversity.edu.n

References

[1] Osabohien R, Matthew O, Aderounmu B, Olawande T. Greenhouse gas emissions and crop production in West Africa: Examining the mitigating potential of social protection. International Journal of Energy Economics and Policy. 2019; **9**(1):57-66

[2] United State Institute of Agriculture and National Institute of Agriculture. Crop Production. 2019. Available from: https://nifa.usda.gov/topic/crop-production

[3] International Atomic Energy Agency. Crop Nutrition. 2018. Retrieved from: https://www.iaea.org/topics/crop-nutrition

[4] Obasi PC, Henri-Ukoha A, Ukewuihe IS, Chidiebere-Mark NM. Factors affecting agricultural productivity among arable crop farmers in Imo state Nigeria. American Journal of Experimental Agriculture. 2013;**3**(2): 443-454

[5] Awodun MA, Otaru MS, Ojeniyi SO, Akintunde AY, Obigbesan GO, Kiru SK, et al. Effect of inorganic fertilizer and foliage of Azadirachta and Parkia species on the productivity of maize. Asian Journal of Agricultural Research, 2000;**1**(1):35-44

[6] Ayeni FA, Sánchez B, Adeniyi BA, Clara G, Margolles A, Ruas-Madiedo P. Evaluation of the functional potential of Weissella and Lactobacillus isolates obtained from Nigerian traditional fermented foods and cow's intestine. International Journal of Food Microbiology. 2011;**147**(2): 97-104

[7] Nottidge DO, Ojeniyi SO, Aswalam DO. A comparative effect of plant residue and NPK fertilizer on nutrient status and yield of maize in a humid. Soil Nigerian Journal of Soil Science. 2005;**15**:1-8

[8] Enete AA, Amusa TA. Challenges of agricultural adaptation to climate change in Nigeria: A synthesis from the literature. Field Actions Science Reports. The Journal of Field Actions. 2010;4. Available from: https://journals.openedition.org/factsreports/678

[9] Yobannes H. A review on relationship between climate change and agriculture. Journal of Earth Science and Climate Change. 2016;**7**(2): 335

[10] Ufiobor KA. Nigerian Agriculture and Sustainability: Problems and Solutions. 2017. https://www.theseus.fi/handle/10024/132525

[11] Nwajiuba CU, Amazu GO, Nwosu CS, Onyeneke RU. Motivation factors and constraints to the growth of small scale food processing enterprises in Owerri metropolis, Imo State, Nigeria. International Journal of Entrepreneurship and Small Business. 2013;**19**(4):488-497

[12] Nations Encyclopedia. 2019. Retrieved from: https://www.nationsencyclopedia.com/Africa/Nigeria-AGRICULTURE.html

[13] Food and Agricultural Organisation of the United Nations. 2017. Retrieved from: http://www.fao.org/faostat/en/#data/EE

[14] International Food Policy Research Institute. 2010. Retrieved from: http://nssp.ifpri.info/files/2010/11/nssp-report-9_final.pdf

[15] Ayoola GB, Chude VO, Abdulsalam AA. Towards a Fertilizer RegulatoryPolicy for Nigeria: An Inventorization of the Fertilizer Sector. 2002

[16] Ukojie J, Yusuf R. Organic fertilizer: The underestimated component in Agricultural Transformation Initiatives for Sustainable Small Holder Farming in Nigeria. Ethiopian Journal of Environmental Studies and Management. 2013;**6**:794-801

[17] Osabohien R, Afolabi A, Godwin A. An econometric analysis of food security and agricultural credit facilities in Nigeria. The Open Agriculture Journal. 2018;**12**:227-239

Chapter 10

Enhancing Soil Properties and Maize Yield through Organic and Inorganic Nitrogen and Diazotrophic Bacteria

Arshad Jalal, Kamran Azeem,
Marcelo Carvalho Minhoto Teixeira Filho and Aeysha Khan

Abstract

In arid and semiarid ecosystems, low organic matter is an important threat to soil fertility, crop productivity, and economic returns. Sustainable approaches are required to build organic matter in such soils to improve nutrient use efficiency and food security. Therefore, we conducted an experiment on spring maize to test with and without diazotrophic bacteria (BM) (*Azotobacter chroococcum* and *Azospirillum brasilense*) on crop productivity and soil properties when applied with organic (farm yard manure FYM) and inorganic (commercial fertilizer) nitrogen (N) sources (with percentile of 0, 25, 50, 75, and 100%) in 2014. The analysis of the study showed that the application of BM and organic and inorganic N ratio were significant and have a positive effect in crop yield and soil properties. BM with a 50:50 ratio of organic and inorganic N was improved biological yield (kg ha^{-1}), grain yield (kg ha^{-1}), stover nitrogen (%), and grain nitrogen (%). However, soil organic matter (%) and soil total nitrogen (%) were enhanced with the application of BM with 100% organic source. Soil bulk density (g cm^{-3}) was significantly reduced by BM with 100% organic. From overall results, it is concluded that the application of beneficial microbes and organic and inorganic N positively improved maize yield and quality and soil health in Peshawar valley.

Keywords: plant growth-promoting bacteria, nitrogen fertilization, soil quality, grain yield

1. Introduction

Maize (*Zea mays* L.) is the third most important cereal crop after wheat and rice, while in the farming system of Khyber Pakhtunkhwa, it ranks third after wheat and rice in its importance [1, 2]. It is an exhaustive and multipurpose cereal crop that provides food for human, feed for animals, and raw material for the industries [3]. It has greater nutritional value as it contains about 72% starch, 10.4% proteins, and 4.5% fats, minerals, and non-cholesterol oil [2, 4].

To mitigate the problem of low yield and contamination in an eco-friendly way is to use effective microorganisms (EM) [5] also known as beneficial microorganism (BM). Microorganisms economically support the farmer community by improving the soil activities [6] and assimilate accumulation in the final product of the production which in turn maintains the balance of organic and inorganic mechanisms of the soil and plant [7]. Many researches reflected advantageous effects of BM on soil physicochemical status [8, 9]. Beneficial microorganism increases the decomposition rate of organic fertilizer and increases nutrients availability [10]. Beneficial microorganism also promotes soil fertility, crop growth, and yield [11]; also improves soil health, soil quality, yield, and quality [12] of various physiological attributes; and regulates various metabolites and atmospheric nitrogen [13].

Nitrogen significantly improves crop productivity [14]. Fertilizers are usually applied to soil for increasing or maintaining crop yields to meet the increasing demand of food [15, 16]. The application of inorganic fertilizers results in higher soil organic matter (SOM) accumulation and biological activity due to increased plant biomass production and organic matter return to the soil in the form of decaying roots, litter, and crop residues [17, 18]. The addition of SOM enhances soil organic carbon (SOC) content, which is an important indicator of soil quality and crop productivity [19]. Chemical fertilizer applications could also affect soil physical properties directly or indirectly such as aggregate stability, water holding capacity, porosity, infiltration rate, hydraulic conductivity, and bulk density due to increases in SOM and SOC content [20, 21]. In turn, the formation of stable aggregates enhances physical protection of SOM against microbial decomposition [22]. Some fertilizer additions also affect the chemical composition of soil solution which can be responsible for dispersion/flocculation of clay particles and thus affect the soil aggregation stability [21, 23]. We can see the plant and microbial interaction in the retention and improvement of N cycle for better plant growth (**Figure 1**).

The present study aimed to investigate the effect of organic and inorganic nitrogen ratios along with the effect of beneficial diazotrophic bacteria on maize yield and quality and physio-chemical properties of soil.

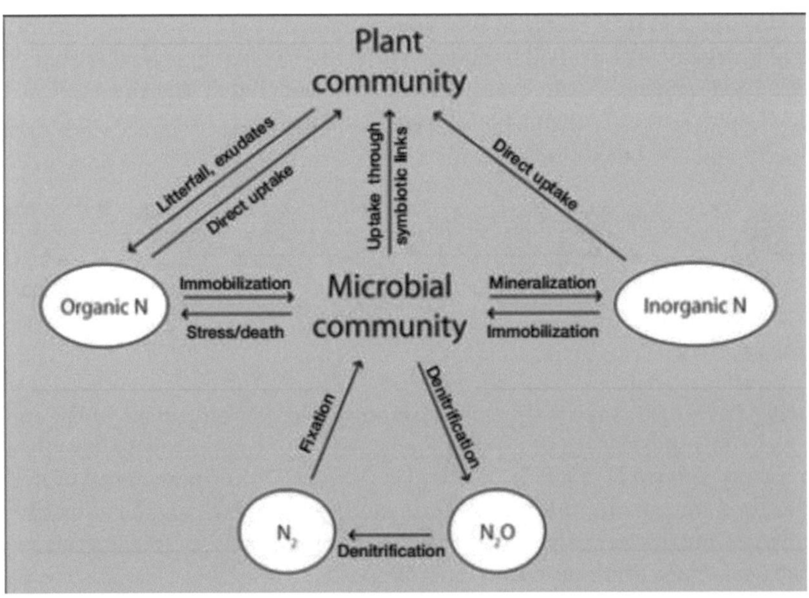

Figure 1.
Plants and microbial interaction in the retention and improvement of N cycle for better plant growth [24].

2. Materials and methods

2.1 Experimental site

The impact of beneficial microbes on enhancing efficiency of organic and inorganic N fertilizers sources was studied on spring maize cropping system in the year of 2014 at Agronomy Research Farm, The University of Agriculture Peshawar, Pakistan. The research area is geographically located at 17°, 35′ N and 35°, 41′ W and altitude of 450 m above sea level. The soil of the experimental farm was silt loam, well drained, and fine textured. The experimental site has a semiarid subtropical continental climate with a mean annual rainfall of about 550 mm. The soil is deficient in total N (<0.5%) and AB-DTPA extractable P (<4.0 mg kg^{-1} soil) but has adequate AB-DTPA extractable K (>100 mg kg^{-1} soil) with a pH of 7.60 and organic matter content <1% (**Table 1**). Rainfall and temperature data were collected from the weather station of Agronomy Research Farm and summarized in **Figure 2**. In addition to rainfall, crop water requirement was fulfilled by supplying water through surface irrigation according to crop requirement.

2.2 Materials and treatments

The experimental field was irrigated before sowing for weed germination and then plowed with cultivator to prepare a fine seed bed for sowing. The experiment consisted of two factors, i.e., beneficial diazotrophic bacteria (*Azotobacter chroococcum* and *Azospirillum brasilense*) (with BM and without BM) and organic (FYM) and inorganic (urea commercial fertilizer) N ratios (0:100, 25:75, 50:50, 75:25 and 100:0). The recommended doses of phosphorus (90 kg ha^{-1}) and potassium (60 kg ha^{-1}) were applied at the time of seed bed preparation from the sources of DAP and SOP. The fertilizer of nitrogen was applied in two equal splits, and organic N was applied 4 weeks before sowing.

2.3 Experimental design

The experiment was conducted at three factorial randomized complete block designs (RCBD) with three replications. The size of plots was 4.2 × 4 m. Row-to-row

Property	Unit	Data
Clay	%	2.8
Silt	%	50
Sand	%	47.2
Textural class	—	Silty loam
pH (1:5)	—	7.60
EC (1:5)	dS m^{-1}	0.18
Organic matter	%	0.39
Total nitrogen	%	0.06
Phosphorus	ppm	2.86
Potassium	Ppm	120.48
Mineral nitrogen	mg kg^{-1}	35

Table 1.
Pre-sowing physicochemical properties of soil (0–0.30 m depth).

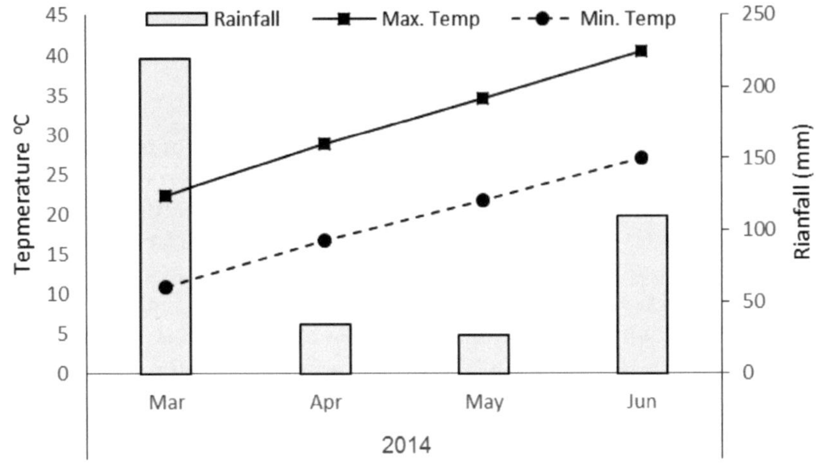

Figure 2.
Weather data of spring maize growing season from March to June, 2014.

distances for maize crop were 0.70 m, whereas plant-to-plant distances were 0.20 m. Each plot had six rows. There were 30 plots having treatment combination of two beneficial diazotrophic bacteria and five organic and inorganic source ratios.

2.4 Observations recorded

Biological yield data was recorded by harvesting four central rows in each plot, sundried and weighed by electronic balance whereas harvested central rows were threshed individually through electric thresher and weighed through electronic balance to obtain grain yield and then converted into kg ha^{-1} by the following formula:

$$\text{Grain yield } \left(\text{kg ha}^{-1}\right) = \frac{\text{grain yield in four central rows}}{\text{row} - \text{row distance (m)} \times \text{row length (m)} \times \text{no.of rows}} \times 10,000 \qquad (1)$$

Organic matter in soil was determined by the modified method of Nelson and [25]. The nitrogen content in soil, stover, and grains were determined by following Kjeldahl method according to the proposed methodology of Bremner and Mulvaney [26].

2.5 Statistical analysis

The data recorded was analyzed statistically using analysis of variance techniques appropriate for randomized complete block design. Statistical analysis was done with Statistic-X software. Means were compared using LSD test at 0.05 level of probability, when the F-values was significant [27]. The possible interactions were graphically made using a software of Microsoft Excel 365.

3. Results and discussion

3.1 Biological and grain yield as influenced by organic and inorganic N with BM

Beneficial microbes significantly influenced biological yield (**Table 2**). Highest biological yield (11,708 kg ha^{-1}) has been observed with the application of

Beneficial microbes	Biological yield (kg ha^{-1})	Grain yield (kg ha^{-1})	Soil organic matter (%)	Bulk density (g cm^{-3})
Without BM	10,733 b	3662 b	0.87 b	1.19 a
With BM	11,708 a	3803 a	1.19 a	1.17 b
LSD	602.29	79.52	0.08	0.02
Organic and inorganic ratios				
0:100	10,961 b	3592 b	0.87 c	1.23 a
25:75	11,401 ab	3793 ab	0.94 bc	1.22 ab
50:50	12,092 a	3907 a	1.02 b	1.19 b
75:25	10,621 b	3684 b	1.10 ab	1.15 c
100:0	11,027 ab	3686 b	1.22 a	1.09 d
LSD	952.3	125.7	0.13	0.03
Interaction				
BM × R	**Figure 3**	ns	**Figure 4**	ns

Mean values of the different categories in each column with different letters discloses significant differences ($p \leq 0.05$) using LSD test.

Table 2.
Biological yield and grain yield of spring maize and soil organic matter and soil bulk density as influenced by beneficial microbes and organic and inorganic ratios.

beneficial diazotrophic bacteria as compared to without beneficial microbes. This may be due to diazotrophic bacteria increase the speed of decomposition and mineralization that improve nutrients' availability to the crop and total dry matter production [28]. Similarly, the application of organic and inorganic fertilizers also significantly affects the biological yield, and greater biological yield (12,092 kg ha^{-1}) was achieved with the application of 50:50 N ratio of organic and inorganic fertilizers. Lower biological yield (10,961 kg ha^{-1}) was attained with the application of 100% N from inorganic source. It might be due to the reason that nitrogen from organic sources are slow release, while inorganic nitrogen is readily available to plant which may not be available at later stages. The combined application of N from inorganic source (urea) and organic source in a ratio of 75:25 improved grain yield, straw yield, and biological yield, whereas 50:50 N ratio increased uptake of nitrogen [29]. Biological and grain yield was significantly improved with the application of 50% nitrogen from inorganic sources in combination with the application of 25% N from FYM and 25% N from poultry manure [30]. The applications of organic and inorganic N fertilizers significantly enhanced biological yield and grain yield [31, 32]. The application of organic and inorganic nitrogen 50% from urea and 50% from FYM or 50% poultry manure significantly enhanced biological yield, grain yield, and harvest index (%) [28].

The graph trend showed that biological yield increased with organic and inorganic nitrogen ratio from 0:100 to 50:50, whereas a decreased trend in biological yield was observed from 50:50 to 100:0 with both beneficial microbes (**Figure 3**). This might be due to the fact that beneficial microbes rapidly decomposed organic matter, provided nutrients, and increased availability of nitrogen from both organic and inorganic sources [28].

Beneficial diazotrophic bacteria significantly increased the grain yield (**Table 2**). Highest grain yield (3803 kg ha^{-1}) has been noted with the application of beneficial microbes as compared to without beneficial microbes. This may be due to the reason that beneficial microbes increase decomposition and mineralization and improve nutrients availability for more total dry matter production [28].

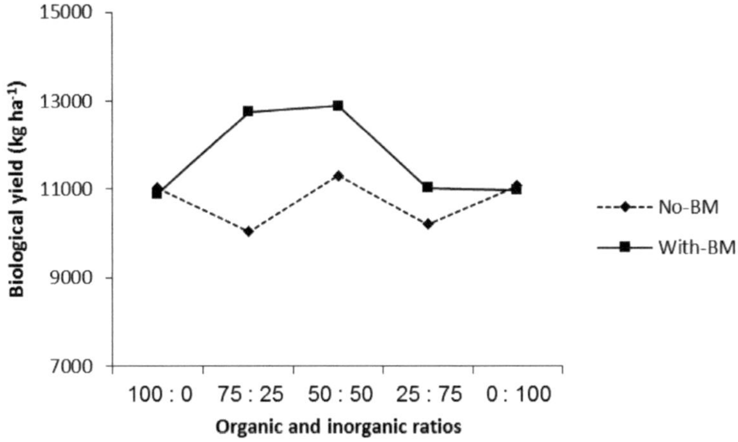

Figure 3.
Biological yield as affected by beneficial microbes and organic and inorganic ratios.

Organic and inorganic fertilizer significantly influenced grain yield, and higher grain yield (3907 kg ha^{-1}) was achieved with the application of 50:50 N organic and inorganic ratio, whereas lower grain yield (3592 kg ha^{-1}) was observed with the application of 100% N from inorganic. It might be due to the reason that N from organic sources is slow release, whereas inorganic nitrogen is readily available to plant that may function in vegetative growth. The application of organic and inorganic N with a ratio of 75:25 prominently improved yield and yield indices, whereas a 50:50 ratio increases nitrogen uptake [29]. Biological and grain yield was significantly higher with the application of 50% nitrogen from inorganic sources with the application of 25% N from FYM and 25% N from poultry manure [30]. The application from organic and inorganic N fertilizer significantly influenced biological yield and grain yield [31, 32]. The application of organic and inorganic nitrogen 50% from urea and 50% from FYM or 50% poultry manure significantly enhanced biological yield, grain yield, and harvest index % [28].

3.2 Soil nitrogen analysis as influenced by organic and inorganic N with BM

Beneficial microbes significantly affected soil organic matter (**Table 2**). Highest soil organic matter (1.19%) has been perceived with the application of beneficial microbes as compared to without diazotrophic bacteria. This may be due to the beneficial microbe increases the speed of decomposition and increase mineralization and produced more exudes [28]. Organic and inorganic ratios significantly affected soil organic matter. More soil organic matter (1.22%) was achieved with the application of a 100:0 ratio of organic and inorganic fertilizer, whereas less soil organic matter (0.87%) was perceived by the application of 100% from inorganic fertilizer. It might be due to beneficial microbe rate of decomposition of organic fertilizer which improves soil organic matter and soil organic carbon. The combined application of organic and inorganic fertilizers improved soil organic matter and total nitrogen [33]. Organic manure linearly increased soil organic matter [34]. The integration of organic with inorganic fertilizer significantly improved crop production and N, P, K, soil pH, soil EC, and organic matter [35, 36]. Organic sources improved soil nutrient, soil organic matter, and soil organic carbon [37]. The graph trend showed that soil organic matter increased linearly with the increased of organic and inorganic ratio from 0:100 to 100:0 in both beneficial and without beneficial microbes (**Figure 4**). This might be due to organic matter was applied

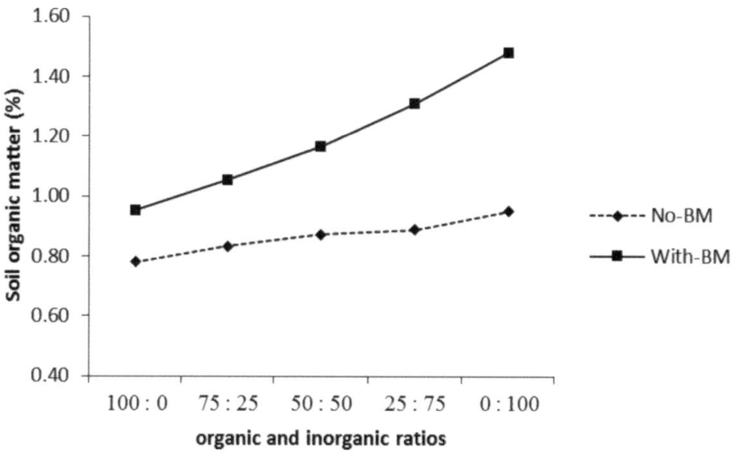

Figure 4.
Soil organic matter as affected by beneficial microbes and organic and inorganic ratios.

100% from organic sources, and beneficial microbes rapidly decomposed organic matter, thus increasing mineralization and soil organic matter [28].

Beneficial microbes significantly affected soil bulk density as highest soil bulk density (1.19 g cm^{-3}) has been perceived without the application of beneficial microbes as compared to with beneficial microbes (**Table 2**). This may be due to beneficial microbes increase the speed of decomposition and increase mineralization and provide nutrients, thus decreasing bulk density; these results are in line with Muhammad et al. [28]. Organic and inorganic ratios significantly affected soil bulk density. Soil bulk density (1.23 g cm^{-3}) was recorded with the application of a 0:100 ratio of organic and inorganic, whereas less soil bulk density (1.09 g cm^{-3}) was determined by the application of 100% of organic. It might be due to the integration of organic and inorganic fertilizer which improved soil bulk density. Bulk density decreased with the application of FYM and poultry manure to soil [38]. Bulk density linearly decreased with soil organic matter [34]. The integration of organic with inorganic fertilizer significantly decreased bulk density of soil [35, 36].

Beneficial diazotrophic bacteria significantly affected soil nitrogen content (**Table 3**). Highest soil nitrogen content (0.39%) has been received with the application of beneficial microbes as compared to without beneficial microbes. It may be due to beneficial microbe increases the speed of decomposition and increases mineralization and provides nutrients, thus more nitrogen in soil; these results are in line with Muhammad et al. [28]. Organic and inorganic ratios significantly affected soil nitrogen content. More soil nitrogen content (0.47%) was achieved with the application of 50:50 ratio of organic and inorganic, whereas less soil nitrogen content (0.24%) was recorded by the application of 100% from inorganic. It might be due to the decomposition of organic matter is slow and the slow release of nutrients; therefore, plots of organic sources have higher N, P, and K than plots having inorganic fertilizer. Organic sources improved soil nutrient and organic carbon [37]. Soil mineral nitrogen increased (22.4%) with the addition of organic fertilizers like FYM, poultry, and legume residues [39]. For better crop growth and sustainability, addition of organic matter is best source of nutrient availability [34]. The integration of organic with inorganic fertilizer significantly improved crop production and N, P, K, soil pH, soil EC, and organic matter [35, 36].

The graph trend showed that soil organic matter increased linearly with the increased of organic and inorganic ratio from 0:100 to 100:0 in both beneficial and without beneficial microbes (**Figure 5**). This might be due to organic matter was

Beneficial microbes	Soil nitrogen content (%)	Stover nitrogen content (%)	Grain nitrogen content (%)
Without BM	0.34 b	0.88 b	1.74 b
With BM	0.39 a	1.10 a	1.91 a
LSD	0.01	0.08	0.01
Organic and inorganic ratios			
0:100	0.24 e	0.89 b	1.56 c
25:75	0.30 d	0.98 ab	1.81 b
50:50	0.38 c	1.11 a	2.01 a
75:25	0.44 b	0.99 ab	1.97 ab
100: 0	0.47 a	0.98 ab	1.78 b
LSD	0.02	0.13	0.16
Interaction			
BM × R	**Figure 5**	ns	**Figure 6**

Mean values of the different categories in each column with different letters discloses significant differences ($p \leq 0.05$) using LSD test.

Table 3.
Soil nitrogen content, stover, and grain nitrogen content of spring maize as influenced by beneficial microbes and organic and inorganic ratios.

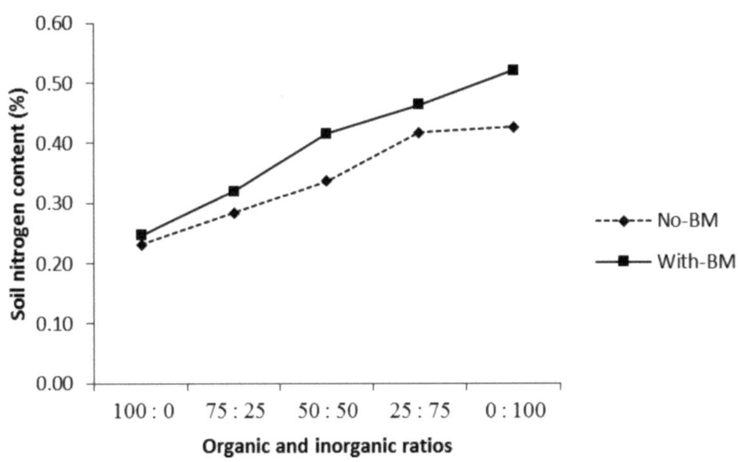

Figure 5.
Soil nitrogen content as affected by beneficial microbes and organic and inorganic ratios.

applied 100% from organic sources, and beneficial microbes rapidly decomposed organic matter, thus increasing mineralization and soil organic matter [28].

3.3 Plant nitrogen analysis as influenced by organic and inorganic N with BM

Beneficial diazotrophic bacteria significantly affected stover nitrogen content (**Table 3**). Highest stover nitrogen content (1.1%) has been perceived with the application of beneficial microbes as compared to without beneficial microbes. It can be due to beneficial microbes increase the speed of decomposition and increase mineralization and provide nutrients for crop to achieved more total nutrient

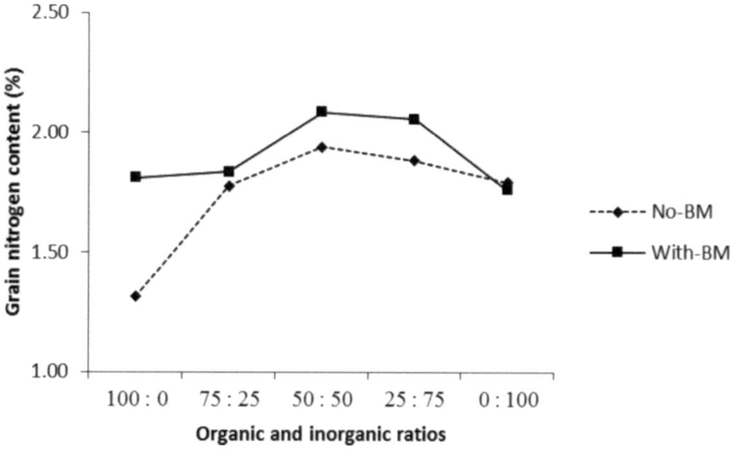

Figure 6.
Grain nitrogen content as affected by beneficial microbes and organic and inorganic ratios.

production [28]. Our results indicated that organic and inorganic N ratios significantly influenced N stover content in maize crop. The application of a 50:50 ratio of organic and inorganic nitrogen resulted in higher stover N content (1.11%), whereas lower stover nitrogen content (0.89%) was attained with the application of 100% N from inorganic source. It might be due to the reason that inorganic fertilizer was quickly available, while organic fertilizer was slowly available to crop. The N, P, and K concentration in straw was significantly increased with the combined application of 10 t N ha^{-1} from poultry manure (PM) and 200 kg N ha^{-1} from NPK as compared to control [40]. N, P, and K uptake by straw and grains was significantly influenced by organic and inorganic fertilizer [41–43]. The application of chemical fertilizers, FYM, green manures, and compost to the soil resulted in improved uptake of N, P, and K [44, 45].

Beneficial diazotrophic bacteria significantly increase grain nitrogen content (**Table 3**). Highest grain nitrogen content (1.91%) has been perceived with the application of beneficial microbes as compared to without beneficial microbes. This may be due to beneficial microbes increase the speed of decomposition and increase mineralization, and more nitrogen content was transferred to grain [28]. Organic and inorganic ratios significantly improved grain nitrogen content, and higher grain nitrogen content (2.01%) was achieved with the application of a 50:50 organic and inorganic N ratio, whereas less grain nitrogen content (1.56%) was achieved with the application of 100% inorganic fertilizer. It may be due to the quick availability of inorganic fertilizer, whereas organic fertilizer is slowly available to the crop. Nitrogen and phosphorus are higher in grains than straw, while potassium content was higher in straw as compared to grains. N, P, and K content was significantly improved by organic fertilizer both in straw and grain [46]. Growth, yield, and NPK concentrations were significantly increased with integrated organic and inorganic fertilizers [3]. Macro- and micronutrients in the grains and straw of wheat were significantly improved with the application of FYM as inorganic N fertilizer [47]. Higher N, P, and K uptake by crop was observed with organic N sources [48].

The graph showed that grain nitrogen increased with organic and inorganic N ratio from 0:100 to 50:50. This trend was declined with N ratio from 50:50 to 100:0 for both beneficial microbes (**Figure 6**). This might be due to the reason that beneficial microbes rapidly decomposed organic matter, provided nutrients, and also increased availability of nitrogen from both organic and inorganic sources [28].

4. Conclusion

High maize yield and nitrogen content was better observed with the application of beneficial microbes and organic and inorganic nitrogenous fertilizer with a ratio of 50:50. Soil bulk density, soil organic matter, and soil total nitrogen were significantly improved with beneficial microbes combinedly apply with organic and inorganic nitrogenous fertilizer in a ratio of 100:0. Therefore, the application of beneficial microbes and organic and inorganic nitrogenous fertilizer with a ratio of 50:50 has been recommended for better nitrogen uptake and higher yield of spring maize in Peshawar valley.

Author details

Arshad Jalal[1], Kamran Azeem[2], Marcelo Carvalho Minhoto Teixeira Filho[1]* and Aeysha Khan[3]

1 Department of Plant Health, Agriculture Engineering and Soils, São Paulo State University (UNESP), São Paulo, Brazil

2 Department of Agricultural Science, The University of Haripur, Pakistan

3 Department of Agricultural Extension Education and Communication, The University of Agriculture Peshawar, Pakistan

*Address all correspondence to: mcm.teixeira-filho@unesp.br

References

[1] Azeem K, Khalil SK, Khan F, Shahenshah S, Qahar A, Sharif M, et al. Phenology, yield and yield components of maize as affected by humic acid and nitrogen. The Journal of Agricultural Science. 2014;**6**:284-286

[2] Imran. Effect of germination on proximate composition of two maize cultivars. Journal of Biology, Agriculture and Healthcare. 2015;**5**(3):123-128

[3] Khaliq T, Mahmood T, Masood A. Effectiveness of farm yard manaure, poultry manure and nitrogen for corn productivity. International Journal of Agriculture & Biology. 2004;**2**:260-263

[4] Aslam M, Iqbal A, Zamir MS, Mubeen M, Amin M. Effect of different nitrogen levels and seed rates on yield and quality of maize fodder. Crop Environment. 2011;**2**(2):47-51

[5] Gomiero T, Pimentel D, Paoletti MG. Environmental impact of different agricultural management practices: conventional vs. organic agriculture. Critical Reviews in Plant Sciences. 2011;**30**(1-2):95-124

[6] Tillman RW, Surapaneni A. Some soil-related issues in the disposal of effluent on land. Australian Journal of Experimental Agriculture. 2002;**42**(3):225-235

[7] Boraste A, Vamsi KK, Jhadav A, Khairnar Y, Gupta N, Trivedi S, et al. Biofertilizers: A novel tool for agriculture. International Journal of Microbiology. 2009;**1**(2):23

[8] Khaliq A, Abbasi MK, Hussain T. Effects of integrated use of organic and inorganic nutrient sources with effective microorganisms (EM) on seed cotton yield in Pakistan. Bioresource Technology. 2006;**97**(8):967-972

[9] Gorski R, Kleiber T. Effect of effective microorganisms (em) on nutrient contents in substrate and development and yielding of rose (rosa × hybrida) and gerbera (*Gerbera jamesonii*). Ecological Chemistry and Engineering. 2010;**17**(4):505-512

[10] Rashid MI, Mujawar LH, Shahzad T, Almeelbi T, Ismail IM, Oves M. Bacteria and fungi can contribute to nutrients bioavailability and aggregate formation in degraded soils. Microbiological Research. 2016;**183**:26-41

[11] Kashyap AS, Pandey VK, Manzar N, Kannojia P, Singh UB, Sharma PK. Role of plant growth-promoting rhizobacteria for improving crop productivity in sustainable agriculture. In: Plant-Microbe Interactions in Agro-Ecological Perspectives. Singapore: Springer; 2017. pp. 673-693

[12] Higa T. What is EM technology? EM World Journal. 2000;**1**:1-6

[13] Kumar A, Verma JP. The role of microbes to improve crop productivity and soil health. In: Ecological Wisdom Inspired Restoration Engineering. Singapore: Springer; 2019. pp. 249-265

[14] Habtegebrial K, Singh BR, Haile M. Impact of tillage and nitrogen fertilization on yield, nitrogen use efficiency of tef (*Eragrostis tef* (Zucc.) Trotter) and soil properties. Soil and Tillage Research. 2007;**94**:55-63

[15] Sainju UM, Ghimire R, Pradhan GP. Nitrogen fertilization I: Impact on crop, soil, and environment. In: Nitrogen in Agricultural Systems. Croatia: IntechOpen; 2019

[16] Singh Brar B, Singh J, Singh G, Kaur G. Effects of long-term application of inorganic and organic fertilizers on soil organic carbon and physical properties in maize–wheat rotation. Agronomy. 2015;**5**(2):220-238

[17] Hati KM, Swarup A, Dwivedi AK, Misra AK, Bandyopadhyay KK. Changes in soil physical properties and organic carbon status at the topsoil horizon of a Vertisol of central India after 28 years of continuous cropping, fertilization and manuring. Agriculture, Ecosystem. & Environment. 2007;**119**(1–2):127-134

[18] Kaur T, Brar BS, Dhillon NS. Soil organic matter dynamics as affected by long-term use of organic and inorganic fertilizers under maize–wheat cropping system. Nutrient Cycling in Agroecosystems. 2008;**81**(1):59-69

[19] Lal R. Soil erosion and the global carbon budget. Environment International. 2003;**29**(4):437-450

[20] Bhattacharyya R, Chandra S, Singh RD, Kundu S, Srivastva AK, Gupta HS. Long-term farmyard manure application effects on properties of a silty clay loam soil under irrigated wheat–soybean rotation. Soil and Tillage Research. 2007;**94**(2):386-396

[21] Haynes RJ, Naidu R. Influence of lime, fertilizer and manure application on soil organic matter content and soil physical conditions: A review. Nutrient Cycling in Agroecosystems. 1998;**51**: 123-137

[22] Zhang M-K, Fang LP. Effect of tillage, fertilizer and green manure cropping on soil quality at an abandoned brick making site. Soil and Tillage Research. 2015;**93**:87-93

[23] Chaplain V, Défossez P, Delarue G, Roger-Estrade J, Dexter AR, Richard G, et al. Impact of lime and mineral fertilizers on mechanical strength for various soil pHs. Geoderma. 2011;**167**: 360-368

[24] de Vries FT, Bardgett RD. Plant–microbial linkages and ecosystem nitrogen retention: Lessons for sustainable agriculture. Frontiers in Ecology and the Environment. 2012;**10** (8):425-432

[25] Nelson DW, Sommers LE. Total carbon, organic carbon, and organic matter. Methods of soil analysis: Part 3 Chemical methods. 1996;**5**:961-1010

[26] Bremner JM, Mulvaney CS. Kjeldhal Method. In method of soil analysis part-2: Chemical & microbiological properties. Madison, Wisconsin: American Society of Agronomy; 1982. pp. 903-948

[27] Jan MT, Shah P, Hollington PA, Khan MJ, Sohail Q. Agriculture research: Design and analysis, a monograph. Pakistan: NWFP Agricultural University Peshawar; 2009

[28] Muhammad H, Khan AZ, Shah Z. The impact of beneficial microbes (BM) on enhancing efficiency of organic and inorganic N fertilizers sources [PhD thesis]. Peshawar: Department of Agronomy, University of Agriculture; 2012

[29] Shah Z, Ahmad MI. Effect of integrated use of farm yard manure and urea on yield and nitrogen uptake of wheat. Journal of Agricultural and Biological Science. 2006;**1**(1):60-65

[30] Shah A, Shafi M, Bakht J, Mohammad W, Shah M, Jan MT, et al. Effect of integrated use of nitrogen on yield and N uptake of maize crop. Pakistan Journal of Botany. 2010;**42**(5): 3633-3638

[31] Ali K, Munsif F, Zubair M, Hussain Z, Shahid M, Din IU, et al. Management of organic and inorganic nitrogen for different maize varieties. Sarhad Journal of Agriculture. 2011;**27**(4):525-529

[32] Ortiz-Monasterio JI, Sayre KD, Rajaram S, McMahom M. Genetic progress in wheat yield and nitrogen use efficiency under four nitrogen rates. Crop Science. 1997;**37**:898-904

[33] Mahmood F, Khan I, Ashraf U, Shahzad T, Hussain S, Shahid M, et al.

Effects of organic and inorganic manures on maize and their residual impact on soil physico-chemical properties. Journal of Soil Science and Plant Nutrition. 2017;**17**(1):22-32

[34] Korsaeth A, Henriksen TM, Bakken LR. Temporal changes in mineralization and immobilisation of N during degradation of plant material: implications for the plant N supply and nitrogen losses. Soil Biology and Biochemistry. 2002;**34**:789-799

[35] Rautaray SK, Ghosh BC, Mittra BN. Effects of fly ash, organic wastes and chemical fertilizers on yield, nutrient uptake, heavy metal content and residual fertility in rice—mustard cropping sequence under acid lateritic soils. Biosource Technology. 2003;**90** (3):275-283

[36] Bekeko Z. Effect of nitrogen and phosphorus fertilizers on some soil properties and grain yield of maize (BH-140) at Chiro, Western Hararghe. Ethiopia. African Journal of Agricultural Research. 2013;**8**(45):5692-5697

[37] Anatoliy GK, Thelen KD. Effect of winter wheat crop residue on no-till corn growth and development. Agronomy Journal. 2007;**99**:549-555

[38] Drinkwater L, Dkletourneau E, Worknch F, Van Bruggen AHC, Shenman C. Fundamental differences between conventional and organic tomato agrosystem in California. Ecological Applications. 1995;**5**: 1098-1112

[39] Kumar K, Goh KM. Management practices of antecedent leguminous and non-leguminous crop residues in relation to winter wheat yield, nitrogen uptake, soil nitrogen mineralization and simple nitrogen balance. European Journal of Agronomy. 2002;**16**(4):295-308

[40] Agbede TM, Oladitan TO, Alagha SA, Ojomo AO, Ale MO. Comparative evaluation of poultry manure and NPK fertilizer on soil physical and chemical properties, leaf nutrient concentrations, growth and yield of yam (*Dioscorea rotundata* Poir) in southwestern Nigeria. World Journal of Agricultural Sciences. 2010;**6**:540-546

[41] Dixit KG, Gupta BR. Effect of farmyard manure, chemical and biofertilizers on yield and quality of rice (*Oryza sativa* L.) and soil properties. Journal of the Indian Society of Soil Science. 2000;**48**:773-780

[42] Sarwar G, Hussain N, Mujeeb F, Schmeisky H, Hassan G. Biocompost application for the improvement of soil characteristics and dry matter yield of *Lolium perenne* (Grass). Asian Journal of Plant Sciences. 2003;**2**(2):237-241

[43] Selvakumari G, Baskar M, Jayanthi D, Mathan KK. Effect of integration of fly ash with fertilizers and organic manures on nutrient availability, yield and nutrient uptake of rice in alfisols. Journal of the Indian Society of Soil Science. 2000;**48**:268-278

[44] Yaduvanshi NPS. Effect of five years of rice-wheat cropping and NPK fertilizers use with and without organic and green manures on soil properties and crop yields in a reclaimed sodic soil. Journal of the Indian Society of Soil Science. 2001;**49**(4):714-719

[45] Singh S, Singh RN, Prasad J, Kumar B. Effect of green manuring, FYM and biofertilizer in relation to fertilizer nitrogen on yield and major nutrient uptake by upland rice. Journal of the Indian Society of Soil Science. 2002;**50** (3):313-314

[46] Kuzyakov Y. Review: Factors affecting rhizosphere priming effects. Journal of Plant Nutrition and Soil Science. 2002;**165**:382-396

[47] El-Ghamry AM, Abd El-Hamid AM, Mosa AA. Effect of farmyard manure

and foliar application of micronutrients on yield characteristics of wheat grown on salt affected soil. American-Eurasian Journal of Agricultural & Environmental Sciences. 2009;**5**: 460-465

[48] Sarwar G, Schmeisky H, Hussain N, Muhammad S, Ibrahim M, Safdar E. Improvement of soil physical and chemical properties with compost application in rice-wheat cropping system. Pakistan Journal of Botany. 2008;**40**(1):275-282

Chapter 11

Microwave Soil Treatment and Plant Growth

Graham Brodie, Muhammad Jamal Khan and Dorin Gupta

Abstract

Crop yield gaps can be partially overcome by soil sanitation strategies such as fumigation; however, there are fewer suitable fumigants available in the marketplace and growing concerns about chemical impacts in the environment and human food chain. Therefore, thermal soil sanitation has been considered for some time and microwave soil treatment has some important advantages over other thermal soil sanitation techniques, such as steam treatment. It is also apparent that microwave soil sanitation does not sterilize the soil, but favors beneficial species of soil biota making more nutrients available for better plant growth. From these perspectives, microwave soil treatment may become an important pre-sowing soil sanitation technology for high value cropping systems, allowing agricultural systems to better bridge the crop yield gap.

Keywords: microwave pasteurization, agriculture, pathogen control, nutrient, production response

1. Introduction

Crop yield gaps are a significant issue for food security and agricultural sustainability. Crop yield gaps are defined as the differences between optimal yield potential and actual crop yield [1]. Yield potential (Yp) is the yield of a crop cultivar when grown in an environment to which it is adapted, with non-limiting water and nutrient supplies, and with pests, weeds, and diseases being effectively controlled [1]. For example, the impact of weeds on crop yield potential has been widely demonstrated [2] and modeled [3–5]. Noling and Ferris [6] demonstrated that nematodes can reduce alfalfa yields by more than 70%. Similarly, fungi can significantly reduce crop yield potential [7, 8]. The impact of various pathogens on crop yield potential can be demonstrated with some simple models.

According to Noling and Ferris [6], the impact of nematode populations on perennial crops, such as alfalfa, can be described by:

$$Y_{loss} = a\left(1 - e^{-bN}\right) \tag{1}$$

where Y_{loss} is the yield loss, a is the maximum yield loss for the system, b is a population sensitivity parameter for the crop (i.e., damage rate), and N is the nematode population. Therefore, the potential crop yield is described by:

$$\mathrm{Y} = \mathrm{Y_o}\left[1 - \mathrm{a}\left(1 - \mathrm{e}^{-\mathrm{bN}}\right)\right] \tag{2}$$

where Y_o is the optimal yield.

In a resource limited environment, the rate of population growth is described by:

$$\frac{dN}{d^{\circ}D} = r\left(\frac{k-N}{k}\right)N \tag{3}$$

where °D is the degree days which are suitable for the growth of the pest or pathogen, k is the maximum sustainable population of the pest or pathogen (i.e., the carrying capacity), and r is the base population growth rate. One Degree Day is determined according to some basis temperature (Tb):

$$^{\circ}D \stackrel{\text{def}}{\equiv} \frac{T_{max} - T_{min}}{2} - T_b > 0.0 \tag{4}$$

Equation (3) can be rearranged to become:

$$\frac{dN}{\left(\frac{k-N}{k}\right)N} = r \bullet d^{\circ}D \tag{5}$$

Integrating both sides of Eq. (5) gives:

$$2\ \tanh^{-1}\left(\frac{2N}{K} - 1\right) = r \bullet {}^{\circ}D + C \tag{6}$$

Therefore, Eq. (6) becomes:

$$N = \frac{K}{2}\left[1 + \tanh\left(\frac{r \bullet {}^{\circ}D}{2} + \frac{C}{2}\right)\right] \tag{7}$$

To evaluate the constant of integration (C), it is appropriate to choose a boundary condition on the problem. It is noted that at the start of any study (i.e., when °D = 0 for this study period), the population will have some starting population value "No." Substituting this into Eq. (7) and setting °D = 0 gives:

$$No = \frac{K}{2}\left[1 + \tanh\left(\frac{C}{2}\right)\right] \tag{8}$$

or:

$$C = 2 \bullet \tanh^{-1}\left(\frac{2 \bullet No}{K} - 1\right) \tag{9}$$

Therefore,

$$N = \frac{K}{2}\left[1 + \tanh\left(\frac{r \bullet {}^{\circ}D}{2} + \tanh^{-1}\left(\frac{2 \bullet No}{K} - 1\right)\right)\right] \tag{10}$$

Using data from Noling and Ferris [6] as a guide, the population of *Meloidogyne hapla* nematodes in their study would increase as shown in **Figure 1**. When these population models are applied to the crop yield model in Eq. (2), the apparent crop yield decline is similar in form to that presented in Noling and Ferris [6], as shown in **Figure 2**.

Different crops require differing numbers of degree days to reach maturity. For example, maize requires between 800 and 2700 degree days while barley requires between 1290 and 1540 degree days. Using the data presented in **Figure 2** to

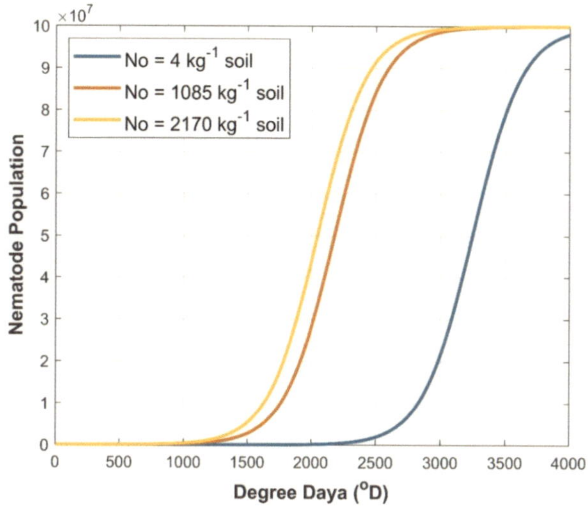

Figure 1.
Population growth in Meloidogyne hapla *nematodes as a function of degree days, based on the initial inoculum of the soil (calculated from Eq. (10)).*

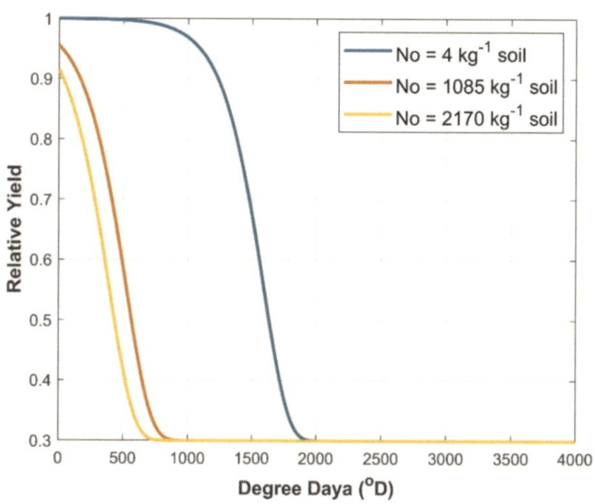

Figure 2.
Crop yield potential for Alfalfa affected by Meloidogyne hapla *nematodes as a function of degree days, based on the initial inoculum of the soil (calculated from Eq. (2)).*

illustrate the importance of the impact of pathogens and pests on crop yield, if a crop requiring 1500 growing degree days to mature is exposed to an initial *Meloidogyne hapla* nematode population of 1085 individuals kg^{-1} of soil, the yield potential would be 0.3 at the end of crop maturation; however, if the crop was exposed to an initial population of only 4 individuals kg^{-1} of soil because of some pre-sowing soil sanitation strategy, the crop yield potential would be approximately 0.7. Therefore, pre-sowing soil sanitation could provide a crop yield increase (compared with untreated soil) of: $\frac{(0.7-0.3)}{0.3} \times 100 = 133\%$.

Although this may appear to be a significant crop yield increase, the pre-sowing soil sanitation is simply bridging a little more of the crop yield gap by treating the soil to remove crop inhibiting organisms before sowing the crop. In fact, the modeling suggests that the crop growing on the sanitized soil may still not have reached its full crop yield potential.

2. Soil sanitation

Many soilborne plant pathogens flourish during the crop growing season and survive between seasons, either in the soil or above-ground, by means of resting structures, such as propagules that are either free or embedded in infected plant debris. Soil sanitation aims to reduce or eliminate the pest population from all sources, thus breaking the continuity of survival in time and space between crops. Soil sanitation (e.g., by fumigation or heating) is a routine procedure in many agricultural systems [9].

3. Fumigation

Soilborne diseases, plant-parasitic nematodes, and weeds can be devastating, and preplant soil fumigation is commonly relied upon to mitigate the risk of crop loss [10]. Methyl bromide has been widely used for soil sanitation in the past; however, because of its ozone depleting impacts it has been included in the 1987 Montreal Protocol as a substance whose use should be reduced and eventually eliminated. Under the Montreal Protocol exemptions were granted for substances (like Methyl Bromide) where no economic alternative existed [11]. Even so, especially in the Strawberry runner industry, alternative treatments have been investigated and found to be wanting [12, 13]. Most alternative treatments involve other fumigants, such as Metam sodium or chloropicrin [14], or thermal processes, such as solarization or applying steam.

Klose et al. [14] showed that weed seeds and soil pathogens exhibit a logistic dose-response to a commercial soil fumigant formulation of 1,3-dichloropropane (1,3-D; 61%) and chloropicrin (33%). It has been shown elsewhere [15] that a more physically meaningful representation of logistic dose responses can be described by:

$$S = a \bullet erfc[b(D - c)] \tag{11}$$

where S is the surviving portion of the population, erfc(x) is the Complementary Gaussian Error Function, D is the fumigant dose (μmol kg^{-1}), and a, b and c are constants that are determined experimentally. Equation (11) is based on an underlying normally distributed population susceptibility to some treatment; therefore, the cumulative effect (mortality) in the population becomes the integral of the normal distribution function, which is described by the Gaussian Error Function, and population survival, which is the whole population minus the mortality rate, is therefore described by the Complementary Gaussian Error Function. Therefore, it is anticipated that the crop yield response to varying doses of pre-sowing soil fumigation treatment should also have a Gaussian Error form, as a function of applied pre-sowing fumigant dose.

Growing concern over the use of excessive chemicals in agriculture, with adverse effects on on-farm and off-farm environments, has prompted a search for alternative soil sanitation options. Soil heating has provided some similar pest and pathogen control to chemicals.

4. Soil heating

The fatal impacts of high temperatures on botanical and zoological specimens have been studied in detail for over a century [16]. In particular, a thoroughly

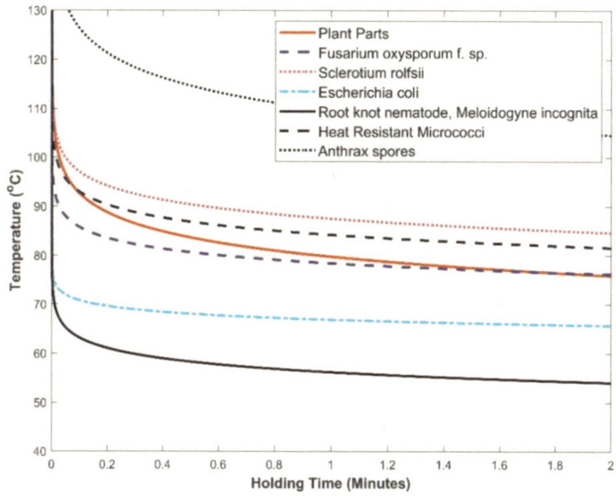

Figure 3.
Lethal temperature/time functions for several important pathogenic organisms.

demonstrated empirical relationship between lethal temperature and temperature holding time has been developed by Lepeschkin [17]:

$$T = 79.8 - 12.8 \bullet \log_{10} Z \tag{12}$$

where T is the lethal temperature (°C), and Z is the lethal temperature holding time, in minutes [16]. Individual relationships for different species of plants and pathogens [9, 17, 18] have been developed over time (**Figure 3**). Ultimately, heat can provide similar lethal effects to chemicals and therefore has been used in soil sanitation processes for some time.

5. Steam treatment

It has been demonstrated that steam soil treatment is as effective as some soil fumigants at reducing pre-sowing soil pathogen loads [19]; however, if the steam is applied to the surface of the soil (i.e., not injected), effective treatment is shallow compared with conventional soil fumigation techniques. This is due to limitations of heat being transferred from the steam into the soil. The governing equation for heat transfer from a hot fluid (air, water or steam) with a temperature of T_f into a solid, such as soil, with an initial temperature of T_s, is expressed as:

$$\frac{q}{A} = h\left(T_s - T_f\right) \tag{13}$$

where q is the heat flow (W), A is the cross sectional area through which the heat passes (m^2), and h is the convective heat flow coefficient of the soil's surface [20]. When studying thermodynamic processes, temperatures are usually expressed in absolute (Kelvin) values.

The convective heat flow coefficient depends on a number of other parameters and conditions [21]. For example, the convective heat flow coefficient for a vertical surface where natural convection achieves turbulent fluid flow conditions over the surface is given by [21]:

$$h = \frac{k}{L}\left\{0.825 + \frac{0.387Ra_L^{1/6}}{\left[1+\left(\frac{0.492}{Pr}\right)^{9/16}\right]^{8/27}}\right\} \tag{14}$$

where k is the thermal conductivity of the heating fluid (W m^{-1} K^{-1}), Pr is the Prandtl number, and L is the characteristic length of the object being heated (m).

The Rayleigh number (Ra_L) in Eq. (14) is also based on a complex relationship between temperature and the physical properties of the fluid. It is given by [21]:

$$Ra_L = \frac{g\beta}{\nu\alpha}(T_s - T_\infty)L^3 \tag{15}$$

where g is the acceleration due to gravity; β is the thermal expansion coefficient of the fluid; υ is the kinematic viscosity of the fluid medium; α is the thermal diffusivity of the fluid medium; and L is the characteristic length of the surface.

Finally, the Prandtl number used in Eq. (14) is a relationship between the fluid's viscous and thermal diffusion rates given by [21]:

$$Pr = \frac{\nu}{\alpha} \tag{16}$$

where v is the kinematic viscosity (m^2 s^{-1}) and α is the thermal diffusivity (m^2 s^{-1}).

Close examination of these equations shows that the convective heat transfer coefficient is dependent on the temperature differential between the fluid and the surface of the soil (see **Figure 4**) and the apparent surface area of the heat transfer interface. Injecting the steam into the soil through hollow tines effectively increases the surface area of the heat transfer interface between the cool soil and hot steam.

Semi-commercial steam soil sanitation systems have been in operation for some time [13, 19]. They are functional, though their application is limited, because they are energy expensive and difficult to use due to their large and heavy operation systems. Soil heat treatment may be better achieved through direct heating of the soil.

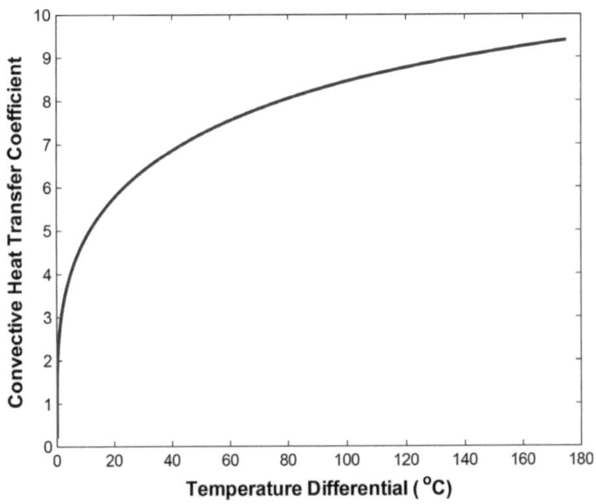

Figure 4.
Convective heat transfer coefficient (h) for air as a function of temperature differential between an object and the air.

6. Microwave soil heating

Microwaves are non-ionizing electromagnetic waves (**Figure 5**) with a frequency of about 300 MHz to 300 GHz and the wavelength range of 1 m to 1 mm [23]. Biological and agricultural systems are electro-chemical in nature [24] and a mixture of organic and dipole molecules, i.e., H_2O, arranged in different geometries [25, 26].

Interest in the study of the interactions of ultra-high frequency electromagnetic energy with complex biological system dates back to the nineteenth century [27]. The interactions of microwave energy with living systems are characterized at atomic, molecular, cellular and subcellular level [24].

The basic consideration in measuring the influence of microwave irradiation on living systems is the determination of the induced electromagnetic field and its spatial distribution. The bio-effects of microwave treatments can be described solely by differences in temperature profile between microwave and conventionally heated systems [28]. The energy of microwave photon at 2.45 GHz is 0.0016 eV [29]. This is not enough energy to break the structure of organic molecules [30]. The basic interactive mechanism of microwave energy with biological system/materials is inducing torsion on polar molecules, i.e., H_2O, Proteins and DNA, by induced electric field [31]. Oscillations in this torsion occur 2.45 billion times/second for 2.45 GHz waves. These oscillations manifest as internal kinetic energy in the material, which is heat.

Microwave (electromagnetic) heating has major advantages over conventional heating techniques. Some of these include: rapid volumetric heating as opposed to surface heating only, precise control, rapid start up and shut down [32], and in the case of soil, having a lighter apparatus than a steam generator to avoid soil compaction issues.

Many of the earlier experiments on plant material focused on the effect of radio frequencies [33] on seeds [27]. In many cases, exposure to low energy densities resulted in increased germination and vigor of the emerging seedlings [34, 35]; however, exposure to higher energy densities usually resulted in seed death [27, 36, 37].

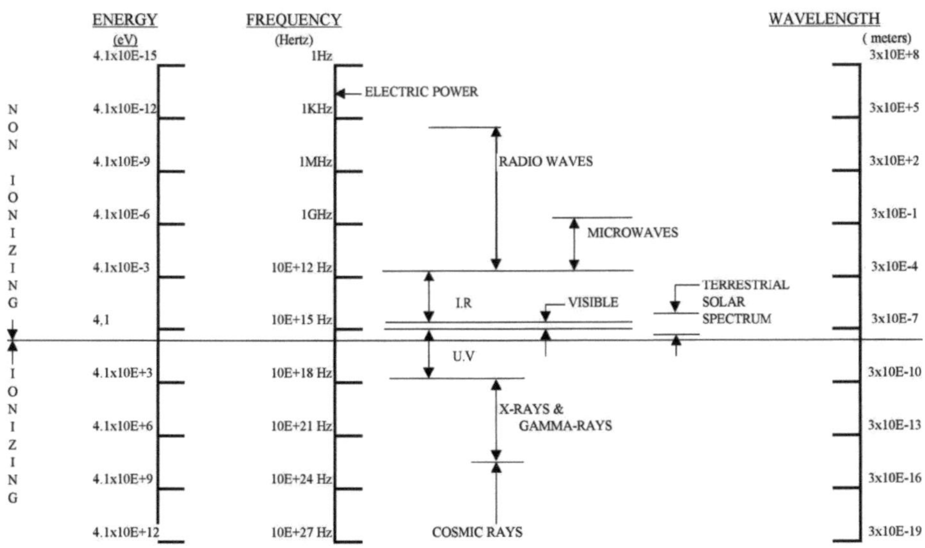

Figure 5.
The electromagnetic spectrum (adapted from [22]).

Davis et al. [38, 39] were among the first to study the lethal effects of microwave heating on weed seeds. They treated seeds, with and without any soil, in a microwave oven and showed that seed damage was mostly influenced by a combination of seed moisture content and the energy absorbed in each seed. In addition, they suggested that both the specific mass and specific volume of the seeds were strongly related to a seed's susceptibility to damage by microwave fields. The association between the seed's volume and its susceptibility to microwave treatment may be linked to the "*radar cross-section*" [40] presented by seeds to propagating microwaves. Large radar cross-sections allow the seeds to intercept, and therefore absorb, more microwave energy.

Ferriss [8] conducted experiments on soil samples with moisture contents between 7 and 37% (wet/dry-weight) and showed that treatment in a microwave oven for 150 seconds eliminated populations of *Pythium*, *Fusarium* and all nematode species, except *Heterodera glycines* in the soil samples. Compared with autoclaving or Methyl bromide (MB) treatment, he found that microwave treatments released less nutrient into the soil solution but had less effect on soil *prokaryotes* and resulted in less recolonization of the soil by *Fusarium* and other fungi after treatment. Similar observations were made by Mattner and Brodie [41] during a preliminary experiment in soils growing strawberry runners at Toolangi, Victoria.

Speir et al. [42] examined the effect of microwave energy on low fertility soil (100 randomly selected cores at a depth of 50 mm), microbial biomass, nitrogen, phosphorus, and phosphatase activity. They reported that an increase in microwave treatment duration (90 seconds) dramatically increased the nitrogen level in the soil by a factor of approximately 10 times (106 μg N g^{-1}) compared with untreated soil (9–10 μg N g^{-1}), but available phosphorus concentration declined as treatment time increased. Furthermore, relevant to soil productivity, Gibson et al. [43], demonstrated that shoot and root growth of birch (*Betula pendula*) significantly increased in microwave irradiated soil. Their experiment evaluated the effect of microwave treatment of soil supplemented with two mycorrhizas on birch seedlings. Shoot growth progressively increased with irradiation duration, with the highest dry shoot weight of 84 mg coinciding with the highest irradiation duration (of 120 seconds) compared to non-irradiated soil which resulted in 25 mg of growth. This result was achieved with no mycorrhizal supplementation. In addition, a recent study reported that microwave (915 MHz; different power × duration) soil treatment increased the dissolved organic carbon (+1.6-fold compared with the control), inorganic phosphorus (+1.2-fold compared with the control), and nitrate content in soil [44]. In addition, they grew the pregerminated seeds of *Medicago truncatula* Gaertn. in microwave treated soil and found that its dry biomass accumulation significantly increased in response to soil heating (75–80°C), compared with the untreated control soils.

Since then there has been ongoing research interest in microwave soil treatment and weed management. **Table 1** lists a subset of the papers that have been published on these and related topics. The consensus from these studies is that: microwave treatment can kill plants; moderate microwave treatment can break dormancy in some hard-seeded species; and high energy microwave treatment can sanitize soil.

Typically, responses of weed seeds and soil biota are both energy and depth dependent, because of the absorption of microwave energy with soil depth. The relationships between applied microwave energy and seed or biota survival at different depths are given by:

$$S = a \bullet \mathrm{erfc}\left[b \bullet \left(\Psi \bullet e^{-2cd} - f\right)\right] \qquad (17)$$

where Ψ is the microwave energy density at the soil surface (J cm^{-2}), d is the depth in the soil (m) and a, b, c, and f are constants to be determined

Paper title	Reference
Douglas- fir tree seed germination enhancement using microwave energy	[45]
Microwave processing of tree seeds	[46]
Increasing legume seed-germination by VHF and microwave dielectric heating	[47]
Effects of low-level microwave radiation on germination and growth rate in corn seeds	[48]
Effects of microwave energy on the strophiole, seed coat and germination of acacia seeds	[35]
The effect of microwave-energy on germination and dormancy of wild oat seeds	[49]
The effect of externally applied electrostatic fields, microwave radiation and electric currents on plants and other organisms, with special reference to weed control	[50]
Control of field weeds by microwave radiation	[51]
Effect of microwave irradiation on germination and initial growth of mustard seeds	[52]
Inhibition of weed seed germination by microwaves	[53]
A possibility of correction of vital processes in plant cell with microwave radiation	[54]
Microwave irradiation of seeds and selected fungal spores	[7]
Response surface models to describe the effects and phytotoxic thresholds of microwave treatments on barley seed germination and vigor	[55]
Energy efficient soil disinfestation by microwaves	[56]
Microwave effects on germination and growth of radish (*Raphanus sativus* L.) seedlings	[57]
Report on the development of microwave system for sterilization of weed seeds: stage I – feasibility	[58]
Design, construction and preliminary tests of a microwave prototype for weed control	[59]
Thermal effects of microwave energy in agricultural soil radiation	[60]
Influence of low-frequency and microwave electromagnetic fields on seeds	[61]
An improved microwave weed killer	[62]
Observations on the potential of microwaves for weed control	[63]
Plant response to microwaves at 2.45 GHz.	[64]
Germination inhibition of undesirable seed in the soil using microwave radiation	[65]
Effect of microwave radiation on seed mortality of rubber vine (*Cryptostegia grandiflora* R. Br.), parthenium (*Parthenium hysterophorus* L.) and bellyache bush (*Jatropha gossypiifolia* L.)	[36]
Effects of microwave treatment on growth, photosynthetic pigments and some metabolites of wheat	[66]
Microwave seed treatment reduces hardseededness in *Stylosanthes seabrana* and promotes redistribution of cellular water as studied by NMR relaxation measurements	[67]
Effect of microwave fields on the germination period and shoot growth rate of some seeds	[68]
Germination of *Chenopodium album* in Response to Microwave Plasma Treatment	[69]
Work conditions for microwave applicators designed to eliminate undesired vegetation in a field	[70]

Table 1.
Literature addressing the application of microwave technology to seed and weed treatment.

experimentally. This is illustrated by the relationships for weed seeds and bacteria in (**Figures 6** and 7).

Unlike in the case of chemical soil fumigants, microwave soil treatment does not sterilize the soil. Although there is a general reduction in soil bacteria after

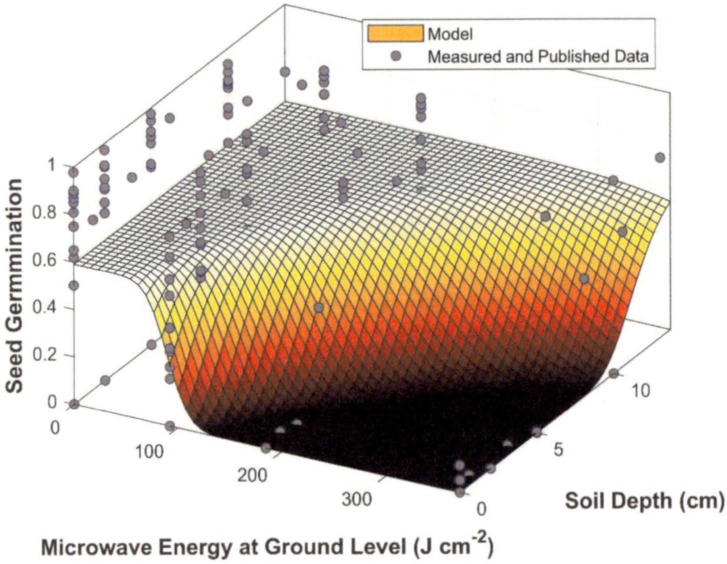

Figure 6.
Response of multiple species of weed seeds as a function of applied microwave energy and soil depth [71].

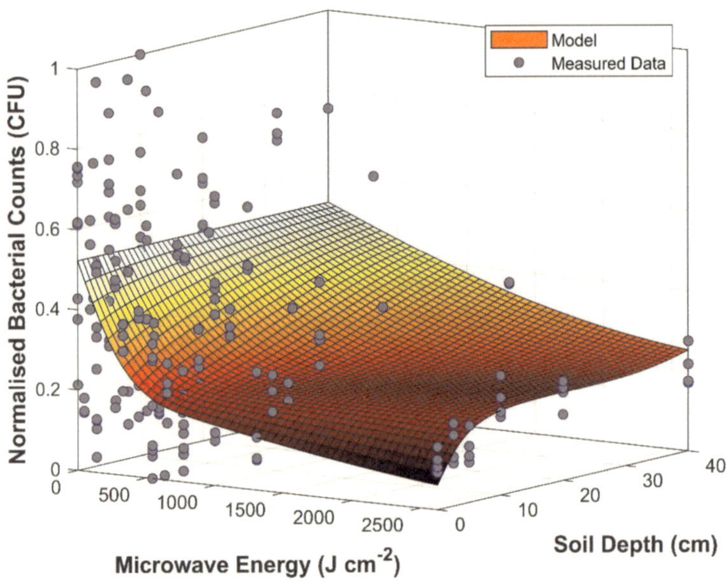

Figure 7.
Response of soil bacteria as a function of applied microwave energy and soil depth [71].

microwave treatment (**Figure** 7), Khan et al. [72] demonstrated that immediately after microwave soil treatments, the relative abundance of *Firmicutes* increased while the relative abundance of *Proteobacteria* decreased significantly. They also showed that the relative abundances of beneficial soil microbes (*Micromonosporaceae*, *Kaistobacter* and *Bacillus*) were significantly higher, as soils recovered from high heating intensities induced by microwave soil treatment, compared with untreated soils.

There is also considerable evidence that microwave soil treatment releases more nitrogen sources in the soil for the crop growth [73]. This may be due to the resilience of nitrifying bacteria and archaea to microwave soil heating. Khan et al. [72] showed that microwave soil treatment did not significantly affect ammonia oxidizing bacteria or ammonia oxidizing archaea. Vela et al. [74] also demonstrated that nitrifying bacteria in the soil were resilient to 40 kJ cm^{-2} of microwave energy at the soil surface; which is 70 times higher than the energy densities used during experimental work undertaken by the current authors.

7. Crop responses

Fully replicated pot and field plot experiments have been undertaken over an extended period of time by the authors to better understand the impact of pre-sowing microwave soil treatment on crop growth. In all cases, the experiments had at least 5 experimental replicates and in many cases, they used 10 experimental replicates. Experiments were undertaken to explore the effect of pre-sowing microwave soil treatments on plant growth and yield of wheat (*Triticum* spp.), rice (*Oryza sativa*), maize (*Zea mays*), canola (*Brassica napus*), processing tomatoes and strawberry runners. In most cases the potted experiments were repeated two or three times and in some cases the field experiments were also repeated. Microwave energy was applied to the soil in pots or in situ using a trailer mounted microwave prototype system with 4 individual 2 kW microwave generators (see **Figure 8**).

The crops were planted within hours of the microwave treatment, once the soil had returned to ambient temperature. Plant growth rate, final plant height, and crop yield showed significant increases with increasing microwave energy (**Table 2**). In the potted trials and in one wheat field trial, hand weeded controls were included in the experiments to determine whether crop growth response was simply due to less weed competition.

Pre-sowing microwave soil treatment was found to have significant beneficial effects on subsequent crop growth. Most crops showed a typical Gaussian Error Function response to increasing microwave soil treatment dosage (**Figure 6**), as would be expected if the pre-sowing soil treatment were acting as a soil fumigant (**Figure 9**).

Figure 8.
Prototype 4 by 2 kW microwave weed killer in a strawberry runner field at Toolangi, Victoria.

Microwave treatment	Control	Hand weeded	Microwave energy (J cm^{-2})			LSD (P = 0.05)	Change from hand weeded/ control (%)
			136	318	545		
Pot trials							
Canola pod yield (g pot^{-1})	0.27^{a}	0.56^{a}	0.36^{a}	1.25^{b}	1.95^{c}	0.55	250%
Wheat grain yield (g pot^{-1})	0.66^{a}	0.67^{a}	0.68^{a}	0.75^{a}	1.25^{b}	0.3	87%
Rice grain yield (g pot^{-1})	40.0^{a}	41.3^{a}	43.3^{a}	59.0^{ab}	64.0^{b}	18.9	55%
Maize (g pot^{-1})	5.3^{a}	6.6^{a}	_	10.3^{ab}	12.8^{b}	4.8	92%
Field trials							
Rice (t ha^{-1}) – Dookie Year 1 (2015/2016)	7.5^{a}	_	_	_	10.1^{b}	2	35
Rice (t ha^{-1}) – Dookie Year 2 – (2016/2017) - crop was cold affected at panicle initiation	2.1^{a}	_	_	_	3.9^{b}	1.3	84
Rice (t ha^{-1}) – Old Coree – (2016/2017)	7.7^{a}	_	_	_	9.1^{b}	1.2	19
Wheat (t ha^{-1})	5.7^{a}	6.6^{ab}	_	_	7.8^{b}	1.4	18
Tomato (t ha^{-1})	64.1^{a}	65.2^{a}	_	_	89.6^{b}	24.7	37
Strawberry runner production (daughter plants m^{-2})	6970^{a}	_	_	_	8445^{b}	670	21

Means with different superscript letters (i.e. a, b, c etc) are statistically different from one another at a probability of 0.05.

Table 2.
Summary of pot and field trial crop yields in response to microwave soil treatment.

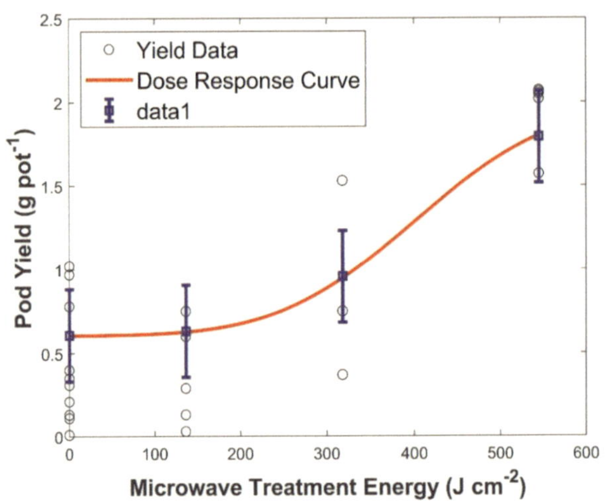

Figure 9.
Canola pod yield response to increasing microwave treatment.

8. Conclusions

Pre-sowing microwave soil treatment acts as a soil sanitation technology and results in significant increases in crop yield, as would be expected from other soil sanitation techniques. Microwave treatment has some major advantages over other soil sanitation techniques in that it is purely thermal in nature and allows immediate access to the site once the soil has cooled to ambient temperatures. Unlike, other thermal treatment systems, such as steam treatment, microwave systems can be light and highly controllable, reducing other impacts on the soil such as compaction.

Also, unlike other soil sanitation techniques, it is evident that microwave treatment does not sterilize the soil, but favors beneficial species of soil biota making more nutrients available for better plant growth. From these perspectives, microwave soil treatment may become an important pre-sowing soil sanitation technology for high-value cropping systems, allowing agricultural systems to better bridge the crop yield gap.

Author details

Graham Brodie*, Muhammad Jamal Khan and Dorin Gupta
The University of Melbourne, Victoria, Australia

*Address all correspondence to: grahamb@unimelb.edu.au

References

[1] Edreiraa JIR, Mourtzinis S, Conley SP, Roth AC, Ciampitti IA, Licht MA, et al. Assessing causes of yield gaps in agricultural areas with diversity in climate and soils. Agricultural and Forest Meteorology. 2017;**247**:170-180

[2] Brodie G. Derivation of a cropping system transfer function for weed management: Part 1 – Herbicide weed management. Global Journal of Agricultural Innovation, Research and Development. 2014;**1**(1):11-16

[3] Cousens R. A simple model relating yield loss to weed density. The Annals of Applied Biology. 1985;**107**(2):239-252

[4] Cousens R, Brain P, O'Donovan JT, O'Sullivan PA. The use of biologically realistic equations to describe the effects of weed density and relative time of emergence on crop yield. Weed Science (USA). 1987;**35**(5):720-725

[5] Schmidt CP, Pannell DJ. Economic issues in management of herbicide-resistant weeds. Research in Agricultural and Applied Economics. 1996;**64**(3):301-308

[6] Noling JW, Ferris H. Nematode-degree days, a density-time model for relating epidemiology and crop losses in perennials. Journal of Nematology. 1987; **19**(1):108-118

[7] Cavalcante MJB, Muchovej JJ. Microwave irradiation of seeds and selected fungal spores. Seed Science and Technology. 1993;**21**(1):247-253

[8] Ferriss RS. Effects of microwave oven treatment on microorganisms in soil. Phytopathology. 1984;**74**(1):121-126

[9] Shlevin E, Saguy IS, Mahrer Y, Katan J. Modeling the survival of two soilborne pathogens under dry structural solarization. Phytopathology. 2003;**93**(10):1247-1257

[10] Walters T, Gigot J, Zasada I. Preplant Soil Fumigation and Alternatives for Berry Production. Mount Vernon Research & Extension Center, Washington, USA: Washington State University; 2011

[11] Carter CA, Chalfant JA, Goodhue RE, Han FM, DeSantis M. The methyl bromide ban: Economic impacts on the California strawberry industry. Review of Agricultural Economics. 2005;**27**(2):181-197

[12] Kabir Z, Fennimore SA, Duniway JM, Martin FN, Browne GT, Winterbottom CQ, et al. Alternatives to methyl bromide for strawberry runner plant production. HortScience. 2005; **40**(6):1709-1715

[13] Samtani JB, Gilbert C, Ben Weber J, Subbarao KV, Goodhue RE, Fennimore SA. Effect of steam and solarization treatments on pest control, strawberry yield, and economic returns relative to methyl bromide fumigation. HortScience. 2012;**47**(1):64-70

[14] Klose S, Ajwa HA, Fennimore SA, Martin FN, Browne GT, Subbarao KV. Dose response of weed seeds ad soilborne pathogens to 1,3-D and chloropicrin. Crop Protection. 2007;**26**: 535-542

[15] Brodie G. Derivation of a cropping system transfer function for weed management: Part 2 – Microwave weed management. Global Journal of Agricultural Innovation, Research and Development. 2016;**3**(1):1-9

[16] Levitt J. Response of Plants to Environmental Stresses. Vol. 1. New York: Academic Press; 1980

[17] Lepeschkin WW. Zur Kenntnis der Einwirkung supamaximaler Temperaturen auf die Pflanze. Berichte.

Deutsche Botanische Gesellschaft. 1912; 30: 713-714

[18] Trevisani M, Mancusi R, Valero A. Thermal inactivation kinetics of shiga toxin-producing *Escherichia coli* in buffalo mozzarella curd. Journal of Dairy Science. 2014;**97**(2):642-650

[19] Rainbolt CM, Samtani JB, Fennimore SA, Gilbert CA, Subbarao KV, Gerik JS, et al. Steam as a preplant soil disinfestant tool in california cut-flower production. HortTechnology. 2013;**23**(2):207-214

[20] Holman JP. Heat Transfer. 8th ed. New York: McGraw-Hill; 1997

[21] Welty JR, Wicks CE, Wilson RE, Rorrer GL. Fundamentals of Momentum, Heat and Mass Transfer. 5th ed. Hoboken, NJ: John Wiley and Sons, Inc.; 2007

[22] Verschaeve L, Maes A. Genetic, carcinogenic and teratogenic effects of radiofrequency fields. Mutation Research. 1998;**410**:141-165

[23] Banik S, Bandyopadhyay S, Ganguly S. Bioeffects of microwave—A brief review. Bioresource Technology. 2003;**87**(2):155-159

[24] Kothari V, Dholiya K, Pate D. Effect of Low Power Microwave on Microbial Growth and Metabolism. Gemany: GRIN Verlag; 2012

[25] Brodie G. The influence of load geometry on temperature distribution during microwave heating. Transactions of the American Society of Agricultural and Biological Engineers. 2008;**51**(4): 1401-1413

[26] Brodie G. Applications of microwave heating in agricultural and forestry related industries. In: Cao W, editor. The Development and Application of Microwave Heating. Rijeka, Croatia: InTech; 2012. pp. 45-78

[27] Ark PA, Parry W. Application of high-frequency electrostatic fields in agriculture. The Quarterly Review of Biology. 1940;**15**(2):172-191

[28] Kozempel M, Cook R, Scullen O, Annous B. Development of a process for detecting non-thermal effects of a microwave energy on microorganisms at low temperature. Journal of Food Processing and Preservation. 2000;**24**: 287-301

[29] Kappe OC, Stadle A, Dallinger D. Microwave in Organic and Medical Chemistry. Weinheim, Germany: John Wiley & Sons; 2005

[30] Yaghmaee P, Durance T. Destruction and injury of *Escherichia coli* during microwave heating under vacuum. Journal of Applied Microbiology. 2005;**98**:498-506

[31] Takashima S, Gabriel C, Sheppard RJ, Grant EH. Dielectric behavior of DNA solution at radio and microwave frequencies (At 200°C). Biophysics Journal. 1984;**46**:29-34

[32] Metaxas AC, Meredith RJ. Industrial Microwave Heating. London: Peter Peregrinus; 1983

[33] Totsche KU, Amelung W, Gerzabek MH, Guggenberger G, Klumpp E, Knief C, et al. Microaggregates in soils. Journal of Plant Nutrition and Soil Science. 2018; **181**(1):104-136

[34] Nelson SO, Stetson LE. Germination responses of selected plant species to RF electrical seed treatment. Transactions of ASAE. 1985;**28**(6):2051-2058

[35] Tran VN. Effects of microwave energy on the strophiole, seed coat and germination of Acacia seeds. Australian Journal of Plant Physiology. 1979;**6**(3): 277-287

[36] Bebawi FF, Cooper AP, Brodie GI, Madigan BA, Vitelli JS, Worsley KJ, et al. Effect of microwave radiation on seed mortality of rubber vine (*Cryptostegia grandiflora* R.Br.), parthenium (*Parthenium hysterophorous* L.) and bellyache bush (*Jatropha gossypiifolia* L.). Plant Protection Quarterly. 2007;**22**(4):136-142

[37] Brodie G, Harris G, Pasma L, Travers A, Leyson D, Lancaster C, et al. Microwave soil heating for controlling ryegrass seed germination. Transactions of the American Society of Agricultural and Biological Engineers. 2009;**52**(1): 295-302

[38] Davis FS, Wayland JR, Merkle MG. Ultrahigh-frequency electromagnetic fields for weed control: Phytotoxicity and selectivity. Science. 1971;**173**(3996): 535-537

[39] Davis FS, Wayland JR, Merkle MG. Phytotoxicity of a UHF electromagnetic field. Nature. 1973;**241**(5387):291-292

[40] Wolf WW, Vaughn CR, Harris R, Loper GM. Insect radar cross-section for aerial density measurement and target classification. Transactions of the American Society of Agricultural and Biological Engineers. 1993;**36**(3):949-954

[41] Mattner S, Brodie G. An economic review of microwave and steam soil treatment as a potential substitute for soil fumigation in the strawberry runner industry. Victorian Strawberry Industry Certification Authority. Toolangi, Australia; 2017

[42] Speir TW, Cowling JC, Sparling GP, West AW, Corderoy DM. Effects of microwave radiation on the microbial biomass, phosphatase activity and levels of extractable N and P in a low fertility soil under pasture. Soil Biology and Biochemistry. 1986;**18**(4):377-382

[43] Gibson BF, Frances MF, Deacon JW. Effects of microwave treatment of soil on growth of birch (*Betula pendula*) seedlings and infection of them by ectomycorrhizal fungi. The New Phytologist. 1988;**108**:189-204

[44] Maynaud G, Baudoin E, Bourillon J, Duponnois R, Cleyet-Marel JC, Brunel B. Short-term effect of 915-MHz microwave treatments on soil physicochemical and biological properties. European Journal of Soil Science. 2019;**70**(3):443-453

[45] Jolly JA, Tate RL. Douglas- fir tree seed germination enhancement using microwave energy. The Journal of Microwave Power. 1971;**6**(2):125-130

[46] Kashyap SC, Lewis JE. Microwave processing of tree seeds. The Journal of Microwave Power. 1974;**9**(2):99-107

[47] Nelson SO, Ballard LAT, Stetson LE, Buchwald T. Increasing legume seed-germination by VHF and microwave dielectric heating. Transactions of ASAE. 1976;**19**(2):369-371

[48] Bigu-Del-Blanco J, Bristow JM, Romero-Sierra C. Effects of low-level microwave radiation on germination and growth rate in corn seeds. Proceedings of the IEEE. 1977;**65**(7):1086-1088

[49] Lal R, Reed WB. The effect of microwave-energy on germination and dormancy of wild oat seeds. Canadian Agricultural Engineering. 1980;**22**(1): 85-88

[50] Diprose MF, Benson FA, Willis AJ. The effect of externally applied electrostatic fields, microwave radiation and electric currents on plants and other organisms, with special reference to weed control. The Botanical Review. 1984;**50**(2):171-223

[51] Vela-Múzquiz R. Control of field weeds by microwave radiation. Acta Horticulturae. 1984;**152**:201-208

[52] Rao YVS, Chakravarthy NVK, Panda BC. Effect of microwave

irradiation on germination and initial growth of mustard seeds. Indian Journal of Agronomy. 1989;**34**(3):378-379

[53] Barker AV, Craker LE. Inhibition of weed seed germination by microwaves. Agronomy Journal. 1991;**83**(2):302-305

[54] Petrov IY, Moiseeva TV, Morozova EV. A possibility of correction of vital processes in plant cell with microwave radiation. In: Proceedings of the IEEE International Symposium on Electromagnetic Compatibility. Cherry Hill, NJ, USA: Institute of Electrical and Electronic Engineers; 1991. pp. 234-235

[55] Stephenson MMP, Kushalappa AC, Raghavan GSV, Mather DE. Response surface models to describe the effects and phytotoxic thresholds of microwave treatments on barley seed germination and vigour. Seed Science and Technology. 1996;**24**(1):49-65

[56] Mavrogianopoulos GN, Frangoudakis A, Pandelakis J. Energy efficient soil disinfestation by microwaves. Journal of Agricultural Engineering Research. 2000;**75**(2): 149-153

[57] Scialabba A, Tamburello C. Microwave effects on germination and growth of radish (*Raphanus sativus L.*) seedlings. Acta Botanica Gallica. 2002; **149**(2):113-123

[58] Advanced Manufacturing Technologies. Report on the Development of Microwave System for Sterilisation of Weed Seeds: Stage I - Feasibility. Wollongong, NSW, Australia: Advanced Manufacturing Technologies; 2003

[59] Zanche Cd, Amista F, Baldoin C, Beria S, Giubbolini L. Design, construction and preliminary tests of a microwave prototype for weed control. Rivista di Ingegneria Agraria. 2003; **34**(2):31-38

[60] Velazquez-Marti B, Gracia-Lopez C. Thermal effects of microwave energy in agricultural soil radiation. International Journal of Infrared and Millimeter Waves. 2004;**25**(7):1109-1122

[61] Kalinin LG, Boshkova IL, Panchenko GI, Kolomiichuk SG. Influence of low-frequency and microwave electromagnetic fields on seeds. Biophysics. 2005;**50**(2):334-337

[62] Vidmar M. An improved microwave weed killer. Microwave Journal. 2005; **48**(10):116-126

[63] Sartorato I, Zanin G, Baldoin C, De Zanche C. Observations on the potential of microwaves for weed control. Weed Research. 2006;**46**(1):1-9

[64] Skiles JW. Plant response to microwaves at 2.45 GHz. Acta Astronautica. 2006;**58**(5):258-263

[65] Velazquez-Marti B, Gracia-Lopez C, Marzal-Domenech A. Germination inhibition of undesirable seed in the soil using microwave radiation. Biosystems Engineering. 2006;**93**(4):365-373

[66] Hamada E. Effects of microwave treatment on growth, photosynthetic pigments and some metabolites of wheat. Biologia Plantarum. 2007;**51**(2):343-345

[67] Anand A, Nagarajan S, Joshi DK, Verma AP-S, Kar A. Microwave seed treatment reduces hardseededness in *Stylosanthes seabrana* and promotes redistribution of cellular water as studied by NMR relaxation measurements. Seed Science and Technology. 2008;**37**(1):88-97

[68] Monteiro JH, Mendiratta SK, Capitão A. Effect of microwave fields on the germination period and shoot growth rate of some seeds. In: Proceedings of the International Conference of Recent Advances in Microwave Theory and Applications. Jaipur, Rajasthan: Institute of Electrical

and Electronic Engineers; 2008. pp. 792-793

[69] Sera B, Stranak V, Sery M, Tichy M, Spatenka P. Germination of *Chenopodium album* in response to microwave plasma treatment. Plasma Science and Technology. 2008;**10**(4):506-511

[70] Velazquez-Marti B, Gracia-Lopez C, de la Puerta R. Work conditions for microwave applicators designed to eliminate undesired vegetation in a field. Biosystems Engineering. 2008; **100**(1):31-37

[71] Brodie GI, Khan MJ, Gupta D, Foletta S. Microwave weed and soil treatment in agricultural systems. Global Journal of Agricultural Innovation, Research and Development. 2018;**5**(2):1-14

[72] Khan MJ, Jurburg SD, He J, Brodie G, Gupta D. Impact of microwave disinfestation treatments on the bacterial communities of no-till agricultural soils. European Journal of Soil Science. (Special Issue Article). 2019:1-12

[73] Khan MJ, Brodie GI, Gupta D, He J. Microwave soil treatment increases soil nitrogen supply for sustained wheat productivity. Transactions of the ASABE. 2019;**62**(2):355-362

[74] Vela GR, Wu JF, Smith D. Effect of 2450 MHz microwave radiation on some soil microorganisms in situ. Soil Science. 1976;**121**(1):44-51

Chapter 12

Fertilizer Use Issues for Smallholder Agriculture in Tropical Africa

Charles S. Wortmann, Anthony O. Esilaba, Kayuki C. Kaizzi, Catherine Kibunja, Keziah W. Ndungu-Magiroi and Nouri Maman

Abstract

Fertilizer is an essential input for wide-scale sustainable intensification of crop productivity in tropical Africa, but its use by smallholders is often financially constrained. Four fertilizer use issues are addressed. Smallholders need high net returns from their investments, with acceptable risk, which can be achieved with good crop-nutrient-rate choices made in consideration of the farmer's financial and agronomic context. Soil acidification, which is affected by crop N supply, is best managed with the use of slightly more acidifying but less costly common N fertilizer, e.g., urea, coupled with lime use compared with the use of more costly but less acidifying N fertilizer such as calcium ammonium nitrate. This chapter addresses the feasibility of tailored fertilizer blends for maximizing farmer profit with respect to the nutrient supply cost, the need for flexibility in nutrient application according to the farmer's context, and the weak justification for tailoring blends based on soil test results. The use of a well-formulated blends is justified in some cases, e.g., for some crops in Rwanda, but the supply of blends does not justify restricting the supply of common fertilizers. Farmers need to be aware that unregulated nontraditional products very often fail to provide the claimed benefits. Fertilizer use, sometimes with timely lime application, can be highly profitable with modest risk with good crop-nutrient-rate choices, adequate free-market fertilizer supply, and avoiding products with unsubstantiated claims.

Keywords: Africa, smallholder, fertilizer, profit, blends, soil acidity, non-traditional products

1. Introduction

Fertilizer use is essential for widespread sustainable increases in crop productivity and for the preservation of the cropland resource base in tropical Africa. Smallholder farmers in tropical Africa generally have severe financial constraints and need high returns to justify an investment, often >100% within a year [1]. Risk needs to be low given the vulnerability of their livelihoods to failed investments. Fertilizer use can have a high probability of high profit with well-informed crop-nutrient-rate choices but also with efficient input supply, favorable credit terms, subsidies, and efficient marketing of the commodity produced [2–4]. The objective of this chapter was to explore four issues affecting the profit potential of fertilizer use by financially constrained smallholder farmers: (1) the choice of fertilizer use options with the greatest potential return on investment, (2) the choice of N source and management of soil acidification, (3) the use of tailored fertilizer blends as alternatives to common straight fertilizers, and (4) the alternative nontraditional products for managing soil productivity. The implications for farm profitability are fundamental to the discussion of these issues.

2. Fertilizer use for maximization of the farmer's profit

Smallholder cropping systems are typically diverse, and each crop or intercrop has some level of profit potential for each nutrient that might be applied [2, 3, 5–13]. Crop-nutrient response functions typically have a diminishing profit-to-cost ratio as the nutrient rate approaches the agronomic optimum. A financially constrained farmer maximizes net returns through optimized choice of crop-nutrient-rate options (**Figure 1**; [4]). In contrast, when fertilizer use is not financially constrained, the profit-oriented farmer targets to apply at the rate at which net returns per hectare are maximized. Fertilizer use decisions can be made by integrating crop-nutrient response functions using linear optimization through computer-based and simple paper decision tools that have been developed for 73 recommendation domains across 15 nations of tropical Africa by the project Optimizing Fertilizer Recommendations in Africa [2, 3, 14].

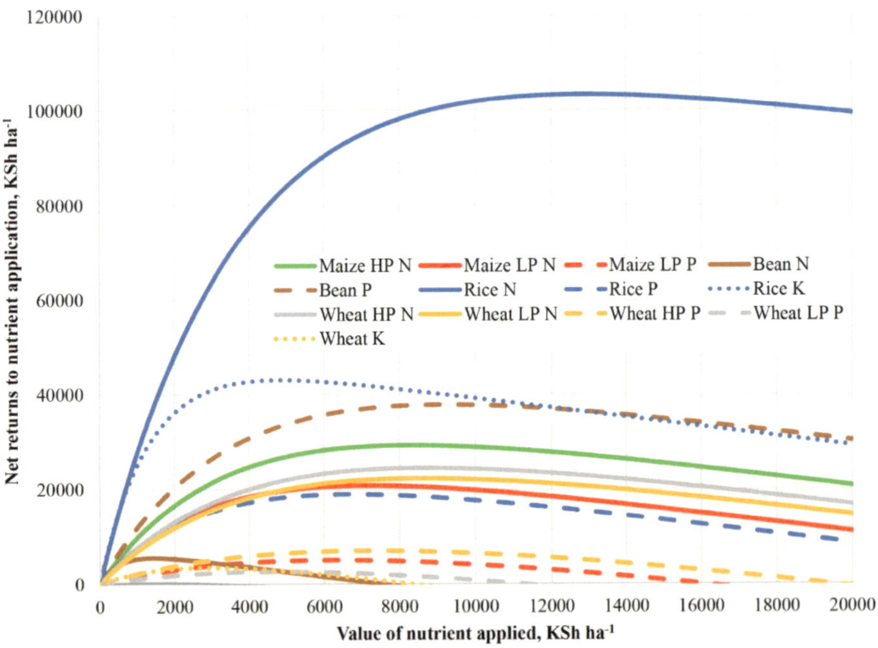

Figure 1.
Net returns in Kenyan shillings (KSh) to investment in nutrient application vary with crop-nutrient-rate choices, exemplified for Central Kenya with fertilizer use costs and on farm commodity values typical in 2016 [4].

3. Management of soil acidification and nitrogen sources

Supply of nitrogen to cropland typically contributes to soil acidification whether the N is supplied through fertilizer, organic materials, biological fixation of atmospheric N, or wet and dry deposition of atmospheric NH_4-N [15]. Soil acidification also occurs with NH_4^+ uptake by plants with subsequent release of cations, mostly H^+. Soil acidification is greater if NO_3^--N is leached from the soil rather than recovered by plants. Soil acidification associated with N sources can be slowed by avoiding excessive application of N and leaching of NO_3^--N and by the use of less acidifying but more costly NO_3^--N fertilizers [16]. Very often, it is most economical to use relatively less expensive but more acidifying NH_4^+-N fertilizers and occasionally amend the soil with lime application rather than using less acidifying fertilizers.

Soil acidification concerns are important in Kenya, for example, especially in some high elevation and high yield potential areas (**Figure 2**). The promoted N fertilizer for these areas is calcium ammonium nitrate (CAN) rather than urea. The chemical composition of CAN varies, but CAN of 27% N contains about 13.5% each of NH_4^+-N and NO_3^--N, and calcium carbonate or calcium-magnesium carbonate (dolomite) may be added to give the fertilizer about 20% calcium carbonate equivalent (CCE). The acidification effect of ammonium nitrate and urea is 3.65 kg CCE for each kg of N applied. If the CCE of CAN is neutralized in the soil, it reduces the net acidification effect of CAN-N by about 22% to about 2.85 kg CCE kg^{-1} N. Therefore, urea is about 28% more acidifying per kg of N compared with CAN. Calcium is supplied by CAN but cannot be credited with economic value to farmers if the yield response to CA is not profitable.

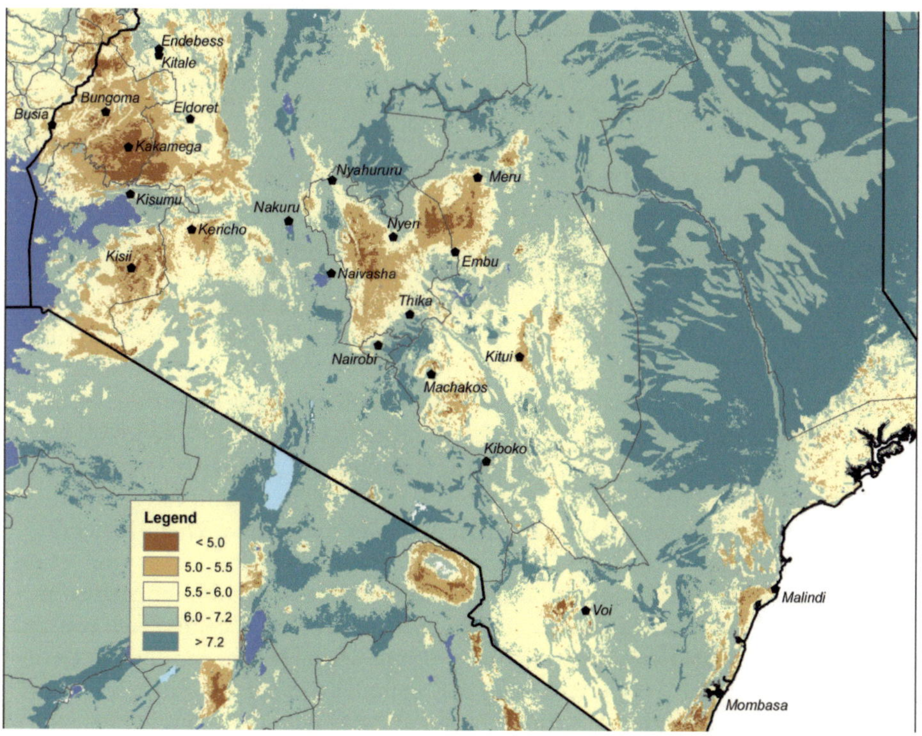

Figure 2.
Soil pH distribution across Kenya determined using AfSIS data.

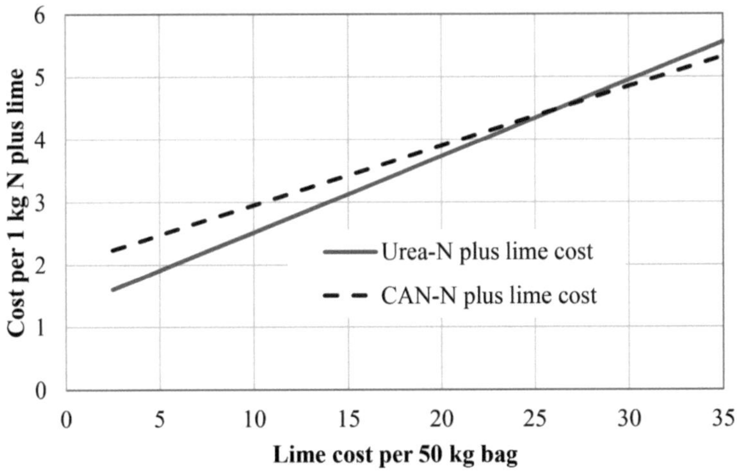

Figure 3.
Comparison of the retail costs of N supply using urea and calcium ammonium nitrate, plus the cost of agricultural lime for neutralizing the fertilizer acidification effects (US\$), based on common fertilizer and lime costs in Kenya in 2016.

The farm-level 2015 costs in western Kenya were 1.3 US\$ kg^{-1} for urea-N, 2.0 \$ kg^{-1} for CAN-N, and 0.17 \$ kg^{-1} for effective CCE of lime. The retail cost of fertilizer N plus lime to neutralize the N effect on soil acidity was \$1.30 + 0.166 × \$ 3.65 = \$1.91 kg^{-1} for urea-N and \$2.00 + 0.166 * \$2.85 = \$2.47 kg^{-1} for CAN-N and 30% more costly for the CAN compared with urea option. The CAN compared with the urea option remains less profitable at these fertilizer prices if the cost for effective CEC of lime is <0.90 \$ kg^{-1} (**Figure 3**).

4. Blended and compound fertilizers

Blended and compound fertilizers are mixtures of common or straight fertilizers. Blended fertilizers are mixtures of common fertilizers which are distinguishable in the mix. Compound fertilizers are formulated by re-granulating the component common fertilizers to have some of each fertilizer in each granule. Hereafter, blended and compound fertilizers are referred to as blends. Common fertilizers often used in dry blends include urea, triple super phosphate (TSP), diammonium phosphate (DAP), and potassium chloride (KCl).

The flexibility in nutrient application with common fertilizers is often important for profit optimization. For example, cereal yield response to N followed by P often has more profit potential than the application of K, secondary nutrients, and micronutrients. The application of several nutrients in a blend can result in increased yield compared to the application of fewer nutrients with the farmer's chosen combination of common fertilizers such as for wheat and maize production in Rwanda, but the profit potential is more often greater with common fertilizers [5, 7, 12, 17]. Blending adds to the cost of nutrient supply, and blends often contain one or more nutrients that have low or no profit potential for the farmer. For example, maize (*Zea mays* L.) yield responses to K included 57% of 164 cases in tropical Africa with increases and 18% with decreases >0.1 Mg ha^{-1} [18]. This indicates an NPK blend may be advantageous in some cases compared to common fertilizers but there are many cases where K use needs to be highly selective. Unfortunately, governmental policy decisions in some countries limit farmer access to some potentially valuable common fertilizers such as TSP, DAP, and KCl. Three concerns of enabling supply of blends while restricting the supply of common fertilizers are addressed here. (1) Claims of profit increases for farmers, with less soil acidification, with the use of blends compared with the judicious use of common fertilizers, are often not true. (2) Tailoring of blends for smallholders in Africa should not be based on soil test information which is scarce and likely to be highly variable across a farm but especially due to the weak basis for interpreting soil test results relative to crop yield response to applied nutrients [19]. (3) Smallholder farmers generally need high return on investment with little risk of failed returns to justify an investment.

5. Blended and compound fertilizers: yield and soil acidification

The acidifying effect of blends depends on their ingredients. Common fertilizers that are generally available on the world market at very competitive prices and with relatively high nutrient content are commonly used to produce blends. The nutrient contents of blends need to be reported, but the constituent fertilizers commonly do not need to be reported. However, a dry NPK blend is expected to contain urea, DAP or mono-ammonium phosphate, and KCl or potassium sulfate. Applying basic algebra for a 17-17-17, for example, it could be composed of 22.5, 37.0, 28.3, and 12.2% of urea, DAP, KCl, and bulking material, respectively, but it could not be 37.0, 37.0, and 28.3% of urea, TSP, and KCl as these total to >100%. The soil acidification effect of the blend depends on the constituent fertilizers. The acidification effect of the urea-DAP-KCl blend could be only slightly reduced by replacing some of the DAP with TSP since TSP is a non-acidifying P source for acid soils or using lime as the 12.2% of bulk material. Therefore, a fertilizer user should not expect a fertilizer blend to be a much less acidifying means for nutrient application compared with judicious application of common fertilizers.

6. Tailoring of blends based on soil test information

Soil test values vary considerably within and across smallholder farming operations with soil texture, depth, and pH generally stronger determinants of crop yield and yield response to nutrients than are soil test results for nutrient availability [20, 21]. However, the probability of maize yield response to N and P is high for agricultural soils not having severe edaphic and other abiotic and biotic constraints. Of 727 N and 672 P yield response functions determined from field maize trials in tropical Africa, yield increases were >0.1 Mg ha^{-1} for 87% of the N functions and 69% of the P functions [18].

Interpretation of soil test results for the estimation of the probability and magnitude of profitable yield response to applied nutrients is generally weak globally for most secondary and micronutrients, with Zn being a possible exception, even where nutrient management is strongly based on field research results. Soil test information has a low or negligible predictive value for crop yield response to applied nutrients in tropical Africa [19]. Soil S tests have been less indicative of crop yield response to S than the use of soil organic matter content and soil texture [22, 23]. Situation-specific interpretations of soil test results for micronutrients have been useful but are unconfirmed for extensive use across geographic and climatic conditions [24]. Hot water extraction of B has been useful in predicting alfalfa yield response to B, but prediction is improved by consideration of soil texture [25]. Different nutrient extraction procedures for soil tests require different interpretations. Mehlich-3 extraction [26] is increasingly used for good reasons but does not correlate well with DTPA extraction for most micronutrients with R^2 values of

0.88 and 0.90 for Zn; 0.42 and 0.63 for Fe; 0.50 and 0.88 for Cu; 0.50 for B; and 0.05 for Mn [27–30]. Therefore, interpretation of Mehlich-3 extraction results is appropriate where crop yield response has been calibrated directly with Mehlich-3 data.

Interpretation of soil test results in terms of probability of profitable yield response to an applied nutrient can be expected to be weak in tropical Africa because crop yield and yield response to inputs in the tropics typically encounter numerous unmitigated constraints that are periodically more constraining than a nutrient deficiency [31, 32]. Each of these constraints not only limits yield but also crop yield response to attempts to mitigate another constraint and ability to predict response. Wendt and Rijpma [33] did not find a relationship between soil test information and crop yield response to applied S, Zn, and B in Malawi for individual fields. Kaizzi et al. [34, 35] did not find a soil test relationship for maize and sorghum yield response to N, P, and K in Uganda. In the analysis of >1100 cases of crop yield response linked to soil test information, Mehlich-3 extracted P and K accounted for <1% of the variation in yield response to application of these nutrients [19]. With more research, interpretation of soil test results for tropical Africa is expected to improve, but soil test results do not provide a practical basis for the tailoring of fertilizer blends in tropical Africa at this time.

7. Blends and the farmer's financial context

The greatest profit/cost potential is likely to be with the application of one or two most limiting nutrients, often N and P for non-legumes and P or P plus another nutrient for legumes [5, 7, 12, 17]. Positive synergistic effects of applying the two most limiting nutrients occur infrequently but tend to account for relatively little yield response compared with the additive effects of individual nutrients e.g., [6, 8–11, 17, 34–37]. Therefore the highest profit/cost ratio can generally be achieved by at least partly alleviating the most limiting nutrient deficiency constraint followed by the second most limiting deficiency.

Farmer profit from fertilizer use may be maximized in some situations through the use of relatively more costly blends compared with common fertilizers such as cited above for wheat and maize in Rwanda [5, 17]. The blends may then at least partly meet the needs for those two most limiting nutrients as well, commonly applied near planting time. Blends should not contain nutrients with inadequately verified yield response unless the added cost to the farmer is minimal as any money that a financially constrained farmer uses for relatively costly fertilizer implies less money available for common fertilizers that may have higher profit potential.

8. Nontraditional materials for crop production

Small bottles of nutrients or other solutions or suspensions are commonly sold in agricultural input shops in Africa with claims that use of small amounts can substitute partly or fully for fertilizer. The price per small bottle, even with a wide profit margin, compared to the price of a 50-kg bag of fertilizer is small, but the nutrient quantity is also very small, and the cost per kg of nutrient may be extremely high. These may contain micronutrients, often as low solubility oxides and carbonates, but the form and solubility are usually not specified. Some such products are sometimes vaguely referred to as bio-fertilizers and bio-stimulants and are mostly unregulated. These may have claims of increased crop growth, yield, or tolerance to insect pests, diseases, or drought or more efficient nutrient cycling. The Compendium of Research Reports on Use of Non-Traditional Materials for Crop Production [38] addresses a fraction of such products that have been marketed in the USA, most of which are no longer available or occur under a different name. Others have been found to be effective for specific situations and have an enduring history of use. No such compendium exists for tropical Africa.

Bio-fertilizers may contain microbes or microbial metabolites claimed to fix atmospheric N, convert insoluble P into soluble forms, or stimulate plant growth. Some products such as *Rhizobium* inoculums for increased symbiotic N fixation with legumes can be very effective in the right situations [39]. A product may contain other N-fixing microbes such as *Azospirillum* and *Azotobacter* which may be effective if they can successfully compete with indigenous *Azospirillum* and *Azotobacter* and the rest of the soil microbial community. The well-targeted use of *Bacillus* and *Pseudomonas* microbes can improve soil P availability but may not compete effectively with the indigenous microbial community [40]. The value of vascular arbuscular mycorrhizal fungi, especially to P and Zn uptake, has been long known, but inoculation very often fails to improve mycorrhizal effectiveness. Such products tend to broadly marketed as effective for all situations, but their use needs to be narrowly targeted to specific conditions. Handling, storage, and application timing and method affect bio-fertilizer effectiveness [41, 42].

Bio-stimulates often are of unknown contents but often contain hormones or humic acid. Hormone application can be effective for specific crops in specific situations, but use across a broad spectrum of production situations is unlikely to be effective for well-adapted crop varieties grown on at least moderately good agricultural soil. Humic acid is important to plant growth but is already abundant in soil. A soil of 3% organic matter may have 1–1.5 Mg ha^{-1} of humic acid in the surface 20-cm soil depth, and adding humic acid at a few kg ha^{-1} has a low probability of increasing yield [35].

9. Conclusion

Fertilizer use is essential for wide-scale sustainable improvement of crop productivity in tropical Africa even though smallholder farmers commonly are severely constrained financially. They require high profit/cost ratios of their investments, with acceptable risk, to gradually reduce the limitations of poverty. Fertilizer use can be highly profitable with good crop-nutrient-rate choices made in consideration of the farmer's financial and agronomic context. Maximizing the profit/cost ratio usually requires adequate access to common fertilizers. Soil acidification is a concern and is a partly an unavoidable consequence of N supply to crops. The most cost-effective means for management of soil acidification often involve avoiding excessive N application and the use of slightly more acidifying but less costly common NH_4^+-N fertilizers coupled with lime use compared with NO_3^--N fertilizers and less lime use. The feasibility of tailored blends has been addressed in consideration of the cost of nutrient supply, the need for flexibility in fertilizer use for maximization of farmer profit, and the weakness of tailoring blends based on soil test results in tropical Africa. However, justification for blends for exceptions such as for wheat and maize in Rwanda should not restrict the supply of common fertilizers. Farmers need to be aware that unregulated products sold in small bottles or packets very often fail to provide the claimed benefits. Fertilizer use, sometimes with timely lime application, can be highly profitable with modest risk if based on good crop-nutrient-rate choices, with adequate fertilizer supply and avoidance of products with unconfirmed claims.

Abbreviations

CAN	calcium ammonium nitrate
CCE	calcium carbonate equivalent
DAP	diammonium phosphate
DTPA	diethylenetriaminepentaacetic acid
TSP	triple super phosphate

Author details

Charles S. Wortmann[1*], Anthony O. Esilaba[2], Kayuki C. Kaizzi[3], Catherine Kibunja[4], Keziah W. Ndungu-Magiroi[5] and Nouri Maman[6]

1 Department of Agronomy and Horticulture, University of Nebraska-Lincoln, Lincoln, NE, USA

2 Kenya Agriculture and Livestock Research Institute, Nairobi, Kenya

3 National Agricultural Research Laboratories, Kampala, Uganda

4 Kenya Agricultural and Livestock Research Organization (KALRO)-Kabete, Nairobi, Kenya

5 KALRO-Kitale, Kitale, Kenya

6 Institut National de Recherche Agronomique du Niger (INRAN), Maradi, Niger

*Address all correspondence to: cwortmann2@unl.edu

References

[1] CIMMYT. From Agronomic Data to Farmer Recommendations: An Economics Training Manual. Mexico, DF: The International Maize and Wheat Improvement Center; 1988

[2] Jansen J, Wortmann CS, Stockton MC, Kaizzi KC. Maximizing net returns to financially constrained fertilizer use. Agronomy Journal. 2013;**105**:573-578. DOI: 10.2134/agronj2012.0413

[3] Kaizzi CK, Mohammed MB, Nouri M. In: Wortmann CS, Sones K, editors. Fertilizer Use Optimization: Principles and Approach. Fertilizer Use Optimization in Sub-Saharan Africa. London, UK: CABI; 2017. pp. 9-19. DOI: 10.1079/9781786392046.0009

[4] Kibunja CN, Ndungu-Magiroi KW, Wamae DK, Mwangi TJ, Nafuma L, Koesch MN, et al. In: Wortmann CS, Sones K, editors. Optimizing Fertilizer Use within the Context of Integrated Soil Fertility Management in Kenya. London, UK: CABI; 2017. pp. 82-99

[5] Cyamweshi AR, Nabahungu LN, Senkoro CJ, Kibunja C, Mukuralinda A, Kaizzi KC, et al. Wheat nutrient response functions for the East Africa highlands. Nutrient Cycling in Agroecosystems. 2017;**111**:21-32. DOI: 10.1007/s10705-018-9912-z

[6] Maman N, Dicko M, Abdou G, Wortmann C. Sorghum and groundnut sole and intercrop nutrient response in semi-arid West Africa. Agronomy Journal. 2017;**109**:1-11. DOI: 10.2134/agronj2017.02.0120

[7] Maman N, MK Dicko G, Abdou ZK, Wortmann C. Pearl millet and cowpea intercrop response to applied nutrients in West Africa. Agronomy Journal. 2017;**109**:2333-2342. DOI: 10.2134/agronj2017.03.0139.

[8] Ndungu-Magiroi KW, Kasozi A, Kaizzi KC, Mwangi T, Koech M, Kibunja CN, et al. Finger millet response to nitrogen, phosphorus and potassium in Kenya and Uganda. Nutrient Cycling in Agroecosystems. 2017;**108**:297-308. DOI: 10.1007/s10705-017-9857-7

[9] Ndungu-Magiroi KW, Wortmann CS, Kibunja C, Senkoro C, Mwangi TJK, Wamae D, et al. Maize bean intercrop response to nutrient application relative to maize sole crop response. Nutrient Cycling in Agroecosystems. 2017;**109**: 17-27. DOI: 10.1007/s10705-017-9862-x

[10] Senkoro CJ, Marandu AE, Ley GJ, Wortmann CS. Maize and pigeon pea sole crop and intercrop nutrient response functions for Tanzania. Nutrient Cycling in Agroecosystems. Mexico, DF: The International Maize and Wheat Improvement Center; 2017;**109**:303-314. DOI: 10.1007/s10705-017-9889-z

[11] Senkoro CJ, Tetteh FM, Kibunja CN, Ndungu-Magiroi KW, Quansah GW, Marandu AE, et al. Cassava yield and economic response to fertiliser in Tanzania, Kenya and Ghana. Agronomy Journal. 2018;**110**:1600-1606. DOI: 10.2134/agronj2018.01.0019

[12] Tarfa BD, Maman N, Ouattara K, Wortmann C. Groundnut and soybean response to nutrient application in West Africa. Agronomy Journal. 2017;**109**:2323-2332. DOI: 10.2134/agronj2017.03.0132

[13] Wortmann CS, Kaizzi CK. Optimization of financially constrained fertilizer use. In: Chatterjee A, Clay D, editors. Soil Fertility Management in Agroecosystems. Madison, WI: ASA, CSSA, and SSSA; 2015. pp. 66-75. DOI: 10.2134/soilfertility.2014.0088

[14] OFRA. Optimizing Fertilizer Recommendations in Africa. 2019. Available from: http://www.agronomy.

unl.edu/OFRA [Accessed: 28 January 2019]

[15] Thomas GW, Hargrove WL. The chemistry of soil acidity. In: Adams F, editor. Soil Acidity and Liming. 2nd ed. Madison, WI: American Society of Agronomy; 1984

[16] Kamprath EJ. In: Adams F, editor. Crop Response to Lime on Soils in the Tropics. In Soil Acidity and Liming. 2nd ed. Madison, WI: ASA, CSSA, SSSA; 1984. pp. 349-368. DOI: 10.2134/agronmonogr12.2ed.c9

[17] Wortmann C, Senkoro C, Cyamweshi AR, Kibunja C, Nkonde D, Munthali M, et al. Maize-nutrient response functions for eastern and southern Africa. Agronomy Journal. 2018;**110**:2070-2079. DOI: 10.2134/agronj2018.04.0268

[18] Wortmann CS, Milner M, Kaizzi KC, Maman N, Cyamweshi RA, Dicko MK, et al. Maize-nutrient response information applied across sub-Saharan Africa. Nutrient Cycling in Agroecosystems. 2017;**107**:175-186. DOI: 10.1007/s10705-017-9827-0

[19] Garba M, Serme I, Wortmann CS. Crop yield response to fertilizer relative to soil properties in sub-Saharan Africa. Soil Science Society of America Journal. 2018;**82**:862-870. DOI: 10.2136/sssaj2018.02.0066

[20] Ugen MA, Wortmann CS. Weed flora composition and soil properties in sub-humid tropical Uganda. Journal of Weed Science and Technology. 2001;**15**:535-543. DOI: 10.1614/0890-037X(2001)015[0535:WFASPI]2.0.CO;2

[21] Vanlauwe B, Tittonell P, Mukalama J. Within-farm soil fertility gradients affect response of maize to fertilizer application in western Kenya. In: Bationo A, Waswa B, Kihaara J, Kimetu J, editors. Advances in Integrated Soil Fertility Management in sub-Saharan Africa: Challenges and Opportunities. Dordrecht, The Netherlands: Springer; 2007. pp. 121-132

[22] Franzen DW. Sulfate-sulfur. In: Recommended Soil Test Procedures for the North Central Region, Ch. 8. North Central Region Research Publication, Columbia MO, USA: Missouri Agricultural Experiment Station; 2015. pp. 8.1-8.6

[23] Shaver T. Corn. In: Shaver T, editor. Nutrient Management for Agronomic Crops of Nebraska. Lincoln NE, USA: University of Nebraska-Lincoln Extension; 2014. pp. 73-81. EC 155

[24] Whitney DA, Missouri A, Wortmann CS, Kirkby RA, Eledu CA, Allen DJ. Micronutrients: zinc, iron, manganese and copper. In: Recommended Soil Test Procedures for the North Central Region. North Central Region Research Publication No. 221, 1998; An Atlas of Common Bean (*Phaseolus vulgaris* L.) production in Africa. Cali, Colombia: Centro Internacional de Agricultura Tropical; 2015

[25] Watson ME. Boron. In: Recommended Soil Test Procedures for the North Central Region. Missouri Agric. Exp. Station; 2015. pp. 10.1-10.4. North Central Reg. Res. Publ. No. 221

[26] Mehlich A. Mehlich-3 soil test extractant: A modification of Mehlich-2 extractant. Communications in Soil Science and Plant Analysis. 1984;**15**:1409-1416. DOI: 10.1080/00103628409367568

[27] Schmisek ME, Cihacek LJ, Swensen LJ. Relationships between the Mehlich-III soil test extraction procedure and standard soil test methods in North Dakota. Communications in Soil Science and

Plant Analysis. 1998;**29**:1719-1729. DOI: 10.1080/00103629809370062

[28] Vocasek FF, Friedericks JB. Soil micronutrient extraction by Mehlich-3 compared to $CaCl_2$-DTPA. Communications in Soil Science and Plant Analysis. 1994;**25**:1583-1593. DOI: 10.1080/00103629409369137

[29] Wendt JW. Evaluation of the Mehlich 3 soil extractant for upland Malawi soils. Communications in Soil Science and Plant Analysis. 2008;**26**:687-702. DOI: 10.1080/00103629509369328

[30] Wortmann M. Comparison of Mehlich 3 extraction with other soil test extraction methods. 2016. unpublished

[31] Wortmann CS, Kirkby RA, Eledu CA, Allen DJ. An Atlas of Common Bean (*Phaseolus vulgaris L.*) production in Africa. Cali, Colombia: Centro Internacional de Agricultura Tropical; 1998. Available from: http://www.ciat.cgiar.org/africa/pdf/atlas_bean_africa/contents.pdf

[32] Wortmann CS, Mamo M, Mburu C, Letayo E, Abebe G, Kayuki KC, et al. Atlas of sorghum (*Sorghum bicolor* (L.) Moench) production in eastern and southern Africa. Univ. Nebraska-Lincoln, Lincoln NE. Agric. Exp. Station SB 1001. 2009. pp. 9.1-9.4

[33] Wendt JW, Rijpma J. Sulphur, zinc, and boron deficiencies in the Dedza Hills and Thiwi-Lifidzi regions of Malawi. Tropical Agriculture. 1997;**74**:81-89

[34] Kaizzi CK, Byalebeka J, Semalulu O, Alou I, Zimwanguyizza W, Nansamba A, et al. Sorghum response to fertilizer and nitrogen use efficiency in Uganda. Agronomy Journal. 2012;**104**:83-90. DOI: 10.2134/agronj2011.0182

[35] Kaizzi CK, Byalebeka J, Semalulu O, Alou I, Zimwanguyizza W, Nansamba A, et al. Maize response to fertilizer and nitrogen use efficiency in Uganda. Agronomy Journal. 2012;**104**:73-82. DOI: 10.2134/agronj2011.0181

[36] Daudu CK, Ugbaje EM, Oyinlola EY, Tarfa BD, Alhaji YA, Amapu IY. et al. Lowland rice nutrient response functions for Nigeria. Agronomy Journal. 2018;**110**:1079-1088. DOI: 10.2134/agronj2017.08.0469

[37] Maman N, Traoré L, Garba M, Dicko MK, Gonda A, Wortmann CS. Maize sole crop and intercrop response to fertilizer in Mali and Niger. Agronomy Journal. 2017;**110**:728-736. DOI: 10.2134/agronj2017.06.0329

[38] ISU. Compendium of Research Reports on Use of Non-traditional Materials for Crop Production. Ames IA, USA: Iowa State Univ. Agron; 2017. Available from: http://extension.agron.iastate.edu/compendium/ [Accessed: 25 January 2019]

[39] Rurangwa E, Vanlauwe B, Giller KE. Benefits of inoculation, P fertilizer and manure on yields of common bean and soybean also increase yield of subsequent maize. Agriculture, Ecosystems and Environment. 2018;**261**:219-229. DOI: 10.1016/j.agee.2017.08.015

[40] Richardson AE, Simpson RJ. Soil microorganisms mediating phosphorus availability update on microbial phosphorus. Plant Physiology. 2011;**156**:989-996. DOI: 10.1104/pp.111.175448

[41] Dobermann A, Wortmann CS, Ferguson RB, Hergert GW, Shapiro CA, Tarkalson DD, et al. Nitrogen response and economics for irrigated corn in Nebraska. Agronomy Journal. 2011;**103**:67-75. DOI: 10.2134/agronj2010.0179

[42] USDA ERA. Fertilizer use and markets. A report of the USDA Economic Research Service. (not dated). Available from: https://data.ers.usda.gov/reports.aspx?ID=17883 [Accessed: 25 January 2019]

Section 3

Sustainable and Advanced Technologies for Crop Production

Chapter 13

Fungal Endophyte-Host Plant Interactions: Role in Sustainable Agriculture

Tamanreet Kaur

Abstract

Fungal endophytes that live inside plant tissues without causing any apparent symptoms in the host plant are important components of plant micro-ecosystems. Endophytic fungi confer profound impacts on their host plants by enhancing their growth, increasing their fitness, strengthening their tolerances to pests and diseases. Moreover, fungal endophytes symbiotic with host plant produce a plethora of bioactive secondary metabolites that are expressed as defensive weapons to protect the host plant against various abiotic stresses. Currently, main focus in endophytic fungi research is associated with the ability of these microorganisms to produce and accumulate biologically active metabolites as these are potent source of novel natural products useful in agriculture sector.

Keywords: fungal endophyte, symbiosis, secondary metabolites, stress, sustainable agriculture

1. Introduction

Over reliance of synthetic pesticides in crop fields from late 1940 to mid-1960s resulted in a number of adverse environmental impacts such as secondary pest outbreak, insect resurgence, effects on non-target organisms, residual problem, environmental pollution, prompted an urgent need for alternative tactics to help make crop protection more sustainable. Biological control using micro-organism has gained much interest, being specific, low relative cost and low risk to ecosystem [1]. Among the various micro-organisms, endophytic fungi can make the chemical intensive crop production system more sustainable as it has ability to enhance plant growth, yield and increase plant fitness by providing biotic and abiotic stress tolerance [2, 3]. Endophytes ("endo" = within, "phyte" = plant) are the microorganisms that inhabit interior of plants especially leaves, stems, roots without causing any apparent harm to the host [4]. These are ubiquitous having rich biodiversity and found in every plant species as nearly 3,00,000 plant species exist on earth with each individual plant host having one or more than one endophytes [5]. Endophytic fungi are considered as plant mutualists as they receive nutrition and protection from host plant while the host plant may benefit from enhanced competitive abilities and increased resistance to herbivores, pathogens and various abiotic stresses [6]. It spends whole or part of their life cycle colonizing inters- and/or intracellularly within the healthy tissues of the host plant without causing visible signs of infection [7, 8]. Moreover, fungal endophytes have gained significant interest in sustainable agriculture due to their great potential to contribute to secondary compounds with unique structure, including alkaloids, benzopyranones, chinones, flavonoids, phenolic acids, quinones, steroids, terpenoids, tetralones, xanthones, etc. [9–11] produced by the fungi or by the plant due to interaction with the fungi. Among the microorganisms, fungal endophytes are the largest group producing secondary metabolites. Fungal toxins produced by these biotic metabolites contribute to plants tolerance towards various biotic and abiotic stresses. Fungal endophytes are known to produce bioactive compounds toxic to insects, nematodes, produces extracellular enzymes (cellulases, proteinase, lipases, esterases) for degradation of dead soil biomass, solubilize insoluble phosphates and produce plant growth-promoting hormones (auxins, cytokinins, gibberellins). Endophyte infected plants manage plant growth under adverse conditions of drought, salinity, temperature and heavy metal stress through different mechanisms. This chapter outlines various approaches for the use of endophytic fungal inoculants to combat various stresses in agricultural fields, thus increasing global crop productivity.

2. Fungal endophyte-host plant association

The association between fungal endophytes and their host plant is due to their unique adaptations which enable the endophytes to harmonize their growth with their host plant [12]. The origin of endophytes is not clear due to complex association between the endophyte and its host plant and the multiplicity of the host's living environment. Exogenous and endogenous are the two hypotheses explaining the origin of endophytes. According to endogenous hypothesis, endophytes are gaged from the mitochondria and chloroplast of the plant, and so it has comparable genetic backgrounds to the host [13], whereas exogenous hypothesis believes that endophytes arrive from outside of the plant and got inserted into the host from root wound, induced channels, or surface [14]. During the long period of co-existence and evolutionary processes, different relationships have been established between endophytic fungi and their host plants ranging from (i) a continuum of mutualism, (ii) antagonism, and (iii) neutralism. As once inside the tissues of a host plant, the endophytic fungi assumed a quiescent (latent) state, either for the whole lifetime of the host plant (neutralism) or for an extended period of time (mutualism or antagonism) until environmental conditions are favorable for endophytic fungi [15]. Endophytes due to its cryptic existence also have its role of decomposers in ecosystem, as they are among the primary colonizers of dead plant tissues [16, 17].

2.1 Fungal endophytes

2.1.1 Transmission

The life history of endophytes in symbiotum with host plant has three modes of reproduction (**Figure 1**). They can either be transmitted (i) vertically from infected plant to offspring via seeds (*Neotyphodium* spp.), (ii) horizontally by sexual spore s from infected individuals (e.g. *Epichloe* spp.) or (iii) mixture of two life cycles [19]. The pure vertical transmission is asexual reproduction of intercellular hyphae of above ground tissues with no symptoms and transmitted vertically via seeds from infected plants to offspring (e.g. *Neotyphodium* spp.). In contrast, the pure horizontal transmission evolves sexual life cycle, relies on the production of contagious

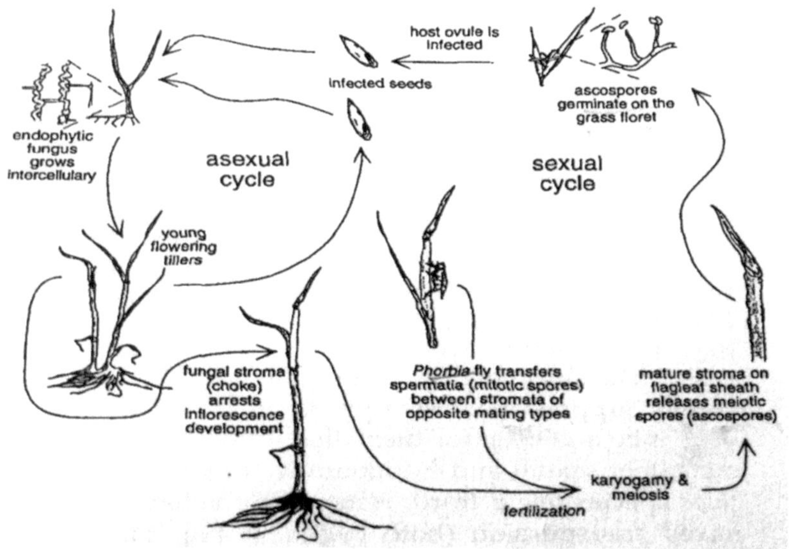

Figure 1.
Asexual and sexual life cycles of Epichloe festucae *symbiotic with* Festuca *spp. [18].*

sexual spores. These spores can only be produced on a fungal structure (stroma) surrounding the grass flag leaf sheath (e.g. some *Epichloe* spp.). Leaves accumulate numerous infections shortly after emergence by means of epiphytic germination of fungal propagules, followed by cuticular penetration or entry through stomata's [20–22] and grow intercellularly within healthy tissues [20, 23]. However, many *Epichloe* spp. use a third mode of reproduction. In this fungi choke some flowering tillers and produce sexual spores leaving majority of tillers uninfected and transmitted asexually via seeds [18]. Endophytes are transmitted vertically (systemic) and horizontally (non-systemic). Vertically transmitted endophytes are mutualistic, whereas those transmitted horizontally depict antagonism to the host [6, 24].

3. Fungal endophytes for sustainable agriculture

In view of escalating pollution and cost due to indiscriminate use of chemical pesticides, diverted researchers interest towards alternative eco-friendly and safe approaches to meet increasing demand of agriculture productivity. Sustainable agriculture requires the use of various strategies to increase or maintain the current rate of food production while minimizing damage to the environment and human health. Symbiotic endophytic fungal associations with crops offer wide range of benefits ranging from the promotion of plant growth to improvements in the tolerance of various biotic and abiotic stresses. Moreover, loss of useful endophytic microbes from crop plants during their domestication and long term cultivation also requires transfer of endophytes from wild relatives of crops to crop species.

4. Fungal endophytes: Biotic stress management

Endophytic fungi have gained importance in the area of agriculture because of their ability to confer resistance to various biotic stress conditions like insect herbivory, nematicidal attack and by aiding plant growth processes.

4.1 Fungal endophytes

4.1.1 Biocontrol agents

Fungal endophytes act as biocontrol agents as they can protect their host plants from pathogens and pests [25, 26]. The mechanism whereby endophytes deter herbivory is through production of antiherbivory/bioactive compounds [27–29] or complex interacting factors of metabolic processes in both the fungus and the plant after infection [26, 30]. These defensive compounds may deter feeding (antixenosis) or reduce insect performance (antibiosis) [31, 32]. Endophytic fungi release the specialized biologically active compounds without any observable damage to their host tissues [33]. Defensive compounds may be categorized into various functional groups: alkaloids, terpenoids, isocoumarin derivatives, quinones, flavonoids, chlorinated metabolites, phenol and phenolic acids and many others [7, 34].

1. **Alkaloids:** Alkaloids are the first reported fungal metabolites to have insecticidal activity. Alkaloids produced by the fungus or by plant in response to fungal infection increase host resistance to herbivores [4, 35]. Endophyte infected grasses contain a variety of alkaloids such as peramines, ergot alkaloids, lolitrems, loline alkaloids and which are absent in non-infected conspecifics [36, 37]. Alkaloids are the first reported fungal metabolites to have insecticidal activity. Most of the alkaloids have been detected in the cultures of grass associated endophytic fungi, such as sexual *Epichloe* spp. and asexual *Neotyphodium* spp. Fungal isolate determines the types of alkaloids produced and plant/fungal genotype interaction can modify the quantities of these alkaloids [38]. The alkaloids from fungal endophytes are categorized into three groups, amines and amides, indole derivatives and pyrrolizidines. Among amines and amides, peramine is toxic to insects without being harmful to mammals [39, 40]. It is a strong feeding deterrent for argentine stem weevil and several other insects [41, 42]. The levels of alkaloids and other toxins may be altered qualitatively depending on the plants physiological state. Ball et al. [43] verified that with plant aging, the amount of peramine decreases in leaves and reaches lower levels during inflorescence phase. The second group of amine and amide alkaloids is ergot alkaloids that also provide significant resistance against insect pests [44]. Feeding experiments with a variety of mammals indicate that ergot alkaloids have significant detrimental effects on mammalian health and reproduction [45, 46]. Among indole derivatives, the lolitrem C and F have been shown to confer resistance against a number of insect species [47]. Other indole derivatives namely chanoclavine, agroclavine and elymoclavine isolated from culture of *Neotyphodium* endophyte [34] were reported to be toxic to some insects and mammals [48]. Among Pyrrolizidines, the saturated aminopyrrolizidine alkaloids as norloline, N-formylloline, N-acetylnorloline, N-acetylloline were exclusively found in endophyte infected grasses of *F. arundinacea* (infected with *Neotyphodium coenophialum*) and *Festuca pratensis* (with *Neotyphodium uncinatum*) [49]. A number of feeding experiments have demonstrated the insecticidal and insect feeding deterrent activities of these lolines [50–52]. Lolines in addition to the well documented effect on insects are also nematicidal [53].

2. **Terpenoids:** Second group of endophytic toxins include terpenoids isolated from some endophytic cultures originating from a variety of host plants. Sesquiterpenes and diterpenes are among the identified terpenoids. Sesquiterpenes as of heptelidic acid and hydroheptelidic acid isolated from *Phyllosticta* sp., an endophytic fungus of balsam fir (*Abies balsamea*) exhibited toxicity to

spruce budworm, *Choristoneura fumiferana* (Clemens) larvae [54]. Two insect toxins, pimarane and diterpene were isolated from an unidentified endophytic fungus symbiotic with needle of *A. balsamea* [54]. Two benzofuran carrying normonoterpene derivatives, toxic to spruce budworm larvae were characterized from an endophytic culture obtained from wintergreen (*Gaultheria procumbens*) [55].

3. **Isocoumarin derivatives:** Toxicity of isocoumarin related metabolites from the conifer endophyte cultures showed toxicity against cells and/ or larvae of spruce budworm [56].

4. **Quinones:** Rugulosin, a metabolite of endophytic fungus *Hormonema dematioides* from balsam fir has been reported to have insecticidal activity [54]. An unidentified endophytic culture isolated from eastern larch (*Larix laricina*) produced a quinone derivative, which was toxic to spruce budworm larvae [55].

5. **Flavonoids:** Among the flavonoids, tricin and related flavone glycosides isolated from endophyte infected blue grass (*Poa ampla*) exhibited toxicity against mosquito larvae [56].

6. **Chlorinated metabolites:** Insecticidal chlorinated metabolite, heptelidic acid chlorohydrins were isolated from cultures of balsam fir needle endophyte *Phyllosticta* spp. [57].

7. **Phenol and phenolic acids:** Phenol and phenolic acids are frequently detected in cultures of endophytes and have pronounced biological activities. Singh et al. [58] purified phenolic compound from ethyl acetate extract of endophytic *Cladosporium* sp. isolated from guduchi (*Tinospora cordifolia*), which induced significant mortality and adversely affected development and survival of tobacco cutworm, *Spodoptera litura* (Fabricius).

Since the 1980s, there is accumulating evidence about factors that influence the outcome of grass–endophyte–insect interactions. Webber [59] was probably the first worker to report plant protection given by fungal endophyte *Phomopsis oblonga* in elm trees (*Ulmus* spp.) against the elm bark beetle, *Physocnemum brevilineum* (Say). Majority of studies for herbivore performance on native grass species symbiotic with endophytic fungi are more consistent showing negative effects including increased mortality [60], reduced mass [61, 62] and decelerated development time [63]. Afkhami et al. [62] reported that bird cherry oat aphid, *Rhopalosiphum padi* (Linnaeus) damaged more endophyte free nodding fescue (*Festuca subverticillata*) than endophyte symbiotic *F. subverticillata*, while positive effect of endophyte infection was reported on eastern lubber grasshopper, *Romalea guttata* (Houttuyn) that preferentially consumed endophyte symbiotic *F. subverticillata* over endophyte free. Similarly increase in growth rate was recorded in third to fifth instars of fall armyworm, *Spodoptera frugiperda* (J.E. Smith) feeding on *N. coenophialum* infected tall fescue [63].

4.2 Fungal endophytes

4.2.1 Nematicidal agents

Fungal endophytes act as nematicidal agents as they are known to produce some compounds which are toxic to nematodes. Diedhiou et al. [64] demonstrated reduced nematicidal activity by an endophytic fungus, *Fusarium oxysporum*, against the plant parasitic nematode *Meloidogyne incognita* in tomato plant. Schwarz et al. [65] reported that several endophytic fungi isolated from above-ground plant organs produced bioactive compound, 3-hydroxypropionic acid (HPA) extracted by bioactivity-guided fractionation of fungal extracts that showed selective nematicidal activity against *M. incognita* with LD50 values of 12.5–15 μg/ml. Similarly, Felde et al. [66] found that combined inoculations of endophytic fungal isolates *Trichoderma atroviride* and *F. oxysporum* is considered an alternative to improve and increase banana yield that reduces the population of burrowing nematode, *Radopholus similis* (Cobb), an important parasitic nematode on banana.

4.3 Fungal endophytes

4.3.1 Phytohormone production

Endophytes can actively or passively regulate the plant growth by solubilization of phosphate, enhance uptake of phosphorus (P), and/ or plant hormones such as auxin, abscisins, ethylene, gibberellic acid (GA), and indole acetic acid (IAA) [67, 68], among these Gibberellic acid is an important phytohormone. The phytohormone GA, a diterpenoid complex, controls the growth of plants, and promotes flowering, stem elongation, seed germination, and ripening [69, 70]. Fungal endophytes *Sebacina vermifera*, *Piriformospora indica, Colletotrichum* and *Penicillium* are distinguished to have better plant growth promoting effects under unfavorable conditions due to their ability to synthesize enzymes and bioactive metabolites [71–73]. Hamayun et al. [69] reported that fungal endophyte, *Cladosporium sphaerospermum* isolated from soybean plant (*Glycine max*) produced gibberellic acid that induced plant growth in rice and soybean. Metabolite pestalotin analogue, isolated from the endophytic *Pestalotiopsis microspora* exhibited significant gibberellin activity in winter-hazel seeds (*Distylium chinense*) and increased their germination rate [74]. Endophytes, *Fusarium tricinctum* and *Alternaria alternata* produced derivatives of plant hormone indole acetic acid that enhanced the plant growth [75]. A study conducted by Johnson et al. [76] on root colonizing endophyte *P. indica* found that association of fungal endophytes with roots modulated phytohormones involved with growth and development of host plant and enhanced nutrient uptake and translocation especially of phosphorus and nitrogen from the soil.

4.4 Fungal endophytes

4.4.1 Agriculturally important enzyme production

Degradation of the dead soil biomass by fungal endophytic is a major step in bringing the utilized nutrients back to the ecosystem that improves soil quality. Endophytic fungi is reported to produce various extracellular hydrolases including cellulase, laccase, pectinase, phosphatase, lipase, xylanase, and proteinase as a resistance mechanism against pathogenic invasion [77] and to obtain nutrition from host as these enzymes break macromolecules such as lignin, sugar-based polymers, proteins, organic phosphate, and carbohydrates to micromolecules that are transported throughout the cells for metabolism and help in host symbiosis process [78]. Sunitha et al. [79] isolated and identified approximately 50 endophytic fungal strains having amylase, laccase, cellulase, pectinase, lipase and protease enzymes. Study conducted by He et al. [80] explained that endophytic fungal species have ability to decompose organic components, including lignin, cellulose, and hemicelluloses that facilitates nutrient cycling. Chathurdevi and Gowrie [81] reported that the endophytic fungi species isolated from medicinal plant *Cardiospermum halicacabum* can support plant growth to overcome the adverse conditions through producing different extracellular enzymes. Fungal chitinases enzyme have vital role in degradation and cycling of carbon and nitrogen from chitin molecule. Chitin molecule is a linear homopolymer of β-1,4N-acetylglucosamine can be obtained from insect's exoskeleton, crustacean's shells, and fungal cell wall. Many fungal endophytes isolated from leaves of trees of Southern India have shown the production of chitinases [82]. An endophyte, *Acremonium zeae*, isolated from maize is reported to produce the extracellular enzyme hemicellulase, which may be used in the bioconversion of lignocellulosic biomass into fermentable sugars [83].

5. Fungal endophytes: Abiotic stress management

Agricultural productivity is significantly threatened by various abiotic stresses. Environmental stresses such as drought, salinity, temperature can collectively cause more than 50% yield losses worldwide [84]. Plants can tolerate abiotic stress by two mechanisms: (i) via activation of response systems directly after exposure to stress [67] (ii) biochemical compounds that are synthesized by fungal endophytes, acts as anti-stress agents [85]. Experimental studies also confirmed that endophytic fungi can help the host plants from environmental stress conditions such as drought, salts, high temperatures and heavy metals and can thus increase the plant growth.

5.1 Drought stress

Among the abiotic stresses, water stress commonly, known as 'drought', is considered as one of the major challenges to crop production worldwide [86]. Drought has a negative impact on the plant growth rate, germination rates, membrane loss of its integrity, repression of photosynthesis, and increase in the productivity of reactive oxygen species [87, 88]. Fungal endophyte infected plants enhance drought tolerance by increased accumulation of solutes in tissues, or by reduced leaf conductance and a slowdown of the transpiration stream, or due to formation of thicker cuticle as compared to non-infected plants [67]. Chippa et al. [89] reported that endophytic, *Neotyphodium* spp. is reported to enhance drought tolerance in grass plant by stomatal and osmoregulations and protect plants in drought and nitrogen starvation. Experimental studies on lavender plants inoculated with *Glomus* spp. showed that these plants accumulated solutes in tissues thereby exhibiting better drought tolerance by improving water contents, N and P contents and root biomass [90, 91]. Moreover, plants harboring endophytes consumes significantly less water and had enhanced biomass than non-symbiotic plants. For instance, endophytes *Chaetomium globosum* and *P. resedanum* isolated from sweet pepper (*Capsicum annuum*) plants enhanced shoot length and biomass of the host plants challenged by drought stress [92, 93]. Similarly, Redman et al. [72] found that inoculation of endophytes *Fusarium culmorum* and *Curvularia protuberata* in drought-affected rice plants resulted in increased biomass than of non-inoculated plants. Fungal endophyte colonization also results in higher chlorophyll content and leaf area in plants under drought stress than non-colonized plant. Higher chlorophyll concentration is related with higher photosynthetic rate [94]. For instance, enhanced photosynthesis rate was recorded in drought stressed *C. annuum* plants colonized by endophytes *C. globosum* [95] and *P. resedanum* [96].

5.2 Salinity stress

High salinity due to extreme climatic conditions and misuse of agricultural land over the past few decades has led to high salinity, which is a limiting factor to global agricultural productivity [97]. Soil salinity is the accumulation of water soluble salts in soil that affects its physical and chemical properties thereby reducing soil's agricultural output [98]. Reactive oxygen species (SOD, CAT, APX) are formed in plants on onset of salt and osmotic stress. Endophytic *Piriformospora indica* induces salt stress tolerance by elevation of antioxidant enzymes [99]. These are involved in the removal of reactive oxygen species either directly or indirectly via regeneration of ascorbate and glutathione in the cell. Experimental studies by Rodriguez et al. [100] reported that constant exposure of non-symbiotic plants dunegrass (*Leymus mollis*) to 500 mmol/l NaCl solution, became severely wilted and desiccated within 7 days and were dead after 14 days. In contrast, symbiotic plants infected with *F. culmorum* showed wilting symptoms only after they were exposed to 500 mmol/l NaCl solution for 14 days.

5.3 Temperature stress (low and high)

High temperature is a major obstacle in crop production that results in major cellular damage such as protein degradation and aggregation [101]. Whereas, low temperature can cause impaired metabolism due to inhibition of enzyme reactions, interactions among macromolecules, changes in protein structure, and modulating cell membrane properties [102]. Endophytic, *Curvularia* spp. is proven to confer thermal tolerance ability plants like tomato, watermelon, and wheat [103]. Herbal plant wooly rosette grass (*Dichanthelium lanuginosum*) that lives in the areas where soil temperatures can reach up to 57°C, the presence of endophytic fungi *Curvularia* sp. protects the plant from temperature stress better than endophyte free plants [104]. Experimental demonstration by Redman et al. [103] showed that grass *D. lanuginosum* survival in soil temperatures ranging between 38 and 65°C is directly linked to its association with the fungus *C. protuberata* and its mycovirus, *Curvularia* thermal tolerance virus (CThTV). Moreover, cold stress tolerance was conferred in germinated seeds of rice under laboratory conditions by *C. protuberata* isolated from *D. lanuginosum* thriving in geothermal soils [72].

5.4 Heavy metal stress

Heavy metal contamination due to increased industrialization has recently received attention because heavy metals cannot be itself degraded [105]. Toxicity by heavy metals can cause the loss of about 25–80% of various cultivated crops. Heavy metals being very toxic to roots of cultivated crop plants can cause poor development of the root system [106]. Endophytic fungi possess metal sequestration or chelation systems that increases tolerances of their host plants to heavy metals via enhancements of antioxidative system thereby changing heavy metal distribution in plant cells and detoxification of heavy metal, thus assisting their hosts to survive in contaminated soil [107, 108]. For instance, dark septate root endophytes (DSEs), *Phialocephala fortinii* can produce the black biopolymer melanin, which can be synthesized from phenolics and binds heavy metals [109] that keep heavy metal ions away from living, plant cells [110]. Siderophores being metal-chelating compounds [111, 112] released from roots into the rhizosphere can be helpful in inhibiting absorption of heavy metals into plant cells as siderophores can form complexes with heavy metals which are not easily absorbed by plant roots. Yamaji et al. [113] recorded that endophytes *P. fortinii* and *Rhizodermea veluwensis* showed an ability to produce siderophores that probably affects heavy metal exclusion in the rhizosphere.

6. Conclusion

Fungal endophytes can be a significant component of sustainable agriculture, being safe, cost-effective, have ability to produce various compounds like phytohormones, defensive compounds, solubilize phosphates, extracellular enzymes, siderophore production, inhibiting plant pathogens, and promoting plant growth. Over the last decade, sharp rise in study of fungal endophytes is seen as they hold huge potential in agricultural sector. However, most of the research on endophytes is still at an experimental level in lab or greenhouse. For permitting the practical use of these endophytes in agriculture it is extremely necessary to encourage field experiments to determine the effectiveness of the endophytes under real world conditions. Simultaneously, it is also necessary to build awareness of this new research field among farmers to improve interactions and collaboration with scientists working in different fields, thereby encouraging the adoption of endophytes in agriculture and maximizing their benefits. If endophytes become feasible in agricultural sector, their practical aspects will also have to be researched so that farmers can learn how to integrate these endophyte species within pre-existing eco-friendly agricultural methods so as to ensure continuity in the approach to sustainability. Moreover, scientific research has to be also focused on use of genetically modified endophytes made by combining endophytes having two or more different ecological roles, such as the suppression of diseases and insect pests to simultaneously improve plant yields and its defensive properties. Thus, optimization of microbial functions to enhance crop production and protection is also required.

Conflict of interest

No conflict of interest is indulged.

Abbreviations

HPA	3-hydroxypropionic acid
P	phosphorus
GA	Gibberellic acid
IAA	indole acetic acid
SOD	superoxide dismutase
CAT	catalase
APx	ascorbate peroxidase
CThTV	*Curvularia* thermal tolerance virus
DSEs	dark septate root endophytes

Author details

Tamanreet Kaur
Department of Zoology, Kanya Maha Vidyalaya, Jalandhar, India

*Address all correspondence to: tamanreetkaur@gmail.com

References

[1] Castillo MA, Moya P, Hernandez E, Primo-Yufera E. Susceptibility of *Ceratitis capitata* Wiedemann (Diptera: Tephritidae) to entomopathogenic fungi and their extracts. BioControl. 2000;**19**:274-282. DOI: 10.1006/bcon.2000.0867

[2] Tanaka A, Tapper BA, Popay A, Parker EJ, Scott B. A symbiosis expressed non-ribosomal peptide synthetase from a mutualistic fungal endophyte of perennial ryegrass confers protection to the symbiotum from insect herbivory. Molecular Microbiology. 2005;**57**:1036-1050. DOI: 10.1111/j.1365-2958.2005.04747.x

[3] Vega FE, Posada F, Aime MC, Pava-Ripoll M, Infante F, Rehner SA. Entomopathogenic fungal endophytes. Biological Control. 2008;**46**:72-82. DOI: 10.1016/j.biocontrol.2008.01.008

[4] Azevedo JL, Maccheroni W Jr, Periera JO, Araujo WL. Endophytic microorganisms: A review on insect control and recent advances on tropical plants. Electronic Journal of Biotechnology. 2000;**3**:40-65. DOI: 10.2225/vol3-issue1-fulltext-4

[5] Smith SA, Tank DC, Boulanger LA, Bascom-Slack CA, Eisenman K, Kingery D, et al. Bioactive endophytes warrant intensified exploration and conservation. PLoS One. 2008;**3**:e3052. DOI: 10.1371/journal.pone.0003052

[6] Saikkonen K, Faeth SH, Helander M, Sullivan TJ. Fungal endophytes: A continuum of interactions with host plants. Annual Review of Ecology and Systematics. 1998;**29**:319-343. DOI: 10.1146/annurev.ecolsys.29.1.319

[7] Tan RX, Zou WX. Endophytes: A rich source of functional metabolites. Natural Product Reports. 2001;**18**:448-459. DOI: 10.1039/B100918O

[8] Hartley SE, Gange AC. Impacts of plant symbiotic fungi on insect herbivores: Mutualism in a multitrophic context. Annual Review of Entomology. 2009;**54**:323-342. DOI: 10.1146/annurev.ento.54.110807.090614

[9] Jalgaonwala RE, Mohite BV, Mahajan RT. Natural products from plant associated endophytic fungi. Journal of Microbiology and Biotechnology Research. 2011;**1**:21-32

[10] Joseph B, Priya RM. Bioactive compounds from endophytes and their potential in pharmaceutical effect: A review. American Journal of Biochemistry and Molecular Biology. 2011;**1**:291-309. DOI: 10.3923/ajbmb.2011.291.309

[11] Pimentel MR, Molina G, Dionisio AP, Marostica MR, Pastore GM. Use of endophytes to obtain bioactive compounds and their application in biotransformation process. Biotechnology Research International. 2011:1-11. Article ID: 566286. DOI: 10.4061/2011/576286

[12] Verma VC, Singh SK, Kharwar RN. Histological investigation of fungal Endophytes in healthy tissues of *Azadirachta indica* A. Juss. Kasetsart Journal (Natural Science). 2012;**46**:229-237

[13] Wen CY. Recent advances and issues on the endophyte. Chinese Journal of Ecology. 2004;**2**:86-91

[14] Li WK. Endophytes and natural medicines. Chinese Journal of Natural Medicines. 2005;**4**:193-199

[15] Sieber TN. Endophytic fungi in forest trees: Are they mutualists? Fungal Biology Reviews. 2007;**21**:75-89. DOI: 10.1016/j.fbr.2007.05.004

[16] Kumaresan V, Suryanarayanan TS. Endophyte assemblage in young,

mature and senescent leaves of *Rhizophora apiculata*: Evidence for the role of endophytes in mangrove litter degradation. Fungal Diversity. 2002;**9**:81-91

[17] Oses R, Valenzuela S, Freer J, Sanfuentes E, Rodriguez J. Fungal endophytes in xylem of healthy chilean trees and their possible role in early wood decay. Fungal Diversity. 2008;**33**:77-86

[18] Bush LP, Wilkinson HH, Schardl CL. Bioprotective alkaloids of grass-fungal endophyte symbioses. Plant Physiology. 1997;**114**:1-7. DOI: 10.1104/pp.114.1.1

[19] Schardl CL, Leuchtmann A, Chung KR, Penny D, Siegel MR. Coevolution by common descent of fungal symbionts (*Epichloe* spp.) and grass hosts. Molecular Biology and Evolution. 1997;**14**:133-143. DOI: 10.1093/oxfordjournals.molbev.a025746

[20] Lebron L, Lodge DJ, Laureano SL, Bayman P. Where is the gate to party? Phytopathology. 2001;**91**:116

[21] Arnold AE. Fungal endophytes in Neotropical trees: Abundance, diversity and ecological interactions [Thesis]. Tucson: University of Arizona; 2002

[22] Arnold AE, Herre EA. Canopy cover and leaf age affect colonization by tropical fungal endophytes: Ecological pattern and process in *Theobroma cacao* (Malvaceae). Mycologia. 2003;**95**:388-398. DOI: 10.1080/15572536.2004.11833083

[23] Arnold AE, Mejia L, Rojas E, Maynard Z, Robbins N, Herre EA. Organismos endofiticos: Microorganismos en plantas. In: Barrantes RM, editor. El uso de Microoganismos Beneficos en la Agricultura Moderna. Costa Rica: Guacimo; 2001. pp. 1-20

[24] Schardl CL, Liu JS, White JF, Finkel RA, An Z, Siegel MR. Molecular phylogenetic relationships of nonpathogenic grass mycosymbionts and clavicipitaceous plant pathogens. Plant Systematics and Evolution. 1991;**178**:27-41. DOI: 10.1007/BF00937980

[25] Bae H, Sicher RC, Kim MS, Kim S-H, Strem MD, Melnick RL, et al. The beneficial endophyte *Trichoderma hamatum* isolate DIS 219b promotes growth and delays the onset of the drought response in *Theobroma cacao*. Journal of Experimental Botany. 2009;**60**:3279-3295. DOI: 10.1093/jxb/erp165

[26] Akello J, Dubois T, Gold CS, Coyne D, Nakavuma J, Paparu P. *Beauveria bassiana* (Balsamo) Vuillemin as an endophyte in tissue culture banana (*Musa* spp.). Journal of Invertebrate Pathology. 2007;**96**:34-42. DOI: 10.1016/j.jip.2007.02.004

[27] Clay K, Schardl C. Evolutionary origins and ecological consequences of endophyte symbiosis with grasses. American Naturalist. 2002;**160**:99-127. DOI: 10.1086/342161

[28] Faeth SH, Hayes CJ, Gardner DR. Asexual endophytes in a native grass: Tradeoffs in mortality, growth, reproduction, and alkaloid production. Microbial Ecology. 2010;**60**:496-504. DOI: 10.1007/s00248-010-9643-4

[29] Torres MS, White JF Jr. Grass endophyte-mediated plant stress tolerance: Alkaloids and their functions. In: Seckbach J, Grube M, editors. Symbiosis and Stress: Joint Ventures in Biology, Cellular Origin, Life in Extreme Habitats and Astrobiology. New York: Springer; 2010. pp. 477-493

[30] Barbosa P, Krischik VA, Jones CG. Microbial Mediation of Plant-Herbivore Interactions. New York: Wiley; 1991. 530p

[31] Braman SK, Duncan RR, Engelke MC, Hanna WW, Hignight K,

Rush D. Grass species and endophyte effects on survival and development of fall armyworm (Lepidoptera: Noctuidae). Journal of Economic Entomology. 2002;**95**:487-492. DOI: 10.1603/0022-0493-95.2.487

[32] Crawford KM, Land JM, Rudgers JA. Fungal endophytes of native grasses decrease insect herbivore preference and performance. Oecologia. 2010;**164**:431-444. DOI: 10.1007/s00442-010-1685-2

[33] Liarzi O, Bucki P, Braun Miyara S, Ezra D. Bioactive volatiles from an endophytic *Daldinia* cf. *concentrica* isolate affect the viability of the plant parasitic nematode *Meloidogyne javanica*. PLoS One. 2016;**11**:e0168437. DOI: 10.1371/journal.pone.0168437

[34] Strobel G, Daisy B, Castillo U, Harper J. Natural products from endophytic microorganisms. Journal of Natural Products. 2004;**67**:257-268. DOI: 10.1021/np030397v

[35] Powell RG, Petroski RJ. Alkaloid toxins in endophyte infected grasses. Natural Toxins. 1992;**1**:163-170. DOI: 10.1002/nt.2620010304

[36] Muller CB, Krauss J. Symbiosis between grasses and asexual fungal endophytes. Current Opinion in Plant Biology. 2005;**8**:450-456. DOI: 10.1016/j.pbi.2005.05.007

[37] Zhang Y, Han T, Ming Q, Wu L, Rahman K, Qin L. Alkaloids produced by endophytic fungi: A review. Natural Product Communications. 2012;7:963-968

[38] Lane GA, Christensen MJ, Miles CO. Coevolution of fungal endophytes with grasses: The significance of secondary metabolites. In: Bacon CW, White JF Jr, editors. Microbial Endophytes. New York: Marcel Dekker; 2000. pp. 341-388

[39] Dew RK, Boissonneault GA, Gay N, Boling JA, Cross RJ, Cohen DA. The effect of the endophyte (*Acremonium coenophialum*) and associated toxin of tall fescue on serum titer response to immunization and spleen cell flow cytometry analysis and response to mitogens. Veterinary Immunology and Immunopathology. 1990;**26**:285-295. DOI: 10.1016/0165-2427(90)90097-c

[40] Rowan DD, Latch GCM. Utilization of endophyte-infected perennial ryegrasses for increased insect resistance. In: Bacon CW, White JF Jr, editors. Biotechnology of Endophytic Fungi of Grasses. Boca Raton: CRC Press; 1994. pp. 169-183

[41] Rowan DD, Dymock JJ, Brimble MA. Effect of fungal metabolite peramine and analogs on feeding and development of argentine stem weevil (*Listronotus bonariensis*). Journal of Chemical Ecology. 1990;**16**:1683-1695. DOI: 10.1007/BF01014100

[42] Rowan DD. Lolitrems, peramine and paxilline: Mycotoxins of the ryegrass/endophyte interaction. Agriculture, Ecosystems and Environment. 1993;**44**:103-122. DOI: 10.1016/0167-8809(93)90041-M

[43] Ball OJP, Barker GM, Prestidge RA, Lauren DR. Distribution and accumulation of the alkaloid peramine in *Neotyphodium lolii*-infected perennial ryegrass. Journal of Chemical Ecology. 1997;**23**:1419-1434. DOI: 10.1023/B:JOEC.0000006473.26175.19

[44] Potter AD, Stokes TJ, Redmond TC, Schardl LC, Panaccione GD. Contribution of ergot alkaloids to suppression of a grass-feeding caterpillar assessed with gene knockout endophytes in perennial ryegrass. Entomologia Experimentalis et Applicata. 2008;**126**:138-147. DOI: 10.1111/j.1570-7458.2007.00650.x

[45] Gadberry MS, Denard TM, Spiers DE, Piper EL. Effects of feeding ergovaline on lamb performance in a heat stress environment. Journal of

Animal Science. 2003;**81**:1538-1545. DOI: 10.2527/2003.8161538x

[46] Parish JA, McCann MA, Watson RH, Hoveland CS, Hawkins LL, Hill NS, et al. Use of nonergot alkaloid-producing endophytes for alleviating tall fescue toxicosis in sheep. Journal of Animal Science. 2003;**81**:1316-1322. DOI: 10.2527/2003.8151316x

[47] Parish JA, McCann MA, Watson RH, Piava NN, Hoveland CS, Parks AH, et al. Use of nonergot alkaloid-producing endophytes for alleviating tall fescue toxicosis in stocker cattle. Journal of Animal Science. 2003;**81**:2856-2868. DOI: 10.2527/2003.81112856x

[48] Ball OJP, Barker GM, Prestidge RA, Sprosen JM. Distribution and accumulation of the Mycotoxin Lolitrem B in *Neotyphodium lolii*-infected perennial ryegrass. Journal of Chemical Ecology. 1997;**23**:1435-1449. DOI: 10.1023/B:JOEC.0000006474.44100.17

[49] Schardl CL, Phillips TD. Protective grass endophytes: Where are they from and where are they going? Plant Disease. 1997;**81**:430-438. DOI: 10.1094/PDIS.1997.81.5.430

[50] Siegel MR, Latch GCM, Bush LP, Fannin FF, Rowan DD, Tapper BA, et al. Fungal endophyte-infected grasses: Alkaloid accumulation and aphid response. Journal of Chemical Ecology. 1990;**16**:3301-3315. DOI: 10.1007/BF00982100

[51] Riedell WE, Kieckhefer RE, Petroski RJ, Powell RG. Naturally occurring and synthetic loline alkaloid derivatives: Insect feeding behavior modification and toxicity. Journal of Entomological Science. 1991;**26**:122-129. DOI: 10.18474/0749-8004-26.1.122

[52] Jensen JG, Popay AJ, Tapper BA. Argentine stem weevil adults are affected by meadow fescue endophyte and its loline alkaloids. New Zealand Plant Protection. 2009;**62**:12-18. DOI: 10.30843/nzpp.2009.62.4800

[53] Bacetty AA, Snook ME, Glenn AE, Noe JP, Nagabhyru P, Bacon CW. Chemotaxis disruption in *Pratylenchus scribneri* by tall fescue root extracts and alkaloids. Journal of Chemical Ecology. 2009;**35**:844-850. DOI: 10.1007/s10886-009-9657-x

[54] Calhoun LA, Findlay JA, Miller JD, Whitney NJ. Metabolites toxic to spruce budworm from balsam fir needle endophytes. Mycological Research. 1992;**96**:281-282. DOI: 10.1016/S0953-7562(09)80939-8

[55] Findlay JA, Li G, Penner PE, Miller JD. Bioactive isocoumarins and related metabolites from conifer endophytes. Journal of Natural Products. 1995;**58**:1759-1768. DOI: 10.1021/np50125a021

[56] Findlay JA, Buthelezi S, Li G, Seveck M, Miller JD. Insect toxins from an endophytic fungus from wintergreen. Journal of Natural Products. 1997;**60**:1214-1215. DOI: 10.1021/np970222j

[57] Ju Y, Sacalis JN, Still CC. Bioactive flavonoids from endophyte-infected blue grass (*Poa ampla*). Journal of Agricultural and Food Chemistry. 1998;**46**:3785-3788. DOI: 10.1021/jf980189m

[58] Singh B, Kaur T, Kaur S, Manhas RK, Kaur A. An alpha-glucosidase inhibitor from an endophytic *Cladosporium* sp. with potential as a biocontrol agent. Applied Biochemistry and Biotechnology. 2015;**175**:2020-2034. DOI: 10.1007/s12010-014-1325-0

[59] Webber J. A natural control of Dutch elm disease. Nature. 1981;**292**:449-451. DOI: 10.1038/292449a0

[60] Brem D, Leuchtmann A. *Epichloe* grass endophytes increase herbivore

resistance in the woodland grass *Brachypodium sylvaticum*. Oecologia. 2001;**126**:522-530. DOI: 10.1007/s004420000551

[61] Bazely DR, Vicari M, Emmerich S, Filip L, Lin D, Inman A. Interactions between herbivores and endophyte-infected *Festuca rubra* from the Scottish islands of St Kilda, Benbecula and Rum. Journal of Applied Ecology. 1997;**34**:847-860. DOI: 10.2307/2405276

[62] Afkhami ME, Rudgers JA. Endophyte-mediated resistance to herbivores depends on herbivore identity in the wild grass *Festuca subverticillata*. Environmental Entomology. 2009;**38**:1086-1095. DOI: 10.1603/022.038.0416

[63] Bultman TL, Bell GD. Interaction between fungal endophytes and environmental stressors influences plant resistance to insects. Oikos. 2003;**103**:182-190. DOI: 10.1034/j.1600-0706.2003.11574.x

[64] Diedhiou PM, Hallmann J, Oerke EC, Dehne HW. Effects of arbuscular mycorrhizal fungi and a non-pathogenic *Fusarium oxysporum* on *Meloidogyne incognita* infestation of tomato. Mycorrhiza. 2003;**13**:199-204. DOI: 10.1007/s00572-002-0215-4

[65] Schwarz M, Kopcke B, Weber RWS, Sterner O, Anke H. 3-Hydroxypropionic acid as a nematicidal principle in endophytic fungi. Phytochemistry. 2004;**65**:2239-2245. DOI: 10.1016/j.phytochem.2004.06.035

[66] Felde ZA, Pocasangre LE, Carnizares Monteros CA, Sikora RA, Rosales FE, Riveros AS. Effect of combined inoculations of endophytic fungi on the biocontrol of *Radopholus similis*. InfoMusa. 2006;**15**:12-18

[67] Malinowski DP, Belesky DP. Adaptations of Endophyte-infected cool-season grasses to environmental stresses: Mechanisms of drought and mineral stress tolerance. Crop Science. 2000;**40**:923-940. DOI: 10.2135/cropsci2000.404923x

[68] Firakova S, Sturdikova M, Muckova M. Bioactive secondary metabolites produced by microorganisms associated with plants. Biologia. 2007;**62**:251-257. DOI: 10.2478/s11756-007-0044-1

[69] Hamayun M, Khan SA, Iqbal I, Na CI, Khan AL, Hwang YH, et al. *Chrysosporium pseudomerdarium* produces gibberellins and promotes plant growth. Journal of Microbiology. 2009;**47**:425-430. DOI: 10.1007/s12275-009-0268-6

[70] Yamaguchi S. Gibberellin metabolism and its regulation. Annual Review of Plant Biology. 2008;**59**:225-251. DOI: 10.1146/annurev.arplant.59.032607.092804

[71] Waller F, Achatz B, Baltruschat H, Fodor J, Becker K, Fischer M, et al. The endophytic fungus *Piriformospora indica* reprograms barley to salt stress tolerance, disease resistance, and higher yield. Proceedings of the National Academy of Sciences of the United States of America. 2005;**102**:13386-13391

[72] Redman RS, Kim YO, Woodward CJ, Greer C, Espino L, Doty SL, et al. Increased fitness of rice plants to abiotic stress via habitat adapted symbiosis: A strategy for mitigating impacts of climate change. PLoS One. 2011;**6**:e14823. DOI: 10.1371/journal.pone.0014823

[73] Hamilton CE, Bauerle TL. A new currency for mutualism? Fungal endophytes alter antioxidant activity in hosts responding to drought. Fungal Diversity. 2012;**54**:39-49. DOI: 10.1007/s13225-012-0156-y

[74] Li X, Guo Z, Deng Z, Yang J, Zou K. A new α-Pyrone derivative from

endophytic fungus *Pestalotiopsis microspora*. Records of Natural Products. 2015;**9**:503-508

[75] Khan AL, Hussain J, Al-Harrasi A, Al-Rawahi A, Lee IJ. Endophytic fungi: Resource for gibberellins and crop abiotic stress resistance. Critical Reviews in Biotechnology. 2015;**35**:62-74. DOI: 10.3109/07388551.2013.800018

[76] Johnson JM, Alex T, Oelmuller R. *Piriformospora indica*: The versatile and multifunctional root endophytic fungus for enhanced yield and tolerance to biotic and abiotic stress in crop plants. Journal of Tropical Agriculture. 2014;**52**:103-122

[77] Leo VV, Passari AK, Joshi JB, Mishra VK, Uthandi S, Ramesh N, et al. A novel triculture system (CC3) for simultaneous enzyme production and hydrolysis of common grasses through submerged fermentation. Frontiers in Microbiology. 2016;**7**:447. DOI: 10.3389/fmicb.2016.00447

[78] Strong PJ, Claus H. Laccase: A review of its past and its future in bioremediation. Critical Reviews in Environmental Science and Technology. 2011;**41**:373. DOI: 10.1080/10643380902945706

[79] Sunitha VH, Devi DN, Srinivas C. Extracellular enzymatic activity of endophytic fungal strains isolated from medicinal plants. World Journal of Agricultural Sciences. 2013;**9**:1-9. DOI: 10.5829/idosi.wjas.2013.9.1.72148

[80] He X, Han G, Lin Y, Tian X, Xiang C, Tian Q, et al. Diversity and decomposition potential of endophytes in leaves of a *Cinnamomum camphora* plantation in China. Ecological Research. 2012;**27**:273-284. DOI: 10.1007/s11284-011-0898-0

[81] Chathurdevi G, Gowrie SU. Endophytic fungi isolated from medicinal plant—A source of potential bioactive metabolites. International Journal of Current Pharmaceutical Research. 2016;**8**:50-56

[82] Rajulu MBG, Thirunavukkarasu N, Suryanarayanan TS, Ravishankar JP, Gueddari NEE, Moerschbacher BM. Chitinolytic enzymes from endophytic fungi. Fungal Diversity. 2011;**47**:43-53. DOI: 10.1007/s13225-010-0071-z

[83] Bischoff KM, Wicklow DT, Jordan DB, de Rezende ST, Liu S, Hughes SR, et al. Extracellular hemicellulolytic enzymes from the maize endophyte *Acremonium zeae*. Current Microbiology. 2009;**58**:499-503. DOI: 10.1007/s00284-008-9353-z

[84] Wang Y, Frei M. Stressed food—The impact of abiotic environmental stresses on crop quality. Agriculture, Ecosystems and Environment. 2011;**141**:271-286. DOI: 10.1016/j.agee.2011.03.017

[85] Schulz B, Boyle C, Draeger S, Rommert A-K, Krohn K. Endophytic fungi: A source of novel biologically active secondary metabolites. Mycological Research. 2002;**106**:996-1004. DOI: 10.1017/S0953756202006342

[86] IPCC. Climate change 2007: Synthesis report. In: Pachauri RK, Reisinger A, editors. Contribution of Working Groups I, II and III to the Fourth Assessment Report of the Intergovernmental Panel on Climate Change. Geneva: IPCC; 2007. p. 104

[87] Granier C, Tardieu F. Reductions in area, cell number and cell area in sunflower leaves subjected to short water deficits with different timings. Variability in responses can be simulated using a simple model of leaf development. Plant Physiology. 1999;**119**:609-619. DOI: 10.1104/pp.116.3.991

[88] Greenberg BM, Huang XD, Gerwing P, Yu XM, Chang P, Wu SS, et al. Phytoremediation of salt impacted

soils: Greenhouse and the field trials of plant growth promoting rhizobacteria (PGPR) to improve plant growth and salt phyto-accumulation. In: Proceeding of the 33rd AMOP Technical Seminar on Environmental Contamination and Response. Ottawa: Environment Canada; 2008. pp. 627-637

[89] Chhipa H, Deshmukh SK. Fungal Endophytes: Rising tools in sustainable agriculture production. In: Jha S, editor. Endophytes and Secondary Metabolites. Swizterland: Springer International Publishing; 2019. pp. 1-24. DOI: 10.1007/978-3-319-76900-4_26-1

[90] Porcel R, Aroca R, Cano C, Bago A, Ruiz-Lozano JM. Identification of a gene from the arbuscular mycorrhizal fungus *Glomus intraradices* encoding for a 14-3-3 protein that is upregulated by drought stress during the AM symbiosis. Microbial Ecology. 2006;**52**:575-582. DOI: 10.1007/s00248-006-9015-2

[91] Marulanda A, Barea JM, Azcon R. Stimulation of plant growth and drought tolerance by native microorganisms (AM fungi and bacteria) from dry environments: Mechanisms related to bacterial effectiveness. Journal of Plant Growth Regulation. 2009;**28**:115-124. DOI: 10.1007/s00344-009-9079-6

[92] Khan AL, Hamayun M, Hussain J, Kang SM, Lee IJ. The newly isolated endophytic fungus *Paraconiothyrium* sp. LK1 produces ascotoxin. Molecules. 2012;**17**:1103-1112. DOI: 10.3390/molecules17011103

[93] Khan AL, Waqas M, Hussain J, Al-Harrasi A, Lee IJ. Fungal endophyte *Penicillium janthinellum* LK5 can reduce cadmium toxicity in *Solanum lycopersicum* (Sitiens and Rhe). Biology and Fertility of Soils. 2014;**50**:75-85. DOI: 10.1007/s11274-013-1378-1

[94] Davies FT, Potter JR, Linderman RG. Drought resistance of mycorrhizal pepper plants independent of leaf P concentration response in gas exchange and water relations. Plant Physiology. 1993;**87**:45-53. DOI: 10.1111/j.1399-3054.1993.tb08789.x

[95] Khan AL, Shinwari ZK, Kim Y, Waqas M, Hamayun M, Kamran M, et al. Role of endophyte *Chaetomium globosum* lk4 in growth of *Capsicum annuum* by production of gibberellins and indole acetic acid. Pakistan Journal of Botany. 2012;**44**:1601-1607

[96] Khan AL, Waqas M, Lee IJ. Resilience of *Penicillium resedanum* LK6 and exogenous gibberellin in improving *Capsicum annuum* growth under abiotic stresses. Journal of Plant Research. 2014;**128**:259-268. DOI: 10.1007/s10265-014-0688-1

[97] Wicke B, Smeets E, Dornburg V, Vashev B, Gaiser T, Turkenburg W, et al. The global technical and economic potential of bioenergy from salt-affected soils. Energy & Environmental Science. 2011;**4**:2669-2681. DOI: 10.1039/C1EE01029H

[98] Hu Y, Schmidhalter U. Limitation of salt stress to plant growth. In: Hock B, Elstner CF, editors. Plant Toxicology. New York: Marcel Dekker Inc; 2002. pp. 91-224. DOI: 10.1201/9780203023884.ch5

[99] Baltruschat H, Fodor J, Harrach BD, Niemczyk E, Barna B, Gullner G, et al. Salt tolerance of barley induced by the root endophyte *Piriformospora indica* is associated with a strong increase in antioxidants. New Phytologist. 2008;**180**:501-510. DOI: 10.1111/j.1469-8137.2008.02583

[100] Rodriguez RJ, Henson J, Van Volkenburgh E, Hoy M, Wright L, Beckwith F, et al. Stress tolerance in plants via habitat-adapted symbiosis. The ISME Journal. 2008;**2**:404-416. DOI: 10.1038/ismej.2007.10

[101] Hussain SS, Mehnaz S, KHM S. Harnessing the plant microbiome for improved abiotic stress tolerance. In: Egamberdieva D, Ahmad P, editors. Plant Microbiome: Stress Response. Singapore: Springer Nature; 2018. pp. 21-43. DOI: 10.1007/978-981-10-5514-0_2

[102] Andreas T, Christophe C, Essaid AB. Physiological and molecular changes in plants at low temperatures. Planta. 2012;**235**:1091-1105. DOI: 10.1007/s00425-012-1641-y

[103] Redman RS, Sheehan KB, Stout RG, Rodriguez RJ, Henson JM. Thermotolerance enerated by plant/fungal symbiosis. Science. 2002;**298**:1581-1581

[104] Stierle A, Strobel G, Stierle D. Taxol and taxane production by *Taxomyces andreanae*, an endophytic fungus of pacific yew. Science. 1993;**260**:214-216. DOI: 10.1126/science.8097061

[105] Kidd P, Barcelo J, Bernal MP, Navari-Izzo F, Poschenrieder C, Shilev S, et al. Trace element behavior at the root-soil interface: Implications in phytoremediation. Environmental and Experimental Botany. 2009;**67**:243-259. DOI: 10.1016/j.envexpbot.2009.06.013

[106] Singh LP, Gill SS, Tuteja N. Unraveling the role of fungal symbionts in plant abiotic stress tolerance. Plant Signalling and Behaviour. 2011;**6**:175-191. DOI: 10.4161/psb.6.2.14146

[107] Likar M. Dark septate endophytes and mycorrhizal fungi of trees affected by pollution. In: Pirttila AM, Frank AC, editors. Endophytes of Forest Trees. Dordrecht: Springer Science+Business Media; 2011. pp. 189-201. DOI: 10.1007/978-94-007-1599-8

[108] Wang JL, Li T, Liu GY, Smith JM, Zhao ZW. Unraveling the role of dark septate endophyte (DSE) colonizing maize (Zea *mays*) under cadmium stress: Physiological, cytological and genic aspects. Scientific Reports. 2016;**6**:22028. DOI: 10.1038/srep22028

[109] Senesi N, Sposito G, Martin JP. Copper (II) and iron (III) complexation by humic acid-like polymers (melanins) from soil fungi. Science of the Total Environment. 1987;**62**:241-252

[110] Fogarty RV, Tobin JM. Fungal melanins and their interaction with metals. Enzyme and Microbial Technology. 1996;**19**:311-317. DOI: 10.1016/0141-0229(96)00002-6

[111] Bultreys A. Siderotyping, a tool to characterize, classify and identify fluorescent pseudomonads. In: Varma A, Chincholkar S, editors. Microbial Siderophores. New York: Springer; 2007. pp. 67-90. DOI: 10.1007/978-3-540-71160-5_3

[112] Miethke M, Marahiel MA. Siderophore-based iron acquisition and pathogen control. Microbiology and Molecular Biology Reviews. 2007;**71**:413-451. DOI: 10.1128/MMBR.00012-07

[113] Yamaji K, Watanabe Y, Masuya H, Shigeto A, Yui H, Haruma T. Root fungal endophytes enhance heavy-metal stress tolerance of *Clethra barbinervis* growing naturally at mining sites via growth enhancement, promotion of nutrient uptake and decrease of heavy-metal concentration. PLoS One. 2016;**11**:e0169089. DOI: 10.1371/journal.pone.0169089

Chapter 14

Endophytes Potential Use in Crop Production

Fabiana Tonial, Francine Falcão de Macedo Nava, Ana Luisa Gayger and Talita Bernardon Mar

Abstract

The endophytic microorganisms have the potential to improve the yield of agricultural crops. They can be used as biological control, plant growth promoter, or bioremediators. The action of endophytes in controlling phytopathogens, insects, and weeds that harm agriculture may be the result of microbial interactions with other organisms or the production of bioactive metabolites. Also, microorganisms can have the ability to favor plant growth and convert toxic compounds present in the soil. The presence of pollutants in the substrate reduces its quality for plant development, so bioremediation also impacts agricultural production. Therefore, prospecting endophytic microorganisms with agronomic potential may provide sustainable alternatives to increase crop yield.

Keywords: endophyte, agriculture, biological control, plant growth promoter, inoculant, bioremediator

1. Introduction

In order to apply sustainable solutions to problems related to food production, the biotechnological potential of endophytic microorganisms has been prospected in the agronomic area. The use of beneficial microorganisms in agricultural production aims for pest control, improvement of productivity and plant development, and/or recovery of ecological systems. Endophytes play a role in evolution of plant and in resistance of stresses through the production of bioactive metabolites, changes in enzyme metabolism, and gene expression related to resistance [1], and those beneficial effects of various endophytic genera may be the combined [2].

2. Biological control

Biological control of phytopathogens occurs when living microorganisms repress the development of the etiological agent in the plant [3]. Endophytes can act inducing resistance, promoting antibiosis and/or competition in consequence of the mutualistic relation with the plant [4]. These processes can occur independently, but the overlap of mechanisms may also happen [5], like is observed in the association of *Beauveria bassiana* and *Metarhizium brunneum* against the complex of *Fusarium*, the control ocurrs by competition and antibiosis [6].

The physiological definition of resistance is the delay or impediment of entry and/or subsequent activity of the pathogen in the plant [7]. Plants have numerous and efficient defense mechanisms naturally triggered when exposed to elicitors [8] that can be stimulated by the endophytes presence. The plant defense mechanisms are induced after the recognition of molecular patterns associated with pathogens/microbes (PAMPs/MAMPs), or plants' molecular patterns associated with damage (DAMPs) and effectors, by proteins or by nucleotide-binding leucine-rich repeat (NB-LRR) [9]. Endophyte induces systemic resistance on plants providing an alert state, the priming [10, 11]. Priming plants exhibits faster and stronger responses against pathogen attacks because transcription factors and signaling proteins have already accumulated in cells. This defense induction is a consequence of molecular signaling during the establishment of plant-endophyte symbiosis [10]. An example of the host-induced resistance by endophytes is the frequent isolation of *Curtobacterium flaccumfaciens* in plants without symptoms of citrus variegated chlorosis, suggesting that this endophyte has a role in the resistance of the citrus plant [12].

A reprogrammed genetic transcription occurs in plants associated with endophytes. The *Epichloë festucae* symbiosis with ryegrass (*Lolium perenne* var. Lolii) enhances gene expression of jasmonic acid (JA) precursors [13], and the expression of the systemic defense genes HvPr17b and HvHsp70 in barley is associated with the presence of the endophyte *Piriformospora indica* [14]. Further, presence of endophytes may alter pathogenesis-related proteins (PR-proteins) concentration, as chitinase, peroxidase, glucanase and cellulase in cucumber inoculated with *Trichoderma harzianum* [15], lignin and cellulose in *Theobroma cacao* in symbiosis with *Colletotrichum tropicale* [16], and PR2, PR6, PR15, and PR16 in rice with *Bacillus subtilis* [17]. The resistance response induced by symbiosis of plant-endophyte is systemic. Studies have shown that gene expression or protein production related to host defense was evidenced in plant portions distant from those inoculated with *Klebsiella pneumoniae* [18], *Rhizobium etli* [19], and *Pseudomonas fluorescens* [20].

The resistance induction is also related with the activity of defense enzymes, such as phenylalanine ammonia lyase, polyphenol oxidase, superoxide dismutase, peroxidase, ascorbate peroxidase, and guaiacol peroxidase. *Pseudomonas fluorescens* induces resistance related to the activity of lipoxygenase, catalase, aminocyclopropane carboxylate oxidase, and phenylalanine ammonia lyase [20]. *Pseudomonas fluorescens* is also capable to induce systemic resistance in plants by producing 2,4-diacetylphloroglucinol [21].

The vast majority of endophytes are biotrophic [22]. Therefore, it is important to consider that when colonization of the plant by biotrophic endophytes begins, the salicylic acid (SA) route activates defenses, so endophytes need to be able to suppress this defense by specific effectors. The expression of the Ca^{2+}/calmodulin kinase enzyme is capable to suppress the pathway of SA [23]. In addition, the possibility of recruiting gibberellic acid (GA) reduces the proportion of DELLA proteins, altering the salicylic acid and jasmonic acid (JA) signaling [24]. The suppression of the SA stimulates JA route precursors and genes, which increases resistance to chewing insects and necrotrophic fungi and promotes susceptibility to biotrophics [10, 22]. To ensure plant protection against biotrophic fungi and sucking insects, endophytes have the ability to biosynthesize compounds responsible for antibiosis; besides they can also control these organisms through mycoparasitism and competition.

The endophytes are able to biosynthesize secondary metabolites, which are important for plant colonization processes [2] and are toxic to insects, pathogens [10], and algae [25]. These compounds are classified as alkaloids (amines and amides; indole derivatives), steroids, terpenoids (sesquiterpenes, diterpenes, monoterpenes), isocoumarin derivatives, quinones, flavonoids, phenylpropanoids and lignans, peptides, phenol and phenolic acids, aliphatic compounds, and chlorinated metabolites [25]. The antagonistic activity of endophytes associated with antibiosis is described for different cultures, like potato [26, 27] and turmeric rhizome [28].

Alkaloids are an important group of metabolites produced by endophytes; some characterized classes are ergot alkaloids, diterpene indole, pyrrolizidines, and peramine. These compounds have important biological activity (antitumor, antimicrobial), including the reduction of insect performance [10, 13]. The resistance of chickpeas (*Cicer arietinum*) colonized by endophytic *Streptomyces* spp. against *Sclerotium rolfsii* is attributed to the production of phenols and flavonoids by the endophyte [29]. Nematicide compounds such as 4-vinylphenol, methionine, piperine, and palmitic acid were evidenced to have high concentrations in soybean colonized by *Bacillus simplex* [30].

The need for nutritional factors, like carbon, nitrogen, and iron, may also promote biological control. Direct parasitism is a fungus-fungus antagonism, in which one directly attacks another and utilizes its nutrients [31]. This kind of control, independent of a systemic defense response, was observed with the colonization of previously endophyte-free leaves of *Theobroma cacao* that significantly decreases necrosis in the local of inoculation when challenged with *Phytophthora* sp. [32]. Endophyte colonization can directly control a phytopathogen even without inducing defense mechanisms such as PR-proteins, like evidenced by the control of *Trichoderma stromaticum* over *Moniliophthora perniciosa* [33]. A scanning electron microscope showed that the *Trichoderma* endophytes cause deformities in the mycelia of *Pythium aphanidermatum* and *Rhizoctonia solani*, such as hyphal fragmentation, perforation, lysis, and mycelial degeneration [28]. A strain of *Trichoderma harzianum* showed in vitro growth contact points that suggest mycoparasitic activity against *Fusarium solani* [34]. Endophytic and epiphytic fungi isolated from fruits of organic *Olea europaea* were able to inhibit mycelial growth, germination, and sporulation and cause pathogenic hyphae abnormalities of *Colletotrichum acutatum*, particularly at mycelial contact [35]. In addition, endophytic fungi from *Pachystachys lutea*, mainly *Diaporthe* sp. perform antagonistic activity against *Colletotrichum* spp. and *Fusarium oxysporum*, in which contact interactions of the endophyte with the pathogen predominated [36].

Competition and direct parasitism require endophyte-pathogen contact, but those microorganisms have very little to no direct contact with the plant. Because of this, contact mechanisms are not the most important biological control pathway [4].

3. Plant growth promoters

Endophytic bacteria promote plant growth directly or indirectly: directly, producing phytohormones or enzymes [37, 38] and indirectly, contributing to plant nutrient uptake through nitrogen fixation, phosphate solubilization, or iron transformation [39, 40]. For this, the inoculant competes with an adapted indigenous microbiota; therefore, for the colonization of plant, some bacterial characteristics are important, such as motility and polysaccharide production [41–44].

Ethylene and indole-3-acetic acid (IAA) are phytohormones that are involved in almost all aspects of plant growth and development, from seed germination to shoot growth, and they control the response of the plant to stress [45, 46]. Plant growth is promoted by reducing ethylene levels and increasing IAA. Biotic and abiotic stresses result in increased ethylene production in plants, leading to inhibition of root elongation, lateral root development, and root hair formation. Plant-associated microorganisms can increase root growth and budding of plants by reducing ethylene levels [47]. The endophytic bacteria can produce an enzyme called 1-aminocyclopropane-1-carboxylate (ACC) deaminase, which hydrolyzes ACC, an ethylene immediate precursor, relieving stress and improving the growth of plants under disturbed conditions [42, 48, 49]. An inoculum from *Burkholderia phytofirmans* with the gene responsible for producing mutated ACC deaminase was unable to promote root growth of canola. The reintroduction of the ACC deaminase gene restored the microorganism's ability to promote plant growth, highlighting the importance of the enzyme in promoting host plant growth [48]. On the other hand, the IAA is an auxin, a growth hormone that promotes differential cell elongation and functions as the plant growth regulator. Besides being produced by plants, IAA may also be produced by root-associated bacteria, such as *Enterobacter* spp., *Pseudomonas* spp., and *Azospirillum* spp. [50].

Endophytic bacteria can benefit the host by producing cytokines and gibberellins. Corn endophytic bacteria, *Azospirillum lipoferum*, produce gibberellin, which is important in relieving plant stress [51]. Similarly, extracts of two endophytic bacteria from *Gynura procumbens*, *Pseudomonas resinovorans*, and *Paenibacillus polymyxa* presented cytokines [52].

Nitrogen is the most important nutrient for plant growth and productivity. Although abundant in the atmosphere, it is not available to plants. For this, it requires to be transformed by a biological nitrogen fixation (BNF) process in which N_2 is converted to NH_3 by bacteria expressing nitrogenase, such as *Burkholderia* spp., *Azoarcus* sp., *Gluconacetobacter diazotrophicus*, *Herbaspirillum* sp., *Azospirillum brasilense*, and *Paenibacillus* sp. [53–55]. Nitrogen-fixing endophytes outperform rhizosphere microorganisms in this process allowing plants to thrive even in nitrogen-limited soil environments, promoting plant health and growth [56]. Endophytic nitrogen-fixing bacteria can also increase the buildup and the nitrogen fixation rate in plants residing in soils with nitrogen limitation.

Phosphorus is an important micronutrient for the enzymatic reactions of plant physiological processes [57]. Although present in large quantities, most of the soil phosphorus is insoluble and therefore unavailable to the plant. In addition, almost 75% of phosphorus applied as fertilizer forms complexes in the soil, which prevents its absorption by the vegetable [58]. The endophytic bacteria can increase soil phosphorus availability to plants by solubilizing precipitated phosphates through mechanisms of acidification, chelation, ion exchange, and the production of organic acids [59]. They can also increase the availability of phosphorus in the soil by secreting acid phosphatase, which can mineralize organic phosphorus [60]. Furthermore, endophytic bacteria can prevent phosphate adsorption and fixation under phosphate-limiting conditions and assimilate solubilized phosphorus [61]. Studies show that endophytic populations of cactus, strawberry, sunflower, soybean, and other legumes have the ability to solubilize phosphate [62–64]. A study examined the role of phosphate-solubilizing endophytic bacteria in cactus cultivation and observed that inoculated plants grew well without added nutrients and that their growth was comparable to fertilized plants. This indicates that endophytic bacteria provide the limiting nutrient to seedlings [65].

Iron is a component of proteins that control physiological processes such as respiration and transpiration [66]. Generally, it occurs in the ferric insoluble form, unavailable to the plants. The endophytic bacteria produce iron chelators called siderophores that may bind to insoluble ferric ions allowing this nutrient uptake by plants [66–68]. The action of bacterial-produced siderophores has already been correlated with the growth of cultivars such as corn, including shoot and root biomass [69], and on tomato development in hydroponic crops [70].

The ability to promote plant growth by endophytic bacteria may be influenced by host genotype [71]. However, many endophytic bacteria can have a wide range of hosts, such as *B. phytofirmans*, which promote growth of *Arabidopsis thaliana*, grapes, corn, potatoes, grass, tomatoes, and wheat [72–74]. Similarly, the bacterial genotype also influences the capacity and potential of stimulatory effects over host plants. For example, the individual ability of different *B. phytofirmans* strains to promote growth of a single potato cultivar [75] and the plant colonization by different *Salmonella enterica* isolates were observed [76]. Therefore, colonization and growth promotion of plants by endophytic bacteria are active processes controlled by genetic factors of both partners.

4. Bioremediators

The prompt development of agriculture has made it possible to increase the food supply all over the world. However, the intensification of agricultural activities brought serious environmental impacts, which not only affect food security but also have impacts on socioeconomic aspects. These impacts comprise contribution to air pollution, impacts on land, waste of water, loss of biological and ecological diversity, and perturbation of global biogeochemical cycles. The pollutants generated by agricultural activities can affect the global or local scale. An example of global-scale agro-environmental problem is the increase in atmospheric concentrations of the greenhouse gasses (GHG) and carbon dioxide (CO_2) through deforestation and nitrous oxide (N_2O) arising from crop production. Agriculture is the largest water consumer and the main source of nitrate, ammonia, and phosphate pollution. These pollutants affect the local scale; some examples are the salinization of irrigated lands and the buildup of nitrate fertilizer residues in groundwater and surface water [77–81].

Most of the negative environmental impacts generated by the intensification of agricultural activities can be reduced or prevented [77]. The use of new technological approaches, physicochemical- or biological-based, could remove pollutants from nature. Biological-based methods are preferred due to the low cost and because they are less harmful to the environment. Atlas and Pramer [82] defined the term bioremediation as "the use of biological agents to reclaim soils and waters polluted by substances hazardous to human health and/or the environment." In other words, bioremediation is a biological-based method involving the use of living organisms, such as plants or microorganisms (bacteria, fungi, and algae), to remove pollutants from the environment [83].

Degradation of pollutants by a microorganism demands favorable conditions of nutrients, temperature, pH, and oxygen. Bacteria and fungi are commonly used in bioremediation strategies, because they are ubiquitous and capable in withstanding different environmental conditions, so they can be used for a broader range of application. There are two main mechanisms of bioremediation: biosorption and bioaccumulation. Biosorption involves sequestration of pollutants thought binding onto surfaces, such as the cell wall. Bioaccumulation involves transport and accumulation of pollutants in the cells and, in some cases, the transformation of pollutants into less harmful compounds [78, 83]. The degradation of target pollutants can also be achieved by employing nonliving subcellular entities of biological origin as bioremediators [84]. To overcome the instability due to the rapid decline in the inoculated cell amount during its competition with indigenous microorganisms, some authors have proposed solutions. For example, a new strategy for the efficient removal of phenylurea herbicides from contaminated soil uses transgenic plants. Transgenic *Arabidopsis thaliana* plants expressing a bacterial N-demethylase

(PdmAB) that demethylated isoproturon were constructed. The synergistic relationship between the transgenic plant and *Sphingobium* sp., which is capable of mineralizing the intermediate of isoproturon excreted from the transgenic plant in the rhizosphere, is an innovative strategy of treatment [85].

Endophytes can remove pollutants by employing either the biosorption or the bioaccumulation mechanisms [83, 86–90]. They have the ability of decreasing and/or removing contaminants from soil, water, sediments, and air. Endophytic fungi have a great potential to manage toxic pollutants; many studies report those fungi to clean up environmental pollutants, such as white rot fungi like *Phanerochaete chrysosporium* that can degrade pesticides, dyes, and xenobiotics [91, 92]. There are several examples of endophytic microorganisms with promising applications in bioremediation [93]. As an example, symbiotic fungal endophytes from agricultural, coastal, and geothermal native grasses colonized tomato plants and conferred disease, salt, and heat tolerance, respectively. Coastal plant endophyte colonized rice and conferred salt tolerance. In addition, coastal and geothermal plant endophytes conferred drought tolerance to monocot and eudicot hosts [88]. In leguminous plants including soybean, salinity is correlated with poor yield and reduction in plant growth [94]. Basidiomycetous endophytic fungus *Porostereum spadiceum* was reposted to produce six types of gibberellins that reduce the effects of salinity in soybean by modulating endogenous phytohormones of the seedlings [95].

Heavy metals are one example of pollutants generated by agricultural activity that bioremediators can remove. The use of some pesticides and fertilizers can introduce into the environment copper (Cu), and some insecticide and herbicides can contain lead (Pb). Fungi have emerged as potential biocatalysts to access heavy metals and transform them into less toxic compounds [92, 96]. Endophytic fungi isolated of *Portulaca oleracea* growing in metal-contaminated soils increased the biomass *Brassica napus*. The results indicated that the endophytic fungus strain had the potential to remove heavy metals from contaminated water and soils [97]. Bioremediation of Pb-contaminated soil occurs by cultivation of *Solanum nigrum* combined with *Mucor circinelloides* [22]. Endophyte isolates from *Phragmites* also showed potential to metal tolerance and absorption of Cu, Pb, and chromium (Cr) [98].

Phytoremediation is the process that uses plants associated with microorganisms to remediate contaminants from soil, sludge, sediments, wastewater, and groundwater [92, 96]. Plants naturally harbor endophytes that may have natural tolerance and adaptation toward the pollutants. Studies explored the potential of using endophytes associated with plants for removal of pollutants in this process of phytoremediation [86, 88, 96, 99]. Plants growing in metal-contaminated soils accumulate the pollutant consumed directly or indirectly by humans and animals [100, 101]. Besides the human risk, polluted soil slows plant growth and reduces the biomass accumulation, compromising some crop productivity [102, 103]. Endophytic fungi resistant to different metals, including cadmium, lead, zinc (Zn), chromium, manganese (Mn), and cobalt (Co), are associated with plant species present in contaminated sites, indicating that these microorganisms have metal bioremediation potential [83, 97–99, 104, 105]. Chromium toxicity influences a number of processes that can lead to low yield. The accumulation of Cr from industrial activities in soil is a serious threat to some crops [106–108]. To minimize the Cr effects from contaminated soils, it is possible to use plants that harbor endophytic fungi that act as bioremediators. In experiments, strains of *Aspergillus fumigatus*, *Rhizopus* sp., *Penicillium radicum*, and *Fusarium proliferatum* isolated from healthy plants were able to remove Cr from soil and culture media as well as biotransform it from highly toxic hexavalent to least toxic trivalent form, instead of simply storing it. Roots of *Lactuca sativa* colonized by those endophytes restored its normal growth into Cr-contaminated soil, making them potential candidates as biofertilizer in Cr-contaminated soil. Likewise, *Rhizopus* sp. and *F. proliferatum* reduced the translocation of Cr to the leaves, making it safer for human consumption [102]. Other biofertilizer candidates to be used in fields affected with heavy metals are the endophytic *Mucor* sp. MHR-7 that presented tolerance to chromium, manganese, cobalt, copper, and zinc by biotransformation and/or accumulation of those metals in its hyphae. Co-cultivation of MHR-7 reduced in 90% the Cr absorption and promoted growth in mustard cultivation [103].

Studies reported the use of *Mucor* sp. in another remediation strategy called phytoextraction. Phytoextraction refers to the removal of heavy metal from the soil through their uptake by a metal-accumulating plant. One limitation is the long growth cycle of those plants. One strategy is to combine plants with endophytes that promote stress tolerance to toxicity and high biomass accumulation, increasing metal accumulation in plant tissues. Oilseed rape plants combined with *Mucor* sp. strains promoted stress tolerance to Cd and Pb, increasing biomass of plants and reducing the concentrations of those metals in the soil [109]. Similar results were found using the fungal endophyte *Peyronellaea* associated with maize under heavy metal stress [110], and the *Microsphaeropsis* sp. strain isolated from *Solanum nigrum* has also been studied for their biosorption capacity of cadmium [111]. Mercury volatilization and bioaccumulation of this metal in plant tissues mediated by endophytic fungi were demonstrated with *Aspergillus* sp. A31, *Curvularia geniculata* P1, *Lindgomycetaceae* P87, and *Westerdykella* sp. P71 on maize and *Aeschynomene fluminensis* [112].

Similar to metal pollutants, triphenylmethane (TPM) dyes are water-soluble organic compounds extensively used in industrial processes and have adverse effects on living organisms. TPM is phytotoxic for several cultivated plants, such as *Sorghum bicolor*, *Triticum aestivum*, *Vigna radiata*, *Lemna minor*, and *Zea mays* [83]. A *Diaporthe* sp. endophyte presented biosorption and biodegradation potential on TPM dyes. The microorganism removed TPM dyes through biodegradation and biosorption [113]. Other endophytes, *Pleurotus ostreatus, Polyporus picipes*, and *Gloeophyllum odoratum,* also demonstrate potential to remove TPM dye [114, 115].

5. Conclusion

Endophytic microorganisms are inestimable natural resources for solving problems in different areas such as human health, veterinary, industrial and ecological systems, and agronomy. In contrast to current agricultural practices that degrade systems and produce food with high concentrations of various contaminants, endophytes are a sustainable alternative to increase crop productivity. For this, they can be exploited by the ability to control pests, to promote plant growth, and by the bioremediation potential. This is possible because these microorganisms are able to induce resistance mechanisms in the host, release compounds with biological activity, compete for space and nutrients with pathogens, provide nutritional elements present in the soil, stimulate the production of phytohormones and cytokines, and neutralize the presence of pollutants in the system. Ultimately, bioprospecting and the use of endophytes in agriculture are a viable alternative to the need of increased food production with quality and sustainably.

Acknowledgements

The authors would like to acknowledge CAPES (Coordenação de Aperfeiçoamento de Pessoal de Nível Superior), Brazil.

Author details

Fabiana Tonial*, Francine Falcão de Macedo Nava, Ana Luisa Gayger and Talita Bernardon Mar
Passo Fundo University, Passo Fundo, Brazil

*Address all correspondence to: fabianatonial@upf.br

References

[1] Yuan Z, Zhang C, Lin F. Role of diverse non-systemic fungal endophytes in plant performance and response to stress: Progress and approaches. Journal of Plant Growth Regulation. 2010;**29**:116-126. DOI: 10.1007/s00344-009-9112-9

[2] Rosenblueth M, Martínez-Romero E. Bacterial endophytes and their interactions with hosts. Molecular Plant-Microbe Interactions. 2006;**19**:827-837. DOI: 10.1094/MPMI-19-0827

[3] Heimpel GE, Mills N, editors. Biological Control—Ecology and Applications. 1st ed. Cambridge: Cambridge University Press; 2017. DOI: 10.1017/9781139029117

[4] Card S, Johnson L, Teasdale S, Caradus J. Deciphering endophyte behaviour: The link between endophyte biology and efficacious biological control agents. FEMS Microbiology Ecology. 2016;**92**(8):1-19. DOI: 10.1093/femsec/fiw114

[5] Latz MAC, Jensen B, Collinge DB, Jørgensen HJL. Endophytic fungi as biocontrol agents: Elucidating mechanisms in disease suppression. Plant Ecology & Diversity. 2018;**11**:555-567. DOI: 10.1080/17550874.2018.1534146

[6] Jaber LR, Alananbeh KM. Fungal entomopathogens as endophytes reduce several species of *Fusarium* causing crown and root rot in sweet pepper (*Capsicum annuum* L.). Biological Control. 2018;**126**:117-126. DOI: 10.1016/j.biocontrol.2018.08.007

[7] Walters D, Newton A, Lyon G. Induced Resistance for Plant Defence—A Sustainable Approach to Crop Protection. Oxford: Blackwell Publishing; 2007. p. 272

[8] Agrios GN. Plant Pathology. 5th ed. San Diego: Elsevier Academic Press; 2005. p. 952

[9] Jones JDG, Dangl JL. The plant immune system. Nature. 2006;**444**:323-329. DOI: nature05286

[10] Bastias DA, Martínez-Ghersa MA, Ballaré CL, Gundel PE. Epichloë fungal endophytes and plant defenses: Not just alkaloids. Trend in Plant Science. 2017;**22**(11):939-948. DOI: 10.1016/j.tplants.2017.08.005

[11] Iason GR, Taylor J, Helfer S. Community-based biotic effects as determinants of tree resistance to pests and pathogens. Forest Ecology and Management. 2018;**417**:301-312. DOI: 10.1016/j.foreco.2018.01.037

[12] Araújo WL, Marcon J, Maccheroni W Jr, Van Elsas JD, Van Vuurde JW, Azevedo JL. Diversity of endophytic bacterial populations and their interaction with *Xylella fastidiosa* in citrus plants. Applied and Environmental Microbiology. 2002;**68**:4906-4914. DOI: 10.1128/aem.68.10.4906-4914.2002

[13] Bastias DA, Ueno AC, Assefh CRM, Alvarez AE, Young CA, Gundel PE. Metabolism or behavior: Explaining the performance of aphids on alkaloid-producing fungal endophytes in annual ryegrass (*Lolium multiflorum*). Oecologia. 2017;**185**:245-256. DOI: 10.1007/s00442-017-3940-2

[14] Waller F, Mukherjee K, Deshmukh SD, Achatz B, Sharma M, Schäfer P, et al. Systemic and local modulation of plant responses by *Piriformospora indica* and related *Sebacinales* species. Journal of Plant Physiology. 2008;**165**:60-70. DOI: 10.1016/j.jplph.2007.05.017

[15] Yedidia I, Benhamou N, Kapulnik Y, Chet I. Induction and accumulation of PR proteins activity during early stages of root colonization by the mycoparasite *Trichoderma harzianum* strain T-203.

Plant Physiology and Biochemistry. 2000;**38**:863-873. DOI: 10.1016/S0981-9428(00)01198-0

[16] Mejía LC, Herre EA, Sparks JP, Winter K, García MN, Van Bael SA, et al. Pervasive effects of a dominant foliar endophytic fungus on host genetic and phenotypic expression in a tropical tree. Frontiers in Microbiology. 2014;**5**:479. DOI: 10.3389/fmicb.2014.00479

[17] Ji SH, Gururani MA, Chun SC. Expression Analysis of Rice Pathogenesis-related Proteins Involved in Stress Response and Endophytic Colonization Properties of gfp-tagged *Bacillus subtilis* CB-R05. Applied Biochemistry and Biotechnology. 2014;**174**:231-341. DOI: 10.1007/s12010-014-1047-3

[18] Liu D, Chen L, Zhu X, Wang Y, Xuan Y, Liu X, et al. *Klebsiella pneumoniae* SnebYK mediates resistance against *Heterodera glycines* and promotes soybean growth. Frontiers in Microbiology. 2018;**9**:1134. DOI: 10.3389/fmicb.2018.01134

[19] Martinuz A, Schouten A, Sikora RA. Systemically induced resistance and microbial competitive exclusion: Implications on biological control. Phytopathology. 2012;**102**:260-266. DOI: 10.1094/PHYTO-04-11-0120

[20] Cabanás CGL, Schilirò E, Valverde-Corredor A, Mercado-Blanco J. The biocontrol endophytic bacterium *Pseudomonas fluorescens* PICF7 induces systemic defense responses in aerial tissues upon colonization of olive roots. Frontiers in Microbiology. 2014;**5**. DOI: 10.3389/fmicb.2014.00427

[21] Siddiqui IA, Shaukat SS. Suppression of root-knot disease by *Pseudomonas fluorescens* CHA0 in tomato: Importance of bacterial secondary metabolite, 2,4-diacetylpholoroglucinol. Soil Biology and Biochemistry. 2003;**35**:1615-1623. DOI: 10.1016/j.soilbio.2003.08.006

[22] Jacobs S, Zechmann B, Molitor A, Trujillo M, Petutschnig E, Lipka V, et al. Broad-spectrum suppression of innate immunity is required for colonization of *Arabidopsis* roots by the fungus *Piriformospora indica*. Plant Physiology. 2011;**156**:726-740. DOI: 10.1104/pp.111.176446

[23] Molitor A, Kogel KH. Induced resistance triggered by *Piriformospora indica*. Plant Signaling and Behavior. 2009;**4**:215-216. DOI: 10.4161/psb.4.3.7814

[24] Cosme M, Lu J, Erb M, Stout MJ, Franken P, Wurst S. A fungal endophytes helps plants to tolerate root herbivory through changes in gibberellin and jasmonate signaling. New Phytologist. 2016;**211**:1065-1076. DOI: 10.1111/nph.13957

[25] Tan RX, Zou WX. Endophytes: A rich source of functional metabolites. Natural Products Reports. 2001;**18**: 448-459. DOI: 10.1039/b100918o

[26] Sessitsch A, Reitter B, Berg G. Endophytic bacterial communities of field-grown potato plants and their plant-growth-promoting and antagonistic abilities. Canadian Journal of Microbiology. 2004;**50**:239-249. DOI: 10.1139/w03-118

[27] Sturz AV, Kimpinski J. Endoroot bacteria derived from marigolds (*Tagetes* spp.) can decrease soil population densities of root-lesion nematodes in the potato root zone. Plant and Soil. 2004;**262**:241-249. DOI: 10.1023/B:PLSO.0000037046.86670.a3

[28] Vinayarani G, Prakash HS. Fungal endophytes of turmeric (*Curcuma longa* L.) and their biocontrol potential against pathogens *Pythium aphanidermatum* and *Rhizoctonia solani*. World Journal of Microbiology

and Biotechnology. 2018;**34**:9-17. DOI: 10.1007/s11274-018-2431-x

[29] Singh SP, Gaur R. Endophytic *Streptomyces* spp. underscore induction of defense regulatory genes and confers resistance against *Sclerotium rolfsii* in Chickpea. Biological Control. 2017;**104**:44-56. DOI: 10.1016/j.biocontrol.2016.10.011

[30] Kang W, Zhu X, Wang Y, Chen L, Duan Y. Transcriptomic and metabolomic analyses reveal that bacteria promote plant defense during infection of soybean cyst nematode in soybean. BMC Plant Biology. 2018;**18**:86. DOI: 10.1186/s12870-018-1302-9

[31] Viterbo A, Horwitz BA. Mycoparasitism. In: Borkovich KA, Ebbole DJ, editors. Cellular and Molecular of Filamentous Fungi. Washington: ASM Press; 2010. pp. 676-677. DOI: 10.1128/9781555816636.ch42

[32] Arnold AE, Mejía LC, Kyllo D, Rojas EI, Maynard Z, Robbins N, et al. Fungal endophytes limit pathogen damage in a tropical tree. Proceedings of the National Academy of Sciences. 2003;**100**:15649-15654. DOI: 10.1073/pnas.2533483100

[33] Souza JT, Bailey BA, Pomella AWV, Erbe EF, Murphy CA, Bae G, et al. Colonization of cacao seedlings by *Trichoderma stromaticum*, a mycoparasite of the witches' broom pathogen, and its influence on plant growth and resistance. Biological Control. 2008;**46**:36-45. DOI: 10.1016/j.biocontrol.2008.01.010

[34] Amira MB, Lopez D, Mohamed AT, Khouaja A, Chaar H, Fumanal B, et al. Beneficial effect of *Trichoderma harzianum strain* Ths97 in biocontrolling *Fusarium solani* causal agent of root rot disease in olive trees. Biological Control. 2017;**110**:70-78. DOI: 10.1016/j.biocontrol.2017.04.008

[35] Preto G, Martins F, Pereira JA, Baptista P. Fungal community in olive fruits of cultivars with different susceptibilities to anthracnose and selection of isolates to be used as biocontrol agents. Biological Control. 2017;**110**:1-9. DOI: 10.1016/j.biocontrol.2017.03.011

[36] Ribeiro AS, Polonio JC, Costa AT, Santos CM, Rhoden SA, Azevedo JL, et al. Bioprospection of culturable endophytic fungi associated with the ornamental plant *Pachystachys lutea*. Current Microbiology. 2018;**75**:588-596. DOI: 10.1007/s00284-017-1421-9

[37] Glick BR. The enhancement of plant growth by free living bacteria. Canadian Journal of Microbiology. 1995;**41**:109-117. DOI: 10.1139/m95-015

[38] Madhaiyan M, Suresh Reddy BV, Anandham R, et al. Plant growth–promoting *Methylobacterium* induces defense responses in groundnut (*Arachis hypogaea* L.) compared with rot pathogens. Current Microbiology. 2006;**53**:270. DOI: 10.1007/s00284-005-0452-9

[39] Wakelin SA, Warren RA, Harvey PR, et al. Phosphate solubilization by *Penicillium* spp. closely associated with wheat roots. Biology and Fertility of Soils. 2004;**40**:36-43. DOI: 10.1007/s00374-004-0750-6

[40] Rasouli-Sadaghiani M, Malakouti MJ, Khavazi K, Miransari M. Siderophore efficacy of fluorescent pseudomonades affecting labeled iron (^{59}Fe) uptake by wheat (*Triticum aestivum* L.) genotypes differing in Fe efficiency. In: Miransari M, editor. Use of Microbes for the Alleviation of Soil Stresses. New York: Springer; 2014. pp. 121-132. DOI: 10.1007/978-1-4939-0721-2_7

[41] Berninger T, Mitter B, Preininger C. The smaller, the better? The size

effect of alginate beads carrying plant growth-promoting bacteria for seed coating. Journal of Microencapsulation. 2016;**33**:127-136. DOI: 10.3109/02652048.2015.1134690

[42] Santoyo G, Moreno-Hagelsieb G, Orozco-Mosqueda MC, Glick BR. Plant growth-promoting bacterial endophytes. Microbiological Research. 2016;**183**:92-99. DOI: 10.1016/j.micres.2015.11.008

[43] Shahzad R, Khan AL, Bilal S, Asaf S, Lee IJ. Plant growth-promoting endophytic bacteria versus pathogenic infections: An example of *Bacillus amyloliquefaciens* RWL-1 and *Fusarium oxysporum* f. sp. *lycopersici* in tomato. PeerJ. 2017;**5**:e3107. DOI: 10.7717/peerj.3107

[44] Liffourrena AS, Lucchesi GI. Alginate-perlite encapsulated *Pseudomonas putida* A (ATCC 12633) cells: Preparation, characterization and potential use as plant inoculants. Journal of Biotechnology. 2018;**278**:28-33. DOI: 10.1016/j.jbiotec.2018.04.019

[45] Woodward AW, Bartel B. Auxin: Regulation, action, and interaction. Annals of Botany. 2005;**95**:707-735. DOI: 10.1093/aob/mci083

[46] Sun L, Wang X, Li Y. Increased plant growth and copper uptake of host and non-host plants by metal-resistant and plant growth-promoting endophytic bactéria. International Journal of Phytoremediation. 2016;**18**:494-501. DOI: 10.1080/15226514.2015.1115962

[47] Glick BR, Todorovic B, Czarny J, Cheng Z, Duan J, McConkey B. Promotion of plant growth by bacterial ACC deaminase. Critical Reviews in Plant Sciences. 2007;**26**:227-242. DOI: 10.1080/07352680701572966

[48] Sun Y, Cheng Z, Glick BR. The presence of a 1-aminocyclopropane-1-carboxylate (ACC) deaminase deletion mutation alters the physiology of the endophytic plant growth-promoting bacterium *Burkholderia phytofirmans* PsJN. FEMS Microbiology Letters. 2009;**296**:131-136. DOI: 10.1111/j.1574-6968.2009.01625.x

[49] Nikolic B, Schwab H, Sessitsch A. Metagenomic analysis of the 1-aminocyclopropane-1-carboxylate deaminase gene (*acdS*) operon of an uncultured bacterial endophyte colonizing *Solanum tuberosum* L. Archives of Microbiology. 2011;**193**:665. DOI: 10.1007/s00203-011-0703-z

[50] Spaepen S, Vanderleyden J, Remans R. Indole-3-acetic acid in microbial and microorganism-plant signaling. FEMS Microbiology Reviews. 2007;**31**:425-448. DOI: 10.1111/j.1574-6976.2007.00072.x

[51] Cohena AC, Travaglia CN, Bottini R, Piccoli PN. Participation of abscisic acid and gibberellins produced by endophytic *Azospirillum* in the alleviation of drought effects in maize. Botany. 2009;**87**:462-487. DOI: 10.1139/B09-023

[52] Bhore SJ, Ravichantar N, Loh SY. Screening of endophytic bacteria isolated from leaves of *Sambung Nyawa Gynura procumbens* Lour Merr. for cytokinin-like compounds. Bioinformation. 2010;**5**:191-197. DOI: 10.6026/97320630005191

[53] Scherling C, Ulrich K, Ewald D, Weckwerth W. A metabolic signature of the beneficial interaction of the endophyte *Paenibacillus* sp. isolate and in vitro–grown poplar plants revealed by metabolomics. APS Publications. 2009;**22**:1032-1037. DOI: 10.1094/MPMI-22-8-1032

[54] Singh A. Soil salinization and waterlogging: A threat to environment and agricultural sustainability. Ecological Indicators. 2015;**57**:128-130. DOI: 10.1016/j.ecolind.2015.04.027

[55] Ivleva NB, Groat J, Staub JM, Stephens M. Expression of active subunit of nitrogenase via integration into plant organelle genome. PLoS One. 2016;**11**(8):e0160951. DOI: 10.1371/journal.pone.0160951

[56] Hurek T, Reinhold-Hurek B. *Azoarcus* sp. strain BH72 as a model for nitrogen-fixing grass endophytes. Journal of Biotechnology. 2003;**106**: 169-178. DOI: 10.1016/j.jbiotec.2003.07.010

[57] Ahemad M. Phosphate-solubilizing bacteria-assisted phytoremediation of metalliferous soils: A review. 3 Biotech. 2015;**5**:111-121. DOI: 10.1007/s13205-014-0206-0

[58] Ezawa T, Smith SE, Smith FA. P metabolism and transport in AM fungi. Plant and Soil. 2002;**224**:221-230. DOI: 10.1023/A:102025832

[59] Nautiyal CS, Bhadauria S, Kumar P, Lal H, Mondal R, Verma D. Stress induced phosphate solubilization in bacteria isolated from alkaline soils. FEMS Microbiology Letters. 2000;**182**:291-296. DOI: 10.1111/j.1574-6968.2000.tb08910.x

[60] Van Der Heijden MG, Bardgett RD, Van Straalen NM. The unseen majority: Soil microbes as drivers of plant diversity and productivity in terrestrial ecosystems. Ecology Letters. 2008;**11**:296-310. DOI: 10.1111/j.1461-0248.2007.01139.x

[61] Khan KS, Joergensen RG. Changes in microbial biomass and P fractions in biogenic household waste compost amended with inorganic P fertilizers. Bioresource Technology. 2009;**100**:303-309. DOI: 10.1016/j.biortech.2008.06.002

[62] Kuklinsky-Sobral J, Araújo WL, Mendes R, et al. Isolation and characterization of soybean-associated bacteria and their potential for plant growth promotion. Plant and Soil. 2005;**273**:91-99. DOI: 10.1007/s11104-004-6894-1

[63] Dias AC, Costa FE, Andreote FD, Lacava PT, Teixeira MA, Assumpção LC, et al. Isolation of micropropagated strawberry endophytic bacteria and assessment of their potential for plant growth promotion. World Journal of Microbiology and Biotechnology. 2009;**25**:189-195. DOI: 10.1007/s11274-008-9878-0

[64] Puente ME, Li CY, Bashan Y. Endophytic bacteria in cacti seeds can improve the development of cactus seedlings. Environmental and Experimental Botany. 2009;**66**:402-408. DOI: 10.1016/j.envexpbot.2009.04.007

[65] Puente ME, Li CY, Bashan Y. Rock-degrading endophytic bacteria in cacti. Environmental and Experimental Botany. 2009;**66**:389-401. DOI: 10.1016/j.envexpbot.2009.04.010

[66] Ma Y, Rajkumar M, Zhang C, Freitas H. Beneficial role of bacterial endophytes in heavy metal phytoremediation. Journal of Environmental Management. 2016;**174**:14-25. DOI: 10.1016/j.jenvman.2016.02.047

[67] Rajkumar M, Ae N, Freitas H. Endophytic bacteria and their potential to enhance heavy metal phytoextraction. Chemosphere. 2009;**77**:153-160. DOI: 10.1016/j.chemosphere.2009.06.047

[68] Crowley DE, Römheld V, Marschner H, et al. Root-microbial effects on plant iron uptake from siderophores and phytosiderophores. Plant and Soil. 1992;**142**:1-7. DOI: 10.1007/BF00010169

[69] Marques AP, Pires C, Moreira H, Rangel AO, Castro PM. Assessment of the plant growth promotion abilities of six bacterial isolates using *Zea mays*

as indicator plant. Soil Biology and Biochemistry. 2010;**42**:1229-1235. DOI: 10.1016/j.soilbio.2010.04.014

[70] Radzki W, Mañero FG, Algar E, García JL, García-Villaraco A, Solano BR. Bacterial siderophores efficiently provide iron to iron-starved tomato plants in hydroponics culture. Antonie Van Leeuwenhoek. 2013;**104**:321-330. DOI: 10.1007/s10482-013-9954-9

[71] Long HH, Schmidt DD, Baldwin IT. Native bacterial endophytes promote host growth in a species-specific manner; phytohormone manipulations do not result in common growth responses. PLoS One. 2008;**3**:e2702. DOI: 10.1371/journal.pone.0002702

[72] Pillay V, Nowak J. Inoculum density, temperature, and genotype effects on in vitro growth promotion and epiphytic and endophytic colonization of tomato (*Lycopersicon esculentum* L.) seedlings inoculated with a pseudomonad bacterium. Canadian Journal of Microbiology. 1997;**43**:354-361. DOI: 10.1139/m97-049

[73] Sessitsch A, Coenye T, Sturz A, Vandamme P, Barka EA, Salles J, et al. *Burkholderia phytofirmans* sp. nov., a novel plant-associated bacterium with plant-beneficial properties. International Journal of Systematic and Evolutionary Microbiology. 2005;**55**:1187-1192. DOI: 10.1099/ijs.0.63149-0

[74] Sheibani-Tezerji R, Rattei T, Sessitsch A, Trognitz F, Mitter B. Transcriptome profiling of the endophyte *Burkholderia phytofirmans* PsJN indicates sensing of the plant environment and drought stress. MBio. 2015;**6**:e00621-e00615. DOI: 10.1128/mBio.00621-15

[75] Trognitz F, Scherwinski K, Fekete A, Schmidt S, Eberl L, Rodewald J, et al. Interaction between potato and the endophyte *Burkholderia phytofirmans*. In: Abwehrstrategien gegen biotische Schaderreger Züchtung von Hackfrüchten und Sonderkulturen. Tagungsband der 59. Jahrestagung der Vereinigung der Pflanzenzüchter und Saatgutkaufleute Österreichs; 25-27 November 2008; Raumberg-Gumpenstein. Austria. 2008. pp. 63-66

[76] Dong Z, Canny MJ, McCully ME, Roboredo MR, Cabadilla CF, Ortega E, et al. A nitrogen-fixing endophyte of sugarcane stems (a new role for the apoplast). Plant Physiology. 1994;**105**:1139-1147. DOI: 10.1104/pp.105.4.1139

[77] Bruinsma J. World Agriculture: Towards 2015/2030. London: Routledge; 2003

[78] Henis Y. Bioremediation in agriculture: Dream or reality? In: Rosen D, Tel-Or E, Hadar Y, Chen Y, editors. Modern Agriculture and the Environment: Proceedings of an International Conference, held in Rehovot, Israel, 2-6 October 1994, under the auspices of the Faculty of Agriculture, the Hebrew University of Jerusalem. Dordrecht: Springer Netherlands; 1997. pp. 481-489

[79] Parris K. Impact of agriculture on water pollution in OECD countries: Recent trends and future prospects. International Journal of Water Resources Development. 2011;**27**:33-52. DOI: 10.1080/07900627.2010.531898

[80] Tilman D. Global environmental impacts of agricultural expansion: the need for sustainable and efficient practices. Proceedings of the National Academy of Sciences. 1999;**96**:5995-6000. DOI: 10.1073/pnas.96.11.5995

[81] Yu J, Wu J. The Sustainability of agricultural development in China: The agriculture–environment nexus. Sustainability. 2018;**10**:1776. DOI: 10.3390/su10061776

[82] Atlas RM, Pramer D. Focus on bioremediatison. ASM News. 1990;**56**:352-354

[83] Sim CSF, Chen SH, Ting ASY. Endophytes: Emerging tools for the bioremediation of pollutants. In: Bharagava RN, Chowdhary P, editors. Emerging and Eco-Friendly Approaches for Waste Management. Singapore: Springer Singapore; 2019. pp. 189-217

[84] Zaccaria M, Dawson W, Cristiglio V, Reverberi M, Ratcliff LE, Nakajima T, et al. Designing a bioremediator: Mechanistic models guide cellular and molecular specialization. Current Opinion in Biotechnology. 2020;**62**:98-105. DOI: 10.1016/j.copbio.2019.09.006

[85] Yan X, Huang J, Xu X, Chen D, Xie X, Tao Q, et al. Enhanced and complete removal of phenylurea herbicides by combinational transgenic plant-microbe remediation. Applied and Environmental Microbiology. 2018;**84**:e00273-e00218. DOI: 10.1128/AEM.00273-18

[86] Anyasi RO, Atagana HI. Endophyte: Understanding the microbes and its applications. Pakistan Journal of Biological Sciences. 2019;**22**:154-167. DOI: 10.3923/pjbs.2019.154.167

[87] Manzotti A, Jørgensen HJL, Collinge DB, Jensen B. Phytohormones in plant-endophyte interactions: Investigating the role of these compounds in the recruitment of tomato root fungal endophytes. Plant Biology. 2017;**9**:89-105. DOI: 10.1007/978-3-319-65897-1_6

[88] Rodriguez RJ, Henson J, Van Volkenburgh E, Hoy M, Wright L, Beckwith F, et al. Stress tolerance in plants via habitat-adapted symbiosis. The ISME Journal. 2008;**2**:404-416. DOI: 10.1038/ismej.2007.106

[89] Tonial F, Maia BHLNS, Gomes-Figueiredo JA, Sobottka AM, Bertol CD, Nepel A, et al. Influence of culturing conditions on bioprospecting and the antimicrobial potential of endophytic fungi from *Schinus terebinthifolius*. Current Microbiology. 2016;**72**:173-183. DOI: 10.1007/s00284-015-0929-0

[90] Savi DC, Aluizio R, Glienke C. Brazilian plants: An unexplored source of endophytes as producers of active metabolites. Planta Medica. 2019;**85**:619-636. DOI: 10.1055/a-0847-1532

[91] Gadd GM. Geomycology: Biogeochemical transformations of rocks, minerals, metals and radionuclides by fungi, bioweathering and bioremediation. Mycological Research. 2007;**111**:3-49. DOI: 10.1016/j.mycres.2006.12.001

[92] Krishnamurthy YL, Naik BS. Endophytic fungi bioremediation. In: Endophytes: Crop Productivity and Protection. Switzerland: Springer Nature Switzerland AG; 2017. pp. 47-60

[93] Stępniewska Z, Kuźniar A. Endophyticmicroorganisms—Promising applications in bioremediation of greenhouse gases. Applied Microbiology and Biotechnology. 2013;**97**:9589-9596. DOI: 10.1007/s00253-013-5235-9

[94] Dinler BS, Antoniou C, Fotopoulos V. Interplay between GST and nitric oxide in the early response of soybean (*Glycine max* L.) plants to salinity stress. Journal of Plant Physiology. 2014;**171**:1740-1747. DOI: 10.1016/j.jplph.2014.07.026

[95] Hamayun M, Hussain A, Khan SA, Kim H-Y, Khan AL, Waqas M, et al. Gibberellins producing endophytic fungus *Porostereum spadiceum* AGH786 rescues growth of salt affected soybean. Frontiers in Microbiology. 2017;**8**:686. DOI: 10.3389/fmicb.2017.00686

[96] Dixit R, Malaviya D, Pandiyan K, Singh UB, Sahu A, Shukla R, et al. Bioremediation of heavy metals from soil

and aquatic environment: An overview of principles and criteria of fundamental processes. Sustainability. 2015;7:2189-2212. DOI: 10.3390/su7022189

[97] Deng Z, Zhang R, Shi Y, La H, Tan H, Cao L. Characterization of Cd-, Pb-, Zn-resistant endophytic *Lasiodiplodia* sp. MXSF31 from metal accumulating *Portulaca oleracea* and its potential in promoting the growth of rape in metal-contaminated soils. Environmental Science and Pollution Research. 2014;**21**:2346-2357. DOI: 10.1007/s11356-013-2163-2

[98] Sim CSF, Tan WS, Ting ASY. Endophytes from *Phragmites* for metal removal: Evaluating their metal tolerance, adaptive tolerance behaviour and biosorption efficacy. Desalination and Water Treatment. 2016;**57**:6959-6966. DOI: 10.1080/19443994.2015.1013507

[99] Fu X-Y, Zhao W, Xiong A-S, Tian Y-S, Zhu B, Peng R-H, et al. Phytoremediation of triphenylmethane dyes by overexpressing a *Citrobacter* sp. triphenylmethane reductase in transgenic Arabidopsis. Applied Microbiology and Biotechnology. 2013;**97**:1799-1806. DOI: 10.1007/s00253-012-4106-0

[100] Wang S, Shi X. Molecular mechanisms of metal toxicity and carcinogenesis. In: Shi X, Castranova V, Vallyathan V, Perry WG, editors. Molecular Mechanisms of Metal Toxicity and Carcinogenesis. Boston, MA: Springer US; 2001. pp. 3-9

[101] Clemens S, Ma JF. Toxic heavy metal and metalloid accumulation in crop plants and foods. Annual Review of Plant Biology. 2016;**67**:489-512. DOI: 10.1146/annurev-arplant-043015-11230

[102] Bibi S, Hussain A, Hamayun M, Rahman H, Iqbal A, Shah M, et al. Bioremediation of hexavalent chromium by endophytic fungi; safe and improved production of *Lactuca sativa* L. Chemosphere. 2018;**211**:653-663. DOI: 10.1016/j.chemosphere.2018.07.197

[103] Zahoor M, Irshad M, Rahman H, Qasim M, Afridi SG, Qadir M, et al. Alleviation of heavy metal toxicity and phytostimulation of *Brassica campestris* L. by endophytic *Mucor* sp. MHR-7. Ecotoxicology and Environmental Safety. 2017;**142**:139-149. DOI: 10.1016/j.ecoenv.2017.04.005

[104] Sun L, Cao X, Li M, Zhang X, Li X, Cui Z. Enhanced bioremediation of lead-contaminated soil by *Solanum nigrum* L. with *Mucor circinelloides*. Environmental Science and Pollution Research. 2017;**24**:9681-9689. DOI: 10.1007/s11356-017-8637-x

[105] An H, Liu Y, Zhao X, Huang Q, Yuan S, Yang X, et al. Characterization of cadmium-resistant endophytic fungi from *Salix variegata* Franch in three gorges reservoir region, China. Microbiological Research. 2015;**176**:29-37. DOI: 10.1016/j.micres.2015.03.013

[106] Tchounwou PB, Yedjou CG, Patlolla AK, Sutton DJ. Heavy metal toxicity and the environment. In: Luch A, editor. Molecular, Clinical and Environmental Toxicology: Volume 3: Environmental Toxicology. Basel: Springer Basel; 2012. pp. 133-164

[107] Gill RA, Ali B, Cui P, Shen E, Farooq MA, Islam F, et al. Comparative transcriptome profiling of two *Brassica napus* cultivars under chromium toxicity and its alleviation by reduced glutathione. BMC Genomics. 2016;**17**:885. DOI: 10.1186/s12864-016-3200-6

[108] López-Bucio J, Hernández-Madrigal F, Cervantes C, Ortiz-Castro R, Carreón-Abud Y, Martínez-Trujillo M. Phosphate relieves chromium toxicity in *Arabidopsis thaliana* plants by interfering with chromate uptake.

Biometals. 2014;**27**:363-370. DOI: 10.1007/s10534-014-9718-7

[109] Zhu S-C, Tang J-X, Zeng X-X, Wei B-J, Yang S-D, Huang B. Isolation of *Mucor circinelloides* Z4 and *Mucor racemosus* Z8 from heavy metal-contaminated soil and their potential in promoting phytoextraction with Guizhou oilseed rap. Journal of Central South University. 2015;**22**:88-94. DOI: 10.1007/s11771-015-2498-6

[110] Shen M, Liu L, Li D-W, Zhou W-N, Zhou Z-P, Zhang C-F, et al. The effect of endophytic *Peyronellaea* from heavy metal-contaminated and uncontaminated sites on maize growth, heavy metal absorption and accumulation. Fungal Ecology. 2013;**6**:539-645. DOI: 10.1016/j.funeco.2013.08.001

[111] Xiao X, Luo S, Zeng G, Wei W, Wan Y, Chen L, et al. Biosorption of cadmium by endophytic fungus (EF) *Microsphaeropsis* sp. LSE10 isolated from cadmium hyperaccumulator *Solanum nigrum* L. Bioresource Technology. 2010;**101**:1668-1674. DOI: 10.1016/j.biortech.2009.09.083

[112] Pietro-Souza W, de Campos Pereira F, Mello IS, Stachack FFF, Terezo AJ, Cunha CN, et al. Mercury resistance and bioremediation mediated by endophytic fungi. Chemosphere. 2020;**240**:124874. DOI: 10.1016/j.chemosphere.2019.124874

[113] Ting ASY, Lee MVJ, Chow YY, Cheong SL. Novel exploration of endophytic *Diaporthe* sp. for the biosorption and biodegradation of triphenylmethane dyes. Water, Air, & Soil Pollution. 2016;**227**:109. DOI: 10.1007/s11270-016-2810-6

[114] Przystaś W, Zabłocka-Godlewska E, Grabińska-Sota E. Biological removal of azo and triphenylmethane dyes and toxicity of process by-products. Water, Air, & Soil Pollution. 2012;**223**:1581-1592. DOI: 10.1007/s11270-011-0966-7

[115] Gangadevi V, Muthumary J. Isolation of *Colletotrichum gloeosporioides*, a novel endophytic taxol-producing fungus from the leaves of a medicinal plant, *Justicia gendarussa*. Mycologia Balcanica. 2008;**5**:1-4. DOI: 10.1.1.457.1000

Chapter 15

Application Potentials of Plant Growth Promoting Rhizobacteria and Fungi as an Alternative to Conventional Weed Control Methods

Adnan Mustafa, Muhammad Naveed, Qudsia Saeed, Muhammad Nadeem Ashraf, Azhar Hussain, Tanveer Abbas, Muhammad Kamran, Nan-Sun and Xu Minggang

Abstract

Weeds are the plants usually grown on unwanted places and are notorious for causing interruptions in agricultural settings. Remarkable yield losses have been reported in fields infested with weeds worldwide. So far, these weeds cause about 34% of losses to yields of major agricultural crops and pose threats to economic condition of the farmers. Conventionally, weed control was achieved by the use of chemical herbicides and traditional agronomic practices. But these methods are no more sustainable as the magnitude of threats imposed by these conventionally outdated methods such as chemical herbicides is greater than the benefits achieved and their continuous use has disturbed biodiversity and weed ecology along with herbicide resistance in some weeds. Herbicide residues are held responsible for human health hazards as well. Therefore the future of weed control is to rely on alternative approaches which may be biological agents such as bacteria and fungi. This chapter highlights the potentials of using bacterial and fungal biocontrol agents against weeds in farmer fields. Moreover, detailed review on merits and demerits of conventional weed control methods is discussed in this chapter.

Keywords: biological weed control, PGPR, fungi, environment, human health, economic losses

1. Introduction

Agriculture is an approach of deploying natural resources to grow the desired plants. Since the induction of green revolution in the 1950s, the food production has been substantially increased that helped to meet food demands for the ever-increasing world population [1]. Improved irrigation practices, tillage implements, fertilizers, and farm operations were some of the key outputs of green revolution. Nevertheless these practices have paved the way of agricultural sustainability yet there are some concerns associated with these practices as, improved irrigation have given rise to salinity of soils, intensive tillage causes deterioration of soil structure, loss of soil organic carbon and destruction of natural habitats of different flora, higher yielding crop cultivars depleted soil nutrients. With all of the outputs of green revolution, introduction of pests is also acknowledged [2]. Disturbance in agricultural production due to invasion of other living organisms for their own existence is a natural phenomenon which cannot be stopped. These living organisms that survive on others are called as pests which include insects, plant pathogens, nematodes, rodents, and weeds.

Among the agricultural crop pests, weeds are the most potent crop pests reducing crop yields by almost 34% followed by animal pests (18%) and plant pathogens (16%) worldwide [3]. Weeds are unwelcomed plants that interfere with the management of agricultural production systems, compete with the main crops for available nutrient resources and space and reduce growth, yield, and quality of agricultural produce up to a certain extent [4]. Generally, they produce a larger number of seeds, which may remain dormant in the soil seedbank for several decades, having greater plasticity and equipped with specialized seed dispersal mechanisms. Further, they exhibit the ability to invade newly disturbed areas and compete with crops for scarcely available moisture, nutrients, and light [5]. Apart from yield and production losses, they may also provide niches and harbor insects, plant pathogens, and other pests, hence increasing their incidence of attack to the main crop [6]. Weeds are the firstborn problem in agriculture since about 10,000 BC [7] representing the main hindrance in profitable agricultural production under natural resource management. The presence of weeds in natural ecosystems causes various direct and indirect losses, including interference with successful crop production, damage to biodiversity, loss of possibly fruitful land, loss of grazing areas and livestock production, obstruction of navigational and irrigation channels, and reduction of available water in water bodies. Most of the weeds belong to families Poaceae and Asteraceae. A majority of the weeds are terrestrial plants, a few are aquatic weeds and some are parasitic weeds [8]. Globally, reduction losses of wheat yield due to weed infestation are 23% [2]. The economic losses incurred due to this wheat yield reduction amount to Rs. 146 billion [9].

In the light of the abovementioned properties and harmful effects of weeds, it becomes important to control them. Appropriate weed control strategy in arable soils employs both the direct and indirect methods. Direct methods include those with the prime objective of weed control such as mechanical, manual, chemical and biological weed control and indirect being the cultural and preventive practices reducing germination, growth and vigor of weeds [10]. Many practices are available to control and manage weeds in agricultural crops. In ancient times when synthetic herbicides were not introduced, people tried polyculture, crop rotation, and other management practices that have shown sustainability with low inputs [11]. Until recently, weeds were being controlled by manual, mechanical, and chemical methods [12]. However there were drawbacks associated with each of these methods that severely limited their practical use, for example, herbicides cast detrimental effects on environment, humans, and animals [13]. They also cause contamination of water bodies and pollute natural resources like air, soil, and plants, thus destroying nontarget entities such as wildlife [14]. Also due to repeated herbicide applications, there is an increasing trend in herbicide-resistant weed species [15]. Mechanical weeding on the other side requires several repetitions and is only feasible for crops sown in rows; therefore weeds grown near to crop plants and within rows are escaped of control [10]. Similarly, hand weeding needs a huge number of labor and hence cannot be applied on a large scale [10]. Therefore, repeated manual weeding and nonavailability of labor make this method unfeasible and uneconomic [16].

Hence, the prevailing situation demands some weed control measures other than chemicals, and in this context, biological control is gaining much importance around the world. Biological control is a general term used to define the introduction of organisms mostly bacteria and fungi in order to solve one or more problems in the farmer's field [17, 18]. Biological control using bacteria (bacterial herbicides), fungi (mycoherbicides), and viruses has recently gained much attention. Different kinds of fungi showing herbicidal activity are potential candidates of *Phoma* and *Sclerotina* genera. Among the bacteria some members of *Pseudomonas* and *Xanthomonas* depict these attributes.

Broadly speaking the control of weeds using microbes in green areas is a green approach that may reduce costs, decrease dependence on synthetic chemicals, and lower the negative impressions of chemicals on the environment. Microorganisms in the form of bioherbicides can be more selective than synthetic chemicals (herbicides) and target only the desired species [19]. Bioherbicides also lessen the chances of induction of resistance in the target weed species, due to the involvement of a number of mechanisms [20]. Therefore, keeping in view the abovementioned (even more) limitations of conventionally outdated methods necessitates the adoption of newer methods based on biological agents that are environmentally safer, friendly, economic, and feasible. We tried to highlight the need for adoption of innovative methods of weed control with higher efficacy. We then focused on harmful aspects of the judicious use of herbicides that in turn causes threats to environmental quality, food security, and human health followed by future research aims for improvement.

2. Weed control options

About one third of the total costs in field crop production is taken away by the weed management. There exist a variety of weed control strategies that can be applied depending upon various cropping systems [21]. Traditional farming practices generally rely on the application of herbicides and manual weeding. Generally, weed control measures include physical, chemical, and biological methods.

2.1 Physical weed control

Physical approaches of weed control include mechanical (tillage), manual methods, crop rotations, and crop fertilization and are separately discussed with possible limitations.

An increase in the density of weed species has been observed where monocropping was adopted. However due to the diverse nature of crop rotations, the density of such weeds can be tackled for profitable crop production [22]. Using a cover crop in rotation with the main crop is an attractive solution to cope with weed infestations [23, 24]. The integration of cover crops with no-till system has shown significant reduction (78%) in weed density in the USA [25]. The weeds with similar life cycles that match with the crop pose serious threats to crop production. These cover crops when used properly in rotation with the main crop compete with weeds for available nutrients, light, space, and water sources, hence reducing their emergence and numbers [26]. However the ability of cover crops to control weeds is largely governed by the growth habit and performance of the cover crop in a desired area [27]. That makes the use of this method to be only a small scale.

Increasing the competitive ability of crops against weeds is an important aspect to avoid field losses due to weeds and has been seen as a strategy for integrated weed management systems [28]. It can be achieved through manipulating fertilizer timing, rate of fertilizer, and placement methods effectively [29]. Nitrogenous fertilizers have been known to involve in the activation of dormant weed seeds, thus directly affecting specific weed densities. The most agricultural weeds have shown growth rates equal to that of wheat in response to the added nitrogen [30]. However, it is not well known that phosphorus levels of soil affect weed growth and crop as well, but it is a fact that the crop-weed competition is considerably affected by phosphorus fertilizations, for instance, Bansal [31] reported that weed-crop (fenugreek) competition was increased with higher P levels. Similarly, Santos et al. [32] reported that lettuce (*Lactuca sativa* L.) showed a higher competitive ability than the common purslane (*Portulaca oleracea* L.) but not smooth pigweed (*Amaranthus hybridus* L.) with higher P levels than lower levels. Therefore, due to this uncertainty, this method is not widely adopted and acceptable.

Manual weed control methods involve plucking, uprooting, and hoeing with and/or without hand-driven machines [33] and are in use since ancient times. Manual weed control is one of the most efficient methods and is applicable in areas where the labor is easily available. However, immediate availability of labor before the weeds have grown in crops [10], repeated hand weedings [16] and adoptation on only small scale farming are the major limitations of this method to be adopted. Mechanical methods use tillage implements such as cultivators, weeders, and different types of harrows which are being drawn by animals (in the past) or by engines (until recently) around world [34]. Tillage practices in the field affect weed management, weed seed bank in the soil, and soil disturbance patterns. Deep cultivation can be used to burry weeds that germinate in the upper soil layers such as *Phalaris minor* in wheat. However, timely sown wheat in integration with zero tillage has shown significant results in the reduction of *Phalaris minor* infestations, obtaining higher grain yields of wheat [35, 36]. Tillage for weed control is not suitable for all crops and is only limited to crops sown on rows with suitable row-to-row spacing. Weeds that grow in close association with crop plants are not managed properly by this method, and those weeds which are grown within crop rows cause more losses than those sown in between crop rows [10, 37]. Moreover, some weeds may regenerate which are not completely uprooted, and root injury to main crop may occur [38]. However, the use of tillage implements for weed control are associated with adverse environmental impacts such as deterioration of soil structure, disturbed soil biological processes and soil erosion [39], leaching of nutrients which would otherwise be available to plants and eutrophication [40]. Therefore the efficiency of mechanical weed control measures is less than that of chemical weed control [22, 38]. Tillage practices done for weeding aggravate more soil compaction than other tillage operations due to a shorter cover of wheel tracks [38].

2.2 Chemical weed control

The application of synthetic chemicals for crop protection began after the second world war when most of the selective herbicides for broad-leaved weeds were commercialized in 1946 [41]. However, with the advancement in crop protection measures usually at the start of the twentieth century, copper and sulfuric acid containing herbicides were developed [42]. Herbicides are chemical compounds which kill or control weeds and are largely synthesized by crop protection industries nowadays available for almost all cultivated crops. They were rapidly adopted by farming communities as they do not require much labor and hence are not costly; no risks of soil erosion and energy efficiency are further advantages of herbicides [43]. The most widely used chemicals in wheat to control grassy and non-grassy weeds are clodinafop, tralkoxydim, Atlantis (meso-/iodosulfuron), sulfosulfuron, and pinoxaden. However, for the control of broad-leaved weeds, major chemical herbicides are carfentrazone, 2,4-D, and metsulfuron [44]. Herbicides account for 44% of all pesticides worldwide [45]. Nevertheless, chemical methods have controlled the weeds resultantly improving the yields of diverse crops from 10 to 50% [4]. However, the continuous application of such herbicides had led to intraspecific selection of weeds and caused the development of herbicide-resistant biotypes of weeds [46, 47]. Approximately, 300 herbicide-resistant weeds have been reported in 15 families of synthetic herbicides [45, 48, 49] (**Table 1**).

A major portion of applied herbicides falls on nontarget species and soil [50]. Some herbicides like triazines and sulphonylureas may persist in soil long enough to affect the growth of subsequent sensitive crops [38]. Herbicides have also caused toxicity and diseases to exposed animals [51]. Herbicides in soil however may not reduce the population of soil microflora and microfauna but may induce intraspecific and interspecific selections [38].

The magnitude of issues caused by herbicides is much bigger than the outcomes of herbicides (**Figure 1**). Therefore it is a dire need of the hour to move toward some newer methods other than chemicals that can ensure environmental safety and resource conservation and sustain crop production economically.

Herbicide-resistant weeds	Common names	Herbicide (s)
Eichhornia crassipes	Water hyacinth	2,4-D,Glyphosate
Chenopodium album	Common lambsquarters	Triazine
Salsola kali	Russian thistle	Sulfonylurea
Senecio vulgaris	Common groundsel	Triazine (atrazine)
Sesbania exaltata	Hemp sesbania	Glyphosate
Cyperus	Purple nutsedge	Sulfonylureas
Avena fatua	Wild oat	Glyphosate

Table 1.
Some worst weeds that evolved resistance against chemical herbicides.

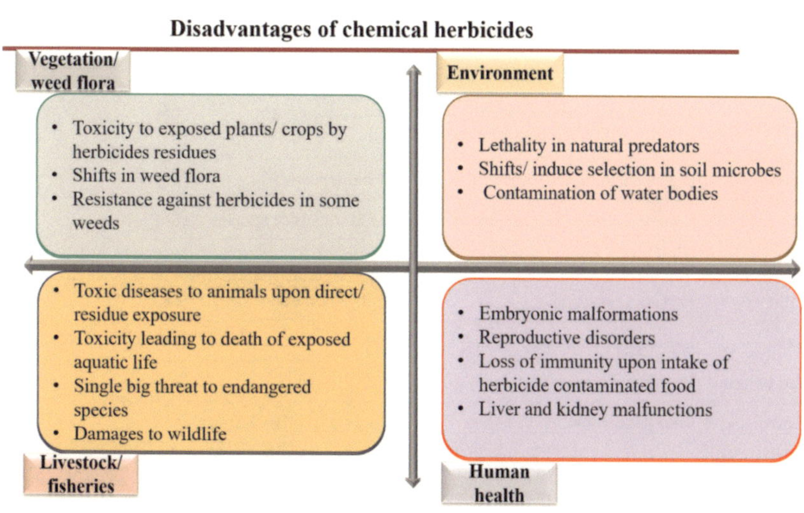

Figure 1.
Disadvantages of herbicides to all life forms. Modified and redrawn from [1].

2.3 Biological weed control

Biological control is the intentional use of biological agents (living organisms) to control plant pathogens or weeds in fields [52, 53]. The application of herbicides for sustaining agricultural production has created so many problems such as contamination of groundwater, destruction to the nontarget species, and induction of resistance against herbicides in a number of weed species [45], and other control methods become even more unsuitable where the land value is small and unaccessible with widespread weed infestations. This situation paved the way of researchers to move toward biological control as an alternative option in weed management. The chemical herbicides can persist in soil for longer periods of time, have limitations for crop rotation, and cause damage to the nontarget organisms [54]. Microbial herbicides on the other hand are more selective and affect only the target species [19]. The other advantage of using microbial agents is the reduced chance of induction of resistance in the target species [20].

Primarily there are two fields of application within the context of biological weed control, viz., the classical and augmentative or inundative. Classical biological control is the introduction and subsequent discharge of a natural enemy of a pest predator with the objective to reduce its virulence without becoming a pest itself [55]. This method is suitable for the control of perennial weeds that grow over a range of large areas such as in the forests, rangelands, along waterways, and roadsides and where reduction in weed competitiveness is required [56]. Several agents might be used in this strategy such as insects, fungi, mites, and different herbivores. The inundative biological control also called as bioherbicide approach is the

Trade name	Microbe(s) involved	Target weed(s)	Representative/initial report reference
BioMal	*Colletotrichum gloeosporioides* f. sp. *nalvae*	Round-leaved mallow	[60]
Casst	*Alternaria cassia*	Sicklepod, coffee senna	[61]
Biochon	*Chondrostereum purpureum*	Woody weeds	[62]
Collego	*Colletotrichum gloeosporioides* f. sp. *aeschynomene*	Northern joint vetch	[61]
Phoma	*Phoma macrostoma*	Broadleaf weeds	[18, 63]
Devine	*Phytophthora palmivora*	Strangle vine	[64]
Camperico poae	*Xanthomonas campestris* pv.	Annual bluegrass	[65]
Hakatak	*Colletotrichum acutatum*	*Hakea sericea*	[17]
Myco-tech	*Chondrostereum purpureum*	Deciduous tree species	[66]
Smolder	*Alternaria destruens*	Dodder	[67]
Dr. Biosedge	*Puccinia canaliculata*	Yellow nutsedge	[68]
Lubao	*Colletotrichum gloeosporioides* f. sp.	Dodder	[61]
Woad warrior	*Puccinia thlaspeos*	Dyer's woad	[69]
Chontrol	*Chondrostereum purpureum*	Alders and other hard woods	[66]
Sarritor	*Sclerotinia minor*	Dandelion	[70]

Table 2.
Successful microbial herbicides (registered) worldwide.

application of mass-produced fungal spores or bacterial cultures in higher concentrations with the objective to eradicate invasive weeds in a managed area [57]. The inundative biocontrol is more related to the agricultural needs and turf management because its implementation is similar to the conventional herbicides as liquid sprays and solid granules [58, 59]. A number of microbial herbicide formulations based on bacteria and fungi have been registered worldwide (**Table 2**).

3. PGPR and stimulation of plant growth

Rhizosphere is the region of the soil surrounded by plant roots and often extended from the surface of roots [94]. This constituency of the soil is much wealthier in bacteria than the contiguous bulk soil [95]. The plant growth-promoting rhizobacteria are the soil bacteria that reside in the rhizosphere and are involved in the stimulation of plant growth through direct and indirect methods [96]. Agricultural production currently relies on the judicious use of synthetic fertilizers [97, 98] that have shown negative environmental impacts due to overuse of these chemical fertilizers [99]. Therefore, the use of PGPR inoculants can be considered as an environmentally sound alternative approach for the sustainable management, decreasing the use of synthetic fertilizers [100–102]. Within the context of PGPR research and their modes of actions, there has been an increasing trend in literature to search for the best PGPR candidate in order to commercialize as bio-fertilizer. Plant growth-promoting rhizobacteria are equipped with a plenty of mechanisms that can result in the promotion of plant growth. For instance, Parmar and Dadarwal [103] suggested the involvement of *fluorescent pseudomonads* to promote nodulation process and increased nitrogen fixation in chickpea [104], in another study, confirmed the ability of *Azospirillum* sp. inoculation on some significant agricultural crops in terms of increased dry weights of the root and shoot. Similarly, [105], who suggested that the foliar application of rhizobacteria in apricot and mulberry causes an increase in total surface area and chlorophyll contents as compared to uninoculated control [106], documented the growth response in wheat after the inoculation with rhizobacteria and revealed that the growth and development of wheat largely depends on the nature of PGPR and environmental factors.

Spaepen et al. [107] reported that various genera of rhizobacteria use tryptophan as a precursor to produce IAA by different pathways. However, the plant pathogenic bacteria only use the indole acetamide pathway to synthesize IAA that causes tumor formation in plants. Swain et al. [108] suggested that cultures of *Bacillus subtilis* when applied on *Dioscorea rotundata* increased the root/stem ratio and number of sprouts as compared to the uninoculated control.

A recent study by Minorsky [109] reported the excellent colonization ability of a PGPR isolate *Pseudomonas fluorescens* (B16) in tomato roots. The positive effects were increased plant height, enhanced flowering, and increased fruit weight. Castro et al. [110] proposed that PGPR stimulates growth and development of crops both by direct and indirect methods. The direct methods of growth promotion may include biological nitrogen fixation, solubilization of mineral phosphorus and iron, production of phytohormones, and synthesis of enzymes and siderophores. Indirect growth promotion occurs through the production of antibiotics and fungal-degrading enzymes and competition for niche exclusion in the rhizosphere [111, 112].

As for the higher uptake of nutrients that is concerned through application of bacterial inoculants, Qin et al. [113] reported the ability of rhizobacteria to dissolve fixed phosphate is related to the rhizosphere acidification. The rhizobium inoculation in soybean plants causes increased availability of phosphorus as compared to non-inoculated plants, hence positively influencing plant growth. Ambrosini et al. [114] suggested that sunflower-associated *Burkholderia* strains were found to be solubilizing $Ca_3(PO4)_2$, hence availing phosphorus for plant use. The management of soil, plant, and environmental interactions evidenced by boosted crop yields is gaining much attention globally. Moreover, agricultural inoculants (cultures) contain plant beneficial bacteria that help plants to meet the demands for nutrients.

4. Bacteria in biological weed control

A number of bacterial species have been studied due to their potential against weed management (**Table 3**). Two major classes of rhizobacteria that show herbicidal activity are *Pseudomonas* and *Xanthomonas* sp. Different rhizobacterial species have been investigated as weed control agents on different crops based on their secondary metabolites [115, 116]. As stated earlier *Pseudomonas* have gained much importance as an agent in biological weed management; there are many strains of this genera, some are plant beneficial [117] and others can have inhibitory effects on plants [118] and so can be applied in biological weed control. Production of extracellular metabolites from these strains is a key mechanism in inhibition of plant growth or germination inhibition [118–120]. However several other mechanisms showing herbicidal activity of bacteria are shown in **Figure 2**.

A strain of *Pseudomonas fluorescens* (D7) isolated from wheat and downy brome rhizosphere has shown inhibitory effects on a number of grassy weeds especially downy brome by virtue of production of a phytotoxin [116, 119, 121]. Kremer et al. [122] tested the phytopathogenic ability of different *fluorescent* and non*fluorescent pseudomonads* which were isolated from the rhizosphere of seven important weeds. About 18% of the strains show phytopathogenecity. However, the ratio of isolates that inhibited seedlings was ranged between 35 and 65%. The mechanism behind is the production of antibiotics, and about 75% of the isolates were active in siderophore production.

Kennedy et al. [121] reported the differential weed inhibition ability of *Pseudomonads* for downy brome and winter wheat. When the culture filtrates were tested on agar, about 8% of the isolates reduced the root growth of downy brome but have no effects on the root growth of wheat. However, under soil application only less than 1% inhibited the growth of downy brome. In the field study, only 0.2% of the total 1000 isolates inhibited the growth of downy brome but increased the growth of winter wheat by 18–35%. Kremer [123] worked with different cover crops associated with deleterious rhizobacteria. Seed bacterization with DRB reduces growth and biomass in weeds associated with cover crops. Adam and Zdor [124] described that rhizobacteria isolated from the rhizosphere of *Abutilon theophrasti* Medik caused growth inhibition of different weeds.

Weissmann and Gerhardson [125] suggested that the application of strain (A153) on *Chenopodium album* suppressed the growth of plants for 10–14 days; however in field conditions, this effect lasts for 2 months. Similarly Weissmann et al. [126] demonstrated excellent growth inhibition ability of a strain (A153) belonging to soil bacteria *Serratia plymuthica* when sprayed on a number of broad-leaved weeds. However, in field experiment this strain showed differential effects on *C. album*, *Stellaria media*, *Polygonum convolvulus*, and *Galeopsis speciosa*. Li and Kremer [127] suggested that the inoculation of *Pseudomonas fluorescens* strain (G2–11) inhibited the growth of *Ipomoea* sp. and *Convolvulus arvensis* weeds and increased the growth of wheat and soybean crops. Zermane et al. [128] in a study stated that *P. fluorescens* has the possible potential to control *Orobanche crenata* and *O. foetida* (*Broomrape*).

DOI: http://dx.doi.org/10.5772/intechopen.86339

Microbe(s) involved	Target weed(s)	Growth condition(s)	Mechanism(s)	Observed effects/comments	References
Pseudomonas fluorescens	Sour cherry	Pot	IAA production	Significant loss in root weight	[71]
Streptomyces chromofuscus cluster	Barnyard grass	Axenic	Antibiosis and H_2S production	ND	[72]
Streptomyces sp. 0H-5093	Reddish	Axenic	Antifungal activity and production of 4-chlorothreonine	Significant growth inhibition	[73]
Streptomyces sp.	Reddish	Axenic	Cellulose inhibition and phthoxazolin A production	Significant growth inhibition due to cellulose inhibition	[74]
Thermoactinomyces sp. A-6019	*Lemna minor*	Axenic	Herbicidal activity and 5′-deoxyguanosine production	ND	[72]
Streptomyces hygroscopicus	*Barnyard grass*	Pot	Antimicrobial and herbicidal activity due to hydantocidine production	Germination inhibition, significant reduction in stem, and leaf structure of weed	[75]
Fusarium and *Rhizoctonia* sp.	Leafy spurge	Greenhouse	Exopolysaccharide and HCN production	Biocontrol activity on leafy spurge leading to significant growth suppression	[76]
Flavobacterium sp.	Sugar beet	Axenic	IAA production	Decreased root elongation and increased shoot to root ratio	[67]
Enterobacter taylorae	Bindweed	Axenic	IAA production	90.5% reduction in root growth, phytotoxic activity	[77]
Pseudomonas fluorescens	Leafy spurge	Field	Auxin production to phytotoxic levels	Reduced cell membrane integrity, inhibited root growth	[60]
Streptomyces saganonensis	Barnyard grass, goose grass, and tufted manna grass	ND	Herbicidine (vi)	Biocontrol activity	[78]
Pseudomonas syringae strain 3366	Corn spurry and fireweed	Pot	Phytotoxin production	Germination inhibition, reduced root, and shoot growth	[79]
Pseudomonas syringae pv. *tagetis*	Annual bluegrass	Field	ND	Greater than 70% weed control	[57]

Microbe(s) involved	Target weed(s)	Growth condition(s)	Mechanism(s)	Observed effects/comments	References
Pseudomonas syringae pv. *phaseolicola*	Kudzu	Greenhouse	ND	ND	[80]
Fusarium tricinctum	Dodder	Field	ND	Effectively controlled dodder at preemergence and postemergence application	[70]
Trichoderma virens	Several weeds	Field	Rhizosphere competence and production of herbicidal compound viridiol	Reduced emergence and seedling growth of different weeds up to a significant extent	[81]
Colletotrichum gloeosporioides f. sp. *malvae*	Round-leaved mallow	Greenhouse	ND	Significant biomass reduction, reduced fresh and dry weight, and inhibited root growth	[82]
Fusarium solani f. sp.	Texas gourd	Field	ND	Greater than 78% mortality, reduced vigor	[83]
Nectria ditissima	Red alder	Field	Infection	ND	[84]
Multiple isolates were screened belonging to *Pseudomonas* spp. and *Xanthomonas* spp.	Jointed goat grass	Axenic and field	ND	Inhibition of weeds by 71% in growth chamber and by 20–74% in different field conditions	[85]
Sclerotinia sclerotiorum	Dandelion	Field	Necrosis and discoloration	80.7% reduction in number of dandelion plants and overall weight reductions	[86]
Pseudomonas putida	Garden asparagus	Pot	Succinic acid and lactic acid production	ND	[87]
Pseudomonas fluorescens and *P. putida*	*Striga hermonthica* (Del.) Benth.	Pot	ND	Significant reduction of weeds and improved biomass of maize	[88]
Collection of multiple rhizobacteria	Leafy spurge	Axenic	Phytotoxin synthesis	30% reduction in leafy spurge growth	[89]
Pseudomonas syringae st. 1 and st. 2	*Polypogon monspeliensis*, *Convolvulus arvensis*, and *Phalaris paradoxa*	Laboratory and field	ND	Reduction in biomass up to 47.5%, 22.8%, and 51.3%. Inhibited 40%, 32.6%, and 46.4% of biomass over control in field conditions	[90]

Microbe(s) involved	Target weed(s)	Growth condition(s)	Mechanism(s)	Observed effects/comments	References
Pseudomonas aeruginosa, *Pseudomonas syringae*, and *Pseudomonas alcaligenes*	Broad-leaved dock, common lambs' quarter	Pot and field	HCN production, IAA production, antibiotic production	Grain yield losses of infested wheat were recovered up to 11.6 to 68% in pot trial, and 17.3 to 62.9% in field trial, respectively	[34]
T. harzianum, *T. pseudokoningii*, *T. reesei*, and *T. viride*	*Avena fatua* L.	Laboratory	ND	Culture filtrates of four *Trichoderma* spp. significantly reduced root, shoot growth, and biomass of *Avena fatua*	[91]
Trichoderma harzianum Rifai, *Trichoderma pseudokoningii* Rifai, *Trichoderma reesei* Simmons, and *Trichoderma viride* Pers	*Phalaris minor* L. and *Rumex dentatus* L.	Laboratory	Synthesis of butanol, n-hexane, chloroform, and ethyl acetate	Original concentration of filtrates reduced root and shoot length and biomass of *Rumex dentatus* significantly, but effect on shoot growth of *Phalaris minor* was not significant	[92]
Trichoderma virens combined with composted chicken manure and rye	Multiple broadleaf and grassweeds	Field	Viridiol (3H)-benzoxazolinone (BOA) and 2,4-dihydroxy-1,4-(2H) benzoxazine-3-one (DIBOA) production	Significant reductions in the emergence of broadleaf and grassweeds and higher reductions in weed biomass was resulted with all treatments as compared to control	[93]

ND = not described.

Table 3.
Features of opportunistic bacteria and fungi in weed control under varying growth conditions.

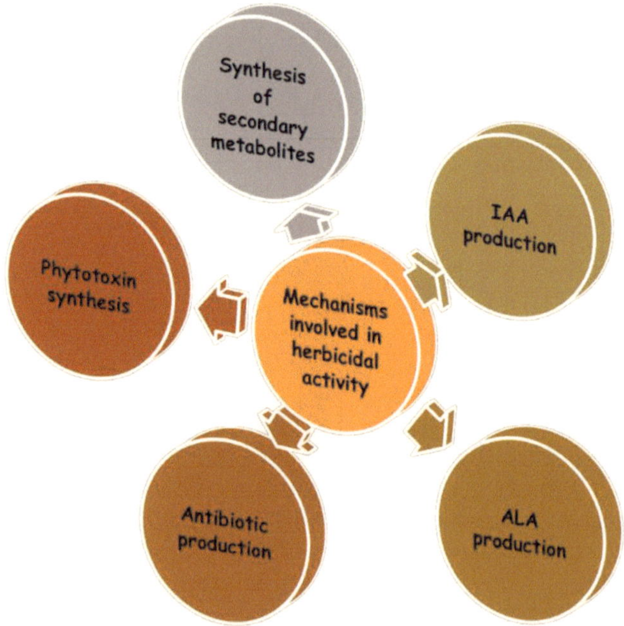

Figure 2.
Possible mechanisms of plant growth-promoting rhizobacteria and fungi involved in herbicidal activity. IAA refers to indole-3 acetic acid, and ALA refers to aminolevulinic acid.

Banowetz et al. [118] tested the germination inhibition activity in various monocot and dicot plants by the application of *P. fluorescens* (strain WH6). The germination inhibition activity was attributed due to the production of a compound called as Germination-Arrest Factor (GAF). Patil [129] screened 15 strains of deleterious rhizospheric bacteria isolated from rhizosphere of different weeds. Among these strains five isolates caused a significant reduction in root and shoot growth of weeds while showing no harmful effects on crop plants. Boyette and Hoagland [130] suggested that *X. campestris* (strain LVA-987) have shown strong growth suppressive effects against horseweed (*Conyza canadensis*). Some of the key herbicidal mechanisms shown by bacteria and fungi are shown in **Figure 2**.

5. Fungi (mycoherbicides) in biological weed control

A list of fungal biological weed control agents is given in **Table 3**. Within the scientific context, three genera of fungi have received worldwide attention to be used in biological weed control. In addition to the abovementioned BioMal and Collego, different other species of genus *Colletotrichum* have been researched extensively. Additionally, *C. truncatum* have been reported to control sesbania (*Sesbania exaltata*) [131] and *C. orbiculare* that has been found to control spiny cocklebur (*Xanthium spinosum*) [63, 132]. It is evident from the literature that these two *Colletotrichum* species produce indole acetic acid [133] which is a phytohormone and derivatives of which show herbicidal activity [134].

Within the genus *Phoma*, three species have a potential against weed control. *P. herbarum* is a fungus that is isolated from lesions of dandelion leaf that have shown control effects of dandelion [135]. *P. macrostoma* has also been studied for weed control due to its inhibitory effects on the dicot plants [18, 136, 137]. *P. macrostoma* strain (94-44B) has been found to control turf associated with broad-leaved weeds in Canada. Mass spectrometric analysis of *P. macrostoma* revealed the production of photobleaching of macrocidins [138] that do not have any inhibitory effects on monocot plants [18]. Despite this macrocidins an anthraquinone pigment in *P. macrostoma* has shown prominent herbicidal effects on some weeds in Central India [139]. The third species under this genus is *Phoma chenopodicola* that is studied widely for its potential against common lamb's quarter [62]. The mechanism behind its virulence against lamb's quarter is the production of diterpene and chenopodolin, a phytotoxic compound isolated from this species [62].

Two species within the genus *Sclerotinia* have been investigated for their herbicidal activity. It is evidenced by the work of Abu-Dieyeh and Watson [140] that *Sclerotinia minor* effectively controlled dandelions in turf management systems. A closely related species of this genus *S. sclerotiorum* has also shown the potential against noxious weeds [141]. Production of oxalic acid has been found by these two species that cause virulence on the host plant [142].

Apart from these three genera, there are other fungal candidates that are registered to control weeds in forest lands and ecosystem managements [143]. A worth mentioning bioherbicide is De Vine containing a fungus *Phytophthora palmivora* [144]. This formulation was registered in 1981 and again in 2006 with the EPA [144].

The mycoherbicide "EcoClear" contains *Chondrostereum purpureum*, a pathogenic fungus which should be applied after the injury to the weeds' branches to retard resprouting [145].

Soil-borne fungi also serve as an important tool in weed management. Their direct application in the soil causes decay of the seeds or emerging seedlings [146]. *Trichoderma virens* is one example that reduces weed populations in horticultural crops [81].

Khattak et al. [147] tested two fungi *Aspergillus* and *Penicillium* for their herbicidal activity against two separate weeds *Silybum marianum* L. and *Lemna minor*. Results showed excellent weed-suppressive characters in the extracts of these fungi.

6. Conclusion and future strategy

Biological control of weeds using bacteria and fungi should be the prime priority for mitigating the negative impressions posed by conventionally adopted weed control methods in order to ensure environmental safety and human health. These biological control agents should be adopted in areas with higher and multiple weed infestations; areas of low value land, where weeds have gotten resistance against herbicides; and areas with lack of labor and where the recommended cultural practices cannot be carried out, for example, restrictions posed by topography and narrow rowed crop cultivations. However, in special cases the combination of biological control agents with other methods could also be a promising approach as an alternative to conventional methods.

The future advancement in biological agents for weed control should be based on advancements in microbial genetics (metagenomics), microbe-plant interactions, and microbial community-level analyses. Further investigations need to be discovered in the future in order to make biological weed control more pragmatic and instrumental. In this context, additional microbe-host relationships containing a match of biological agent and its potential host at greater susceptibility of virulence should be further explored. Since the 1960s a number of formulations have been registered in the world. Formulations that can ensure greater shelf lives, efficacy, and survival of microbial agents should be investigated in the future. Investigations on microbial community structure and function can advance microbial weed control. Traditional methods of microbial community structure solely rely on phenotypic characters; molecular-level characterization should be explored in the future. In a nutshell, fatty acid profiling should be the initial step in targeted weed control. Nucleic acid tools, array pyrosequencing, metagenomics, construction of molecular probes, selection of hyper virulence, genomic studies, and host-microbe interactions should be investigated for the development of innovative weed control methods, reducing reliance on herbicide usage.

Acknowledgements

The authors gratefully acknowledge the Environmental Science Laboratory, Institute of Soil and Environmental Sciences, and Ms. Tayyaba Samreen, (Lecturer) Institute of Soil and Environmental Sciences, University of Agriculture, Faisalabad, Pakistan, for improving the language of this chapter.

Author details

Adnan Mustafa[1,2], Muhammad Naveed[2], Qudsia Saeed[3], Muhammad Nadeem Ashraf[1], Azhar Hussain[4], Tanveer Abbas[2], Muhammad Kamran[5], Nan-Sun[1] and Xu Minggang[1*]

1 National Engineering Laboratory for Improving Quality of Arable Land, Institute of Agricultural Resources and Regional Planning, Chinese Academy of Agricultural Sciences, Beijing, P.R. China

2 Institute of Soil and Environmental Sciences, University of Agriculture, Faisalabad, Pakistan

3 College of Natural Resources and Environment, Northwest Agriculture and Forestry University, Yangling, P.R. China

4 Department of Soil Science, University College of Agriculture and Environmental Sciences, The Islamia University of Bahawalpur, Pakistan

5 Key Laboratory of Arable Land Conservation (Middle and Lower Reaches of Yangtze River), College of Resources and Environment, Huazhong Agricultural University, Wuhan, China

*Address all correspondence to: xuminggang@caas.cn

References

[1] Abbas T, Zahir ZA, Naveed M, Kremer RJ. Limitations of existing weed control practices necessitate development of alternative techniques based on biological approaches. Advances in Agronomy. 2018;**147**: 239-280

[2] Oerke EC. Crop losses to pests. The Journal of Agricultural Science;**144**: 31-43

[3] Jabran K, Mahajan G, Sardana V, Chauhan BS. Allelopathy for weed control in agricultural systems. Crop Protection. 2015;**72**:57-65

[4] Ashiq M, Aslam Z. Chemical control of weeds. In: Weeds and Weedicides. Faisalabad. Pakistan: Department of Agronomy, Ayub Agricultural Research Institute, Faisalabad and University of Agriculture; 2014. pp. 235-256

[5] Rajcan I, Swanton CJ. Understanding maize-weed competition: Resource competition, light quality and the whole plant. Field Crops Research. 2001;**71**: 139-150

[6] Booth BD, Murphy SD, Swanton CJ. Weed Ecology in Natural and Agricultural Systems. Cambridge, UK: CABI Publishing; 2003

[7] Avery AA. Nature's Toxic Tools: The Organic Myth of Pesticide-Free Farming. Center for Global Food Issues; 2006. Available from: www.cgfi.org

[8] Kostov T, Pacanoski Z. Weeds with major economic impact on agriculture in Republic of Macedonia. Pakistan Journal of Weed Science Research. 2007;**13**: 227-239

[9] Razzaq A, Cheema ZA, Jabran K, Hussain M, Farooq M, Zafar M. Reduced herbicide doses used together with allelopathic sorghum and sunflower water extracts for weed control in wheat. Journal of Plant Protection Research. 2012;**52**:281-285

[10] Alam SM. Weeds and Their Ill Effects on Main Crops. Karachi, Pakistan: DAWN media group; 2003. Dawn the internet edition: http:// DAWN.com

[11] Singh HP, Daizy Batish R, Kohli RK. Allelopathic interactions and allelochemicals: New possibilities for sustainable weed management. Critical Reviews in Plant Sciences. 2003;**22** (3–4):239-311

[12] Chauvel B, Guillemin JP, Gasquez J, Gauvrit C. History of chemical weeding from 1944 to 2011 in France: Changes and evolution of herbicide molecules. Crop Protection. 2012;**42**:320-326

[13] Annett R, Habibi HR, Hontela A. Impact of glyphosate and glyphosate based herbicides on the fresh water environment. Journal of Applied Toxicology. 2014;**34**:458-479

[14] Geiger F, Bengtsson J, Berendse F, Weisser WW, Emmerson M, Morales MB, et al. Persistent negative effects of pesticides on biodiversity and biological control potential on European farmland. Basic and Applied Ecology. 2010;**11**: 97-105

[15] Heap I. The International Survey of Herbicide Resistant Weeds. 2015. Available from: www.weedscience.org [Accessed: 16 June 2015]

[16] Rao AN, Johnson DE, Sivaprasad B, Ladha JK, Mortimer AM. Weed management in direct-seeded rice. Advances in Agronomy. 2007;**93**: 153-255

[17] Charudattan R. Biological control of weeds by means of plant pathogens: Significance for integrated weed management in modern agroecology. BioControl. 2001;**46**:229-260

[18] Bailey KL, Pitt WM, Falk S, Derby J. The effects of *Phoma macrostoma* on non-target plant and target weed species. Biological Control. 2011;**58**: 379-386

[19] Bolton H, Elliott LF. Toxin production by a rhizobacterial *Pseudomonas* spp. that inhibits wheat growth. Plant and Soil. 1989;**114**:269278

[20] Crump NS, Ash GJ, Fagan RJ. The development of an Australian Bioherbicide. In: Bishop AC, Boersma M, Bames CD, editors. 12th Australian Week Conference; 1999. p. 235237

[21] Anderson WP. Methods of weed control. In: Weed Science: Principles. 2nd ed. Saint Paul: West; 1983. pp. 65-122

[22] Liebman M, Mohler CL, Staver CP. Ecological Management of Agricultural Weeds. Cambridge: Cambridge University Press; 2001. p. 532

[23] Chauhan BS, Singh RG, Mahajan G. Ecology and management of weeds under conservation agriculture: A review. Crop Protection. 2012;**38**:57-65

[24] Mhlanga B, Cheesman S, Maasdorp B, Muoni T, Mabasa S, Mangosho E, et al. Weed community responses to rotations with cover crops in maize-based conservation agriculture systems of Zimbabwe. Crop Protection. 2015;**69**: 1-8. DOI: 10.1016/j.cropro.2014.11.010

[25] Fisk JW, Hesterman OB, Shrestha A, Kells JJ, Harwood RR, Squire JM, et al. Weed suppression by annual legume cover crops in no tillage corn. Agronomy Journal. 2001;**93**:319-325

[26] Teasdale JR. Cover crops, smother plants, and weed management. In: Hatfield JL, Buhler DD, Stewart, editors. Integrated Weed and Soil Management. Chelsea, MI: Ann Arbor Press; 1998. pp. 247-270

[27] Mhlanga B, Cheesman S, Maasdorp B, Mupangwa W, Thierfelder C. Contribution of cover crops to the productivity of maize-based conservation agriculture systems in Zimbabwe. Crop Science. 2015;**55**: 1791-1805. DOI: 10.2135/cropsci2014.11.0796

[28] Mohler CL. Weed life history: Identifying vulnerabilities. In: Liebman M, Mohler CL, Staver CP, editors. Ecological Management of Agricultural Weeds. Cambridge, UK: Cambridge University Press; 2001. pp. 40-98

[29] O'Donovan JT, Sharma MP, Harker KN, Maurice D, Baig MN, Blackshaw RE. Wild oat (*Avena fatua*) populations resistant to triallate are also resistant to difenzoquat. Weed Science. 1994;**42**: 195-199

[30] Blackshaw RE, Brandt RN, Janzen HH, Entz T. Weed species response to phosphorus fertilization. Weed Science. 2004;**52**:406-412

[31] Bansal C. Efficiency of various weed control methods on growth and yield of fenugreek (*Trigonella foenum-graecum* L.) [Doctoral dissertation]. Gwalior, MP: Rvskvv; 2012

[32] Santos BM, Dusky JA, Stall WM, Shilling DG, Bewick TA. Phosphorus effects on competitive interactions of smooth pigweed (*Amaranthus hybridus*) and common purslane (*Portulaca oleracea*) with lettuce (*Lactuca sativa*). Weed Science. 1998:307-312

[33] Young SL, Pierce FJ, Nowak P. Introduction: Scope of the problem-rising costs and demand for environmental safety for weed control. In: Automation: The Future of Weed Control in Cropping Systems. Dordrecht, The Netherlands: Springer; 2014. pp. 1-8

[34] Abbas T, Zahir ZA, Naveed M, Aslam Z. Biological control of

broad-leaved dock infestation in wheat using plant antagonistic bacteria under field conditions. Environmental Science and Pollution Research. 2017;**24**(17): 14934-11494

[35] Chhokar RS, Sharma RK, Jat GR, Pundir AK, Gathala MK. Effect of tillage and herbicides on weeds and productivity of wheat under rice-wheat growing system. Crop Protection. 2007; **26**:1689-1696

[36] Franke AC, Singh S, McRoberts N, Nehra AS, Godara S, Malik RK, et al. Phalaris minor seed bank studies: Longevity, seedling emergence and seed production as affected by tillage regime. Weed Research. 2007;**47**:73-83

[37] Melander B, Holst N, Rasmussen IA, Hansen PK. Direct control of perennial weeds between crops—Implications for organic farming. Crop Protection. 2012; **40**:36-42

[38] Hakansson S. Soil tillage effects on weeds. In: Weeds and Weed Management on Arable Land: An Ecological Approach. Cambridge: CABI Publishing; 2003

[39] Ahlgren S. Environmental Impact of Chemical and Mechanical Weed Control in Agriculture: A Comparing Study Using Life Cycle Assessment (LCA) Methodology. Sweden: The Swedish Institute for Food and Biotechnology, Gothenburg; 2004. SIK Rapport No. 719-2004

[40] Birkas M, Jolankai M, Gyuricza C, Percze A. Tillage effects on compaction, earthworms and other soil quality indicators in Hungary. Soil and Tillage Research. 2004;**78**:185-196

[41] Cobb AH, Kirkwood RC. Herbicides and Their Mechanisms of Action. Boca Raton, FL: CRC Press; 2000

[42] Hamill AS, Holt JS, Mallory-Smith CA. Contributions of weed science to weed control and management. Weed Technology. 2004;**18**:1563-1565

[43] Ghorbani R, Leifert C, Seel W. Biological control of weeds with antagonistic plant pathogens. Advances in Agronomy. 2005;**86**:191-225

[44] Chhokar RS, Sharma RK, Gill SC. Compatibility of herbicides against grassy weeds in wheat. Indian Journal of Weed Science. 2013;**45**:239-242

[45] Quimby PC, King LR, Grey WE. Biological control as a means of enhancing the sustainability of crop/ land management systems. Agriculture, Ecosystems and Environment. 2002;**88**: 147-152

[46] Bradley KW, Hagood ES Jr. Identification of a Johnsongrass (*Sorghum halepense*) biotype resistant to aryloxyphenoxypropionate and cyclohexanedione herbicides in Virginia. Weed Technology. 2001;**15**:623-627

[47] De Prado R, Osuna MD, Fischer AJ. Resistance to ACCase inhibitor herbicides in a green foxtail (*Setaria viridis*) biotype in Europe. Weed Science. 2004;**52**:506-512

[48] Chhokar RS, Sharma RK, Chauhan DS, Mongia AD. Evaluation of herbicides against (*Phalaris minor*) in wheat in north-western Indian plains. Weed Research. 2006;**46**:40-49

[49] Heap IM. International Survey of Herbicide Resistant Weeds. 2014. Available from: http://www.weedscience.org [Accessed: July 2014]

[50] Crone EE, Marler M, Pearson DE. Non-target effects of broadleaf herbicide on a native perennial forb: A demographic framework for assessing and minimizing impacts. Journal of Applied Ecology. 2009;**46**:673-682

[51] Pimentel D. Environmental and economic costs of the application of

pesticides primarily in the United States. Environment, Development and Sustainability. 2005;7:229-252

[52] Hallet SG. Where are the bioherbicides? Weed Science. 2005;**53**: 404-415

[53] Chutia M, Mahanta JJ, Saikia R, AKS B, Sarma TC. Effect of leaf blight disease on yield of oil and its constituents of *Java citronella* and in vitro control of the pathogen using essential oils. World Journal of Agricultural Sciences. 2006;**2**: 319-321

[54] Guske S, Schulzand B, Boyle C. Biocontrol options for *Cirsium arvense* with indigenous fungal pathogens. Weed Research. 2004;**44**:107-116

[55] Shaw RH, Bryner S, Tanner R. The life history and host range of the Japanese knot weed psyllid, *Aphalara itadori* Shinji: Potentially the first classical biological weed control agent for the European Union. Biological Control. 2009;**49**:105-113

[56] Templeton GE, TeBeest DO, Smith RJJ. Biological weed control with mycoherbicides. Annual Review of Phytopathology. 1979;**17**:301-310

[57] Johnson DR, Wyse DL, Jones KJ. Controlling weeds with phytopathogenic bacteria. Weed Technology. 1996;**10**:621-624

[58] Auld BA, Hetherington SD, Smith HE. Advances in bioherbicide formulation. Weed Biology and Management. 2003;**3**:61-67

[59] Caldwell CJ, Hynes RK, Boyetchko SM, Korber DR. Colonization and bioherbicidal activity on green foxtail by *Pseudomonas fluorescens* BRG100 in a pesta formulation. Canadian Journal of Microbiology. 2012;**58**:1-9

[60] Brinkman MA, Clay SA, Kremer RJ. Influence of deleterious rhizobacteria on leafy spurge (*Euphorbia esula*) roots. Weed Technology. 1999;**13**:835-839

[61] Julien MH, Griffiths MW. Biological control of weeds. In: A World Catalogue of Agents and Their Target Weeds. Wallingford, UK: CAB International; 1998

[62] Cimmino A, Andolfi A, Zonno MC, Avolio F, Santini A, Tuzi A. Chenopodolin: A phytotoxic unrearranged ent-pimaradiene diterpene produced by *Phoma chenopodicola*, a fungal pathogen for *Chenopodium album* biocontrol. Journal of Natural Products. 2013;**76**:1291-1297

[63] Auld BA, Say MM, Ridings HI, Andrews J. Field applications of *Colletotrichum orbiculare* to control *Xanthium spinosum*. Agriculture, Ecosystems and Environment. 1990;**32**: 315-323. DOI: 10.1016/0167-8809(90) 90168-d

[64] Burnett HC, Tucker DPH, Patterson ME, Ridings WH. Biological control of milk weed vine with a race of *Phytophthora citrophthora*. Proceedings of the Florida State Horticultural Society. 1973;**86**:111-115

[65] Imaizumi S, Nishino T, Miyabi K, Fujimori T, Yamada M. Biological control of annual bluegrass (*Poa annua* L.) with a Japanese isolate of *Xanthomonas campestris* pv. poae (JT-P482). Biological Control. 1997;**8**:7-14

[66] Setliff EC. The wound pathogen *Chondrostereum purpureum*, its history and incidence on trees in North America. Australian Journal of Botany. 2002;**50**:645-651. DOI: 10.1071/bt01058

[67] Loper JE, Schroth MN. Influence of bacterial source of indole-3-acetic acid on root elongation of sugar beet. Phytopathology. 1986;**76**:386-389

[68] Phatak SC, Summer DR, Wells HD, Bell DK, Glaze NC. Biological control of

yellow nutsedge with the indigenous rust fungus *Puccinia canaliculata*. Science. 1983;**219**:1446-1447

[69] Lovic BR, Dewey SA, Thomson SV, Evans JO. *Puccinia thlaspeos* a possible biocontrol agent for Dyers woad. Proceedings - Western Society of Weed Science. 1988;**41**:55-57

[70] Hopen HJ, Carusoand FL, Bewick TA. Control of dodder in cranberry vaccinium macrocarpon with a pathogen-based bioherbicide. Acta Horticulturae. 1997;**446**:427-428

[71] Dubeikovsky AN, Mordukhova EA, Kochetkov VV, Polikarpova FY, Boronin AM. Growth promotion of blackcurrant softwood cuttings by recombinant strain *Pseudomonas fluorescens* BSP53a synthesizing an increased amount of indole-3-acetic acid. Soil Biology and Biochemistry. 1993;**25**:1277-1281

[72] Isaac BG, Ayer SW, Letendre LJ, Stonard RJ. Herbicidal nucleosides from microbial sources. The Journal of Antibiotics. 1991;**44**:729-732

[73] Yoshida H, Arai N, Sugoh M, Iwabuchi J, Shiomi K, Shinose M, et al. 4-chlorothreonine, a herbicidal antimetabolite produced by *Streptomyces* sp. OH-5093. The Journal of Antibiotics. 1994;**47**:1165-1166

[74] Tanaka Y, Kanaya I, Takahashi Y, Shinose M, Tanaka H, Omura S. Phthoxazolin A, a specific inhibitor of cellulose biosynthesis from microbial origin. The Journal of Antibiotics. 1993; **46**:1209-1213

[75] Nakajima M, Itoi K, Takamatsu Y, Kinoshita T, Okazaki T, Kawakubo K, et al. Hydantocidin: A new compound with herbicidal activity from *Streptomyces hygroscopicus*. The Journal of Antibiotics. 1991;**44**:293-300

[76] Kremer RJ, Caesar AJ, Souissi T. Soilborne microorganisms of Euphorbia are potential biological control agents of the invasive weed leafy spurge. Applied Soil Ecology. 2006;**32**:27-37

[77] Sarwar M, Kremer RJ. Enhanced suppression of plant growth through production of L-tryptophan-de rived compounds by deleterious rhizobacteria. Plant and Soil. 1995;**172**: 261-269

[78] Cutler HG. Perspectives on discovery of microbial phytotoxins with herbicidal activity. Weed Technology. 1988;**2**:525-532

[79] Norman MA, Patten KD, Gurusiddaiah S. Evaluation of a phytotoxin(s) from *Pseudomonas syringae* for weed control in cranberries. Horticultural Science. 1994;**29**: 1475-1477

[80] Zidack NK, Backman PA. Biological control of kudzu (*Pueraria lobata*) with the plant pathogen *Pseudomonas syringae* pv. phaseolicola. Weed Science. 1996; **44**:645-649

[81] Héraux FMG, Hallett SG, Ragothama KG, Weller SC. Composted chicken manure as a medium for the production and delivery of *Trichoderma virens* for weed control. Horticultural Science. 2005;**40**:1394-1397

[82] Grant NT, Prusinkiewicz E, Mortensen K, Makowski RMD. Herbicide interactions with *Colletotrichum gloeosporiodes* f. sp. malvae, a bioherbicide for round leaved mallow (*Malva pusilla*) control. Weed Technology. 1990;**4**:716-723

[83] Weidemann GJ, Templeton GE. Control of Texas gourd, *Cucurbita texana*, with *Fusarium solani* f. sp. cucurbitae. Weed Technology. 1988;**2**: 271-274

[84] Dorworth CE. Biological control of red alder (*Alnus rubra*) with the fungus

Nectria ditissima. Weed Technology. 1995;**9**:243-248

[85] Kennedy AC, Stubbs TL. Management effects on the incidence of jointed goatgrass inhibitory rhizobacteria. Biological Control. 2007; **40**:213-221

[86] Riddle GE, Burpee LL, Boland GJ. Virulence of *Sclerotinia sclerotiorum* and *S. minor* on dandelion (*Taraxacum officinale*). Weed Science. 1991;**39**: 109-118

[87] Yoshikawa M, Hirai N, Wakabayashi K, Sugizaki H, Iwamura H. Succinic and lactic acids as plant growth promoting compounds produced by rhizospheric *Pseudomonas putida*. Canadian Journal of Microbiology. 1993;**39**:1150-1154

[88] Ahonsi MO, Berner DK, Emechebe AM, Lagoke ST. Selection of rhizobacterial strains for suppression of germination of *Striga hermonthica* (Del.) Benth. seeds. Biological Control. 2002; **24**:143-152

[89] Souissi T, Kremer RJ. A rapid microplate callus bioassay for assessment of rhizobacteria for biocontrol of leafy spurge (*Euphorbia esula* L.). Biocontrol Science and Technology. 1998;**8**(1):83-92

[90] Omer AM, Balah MA. Using of rhizo-microbes as bioherbicides for weeds. Global Journal of Biotechnology and Biochemistry. 2011;**6**(3):102-111

[91] Javaid A, Ali S. Alternative management of a problematic weed of wheat *Avena fatua* L. by metabolites of trichoderma. Chilean Journal of Agricultural Research. 2011;**71**(2):205-211

[92] Javaid A, Ali S. Herbicidal activity of culture filtrates of *Trichoderma* spp. against two problematic weeds of wheat. Natural Product Research. 2011; **25**(7):730-740

[93] Héraux FM, Hallett SG, Weller SC. Combining *Trichoderma virens*-inoculated compost and a rye cover crop for weed control in transplanted vegetables. Biological Control. 2005; **34**(1):21-26

[94] Bringhurst RM, Cardon ZG, Gage DJ. Galactosides in the rhizosphere: Utilization by *Sinorhizobium meliloti* and development of a biosensor. Proceedings of the National Academy of Sciences of the United States of America. 2001;**98**:4540-4545

[95] Hiltner L. Uber neuere Erfahrungen und Probleme auf dem Gebiet der Bodenbakteriologie und unter besonderer Berucksichtigung der Grundungung und Brache. Arbeiten der Deutschen Landwirtschaftlichen Gesellschaft. 1904;**98**:59-78

[96] Ahemad M, Kibret M. Mechanisms and applications of plant growth promoting rhizobacteria: Current perspective. Journal of King Saud University - Science. 2014;**26**:1-20

[97] Wartiainen I, Eriksson T, Zheng W, Rasmussen U. Variation in the active diazotrophic community in rice paddy-nifH PCR-DGGE analysis of rhizosphere and bulk soil. Applied Soil Ecology. 2008;**39**:65-75

[98] Adesemoye AO, Torbert HA, Kloepper JW. Plant growth-promoting rhizobacteria allow reduced application rates of chemical fertilizers. Microbial Ecology. 2009;**58**:921-929

[99] Shenoy VV, Kalagudi GM, Gurudatta BV. Towards nitrogen autotrophic rice. Current Science. 2001; **81**:451-457

[100] Alves BJR, Boddey RM, Urquiaga S. The success of BNF in soybean in Brazil. Plant and Soil. 2004;**252**:1-9

[101] Hungria M, Campo RJ, Souza EM, Pedrosa FO. Inoculation with selected

strains of *Azospirillum brasilense* and *A. lipoferum* improves yields of maize and wheat in Brazil. Plant and Soil. 2010;**331**: 413-425

[102] Hungria M, Nogueira MA, Araujo RS. Co-inoculation of soybeans and common beans with rhizobia and azospirilla: Strategies to improve sustainability. Biology and Fertility of Soils. 2013;**49**:791-801

[103] Parmar N, Dadarwal KR. Stimulation of nitrogen fixation and induction of flavonoid like compounds by rhizobacteria. Journal of Applied Microbiology. 1999;**86**:36-44

[104] Dobbelaere S, Croonenborghs A, Thys A, Ptacek D. Responses of agronomically important crops to inoculation with Azospirillum. Australian Journal of Plant Physiology. 2001;**28**:871-879

[105] Esitken A, Karlidag H, Ercisli S, Turan M, Sahin F. The effect of spraying a growth promoting bacterium on the yield, growth and nutrient element composition of leaves of apricot (*Prunus armeniaca* L. cv. Hacihaliloglu). Australian Journal of Agricultural Research. 2003;**54**:377-380

[106] Khalid A, Arshad M, Zahir ZA. Screening plant growth-promoting rhizobacteria for improving growth and yield of wheat. Journal of Applied Microbiology. 2004;**96**:473-480

[107] Spaepen S, Vanderleyden J, Remans R. Indole-3-acetic acid in microbial and microorganism-plant signaling. In: Unden F, editor. FEMS Microbiology Reviews. New York: Blackwell Publishing Ltd; 2007. pp. 1-24

[108] Swain MR, Naskar SK, Ray RC. Indole-3-acetic acid production and effect on sprouting of yam (*Dioscorea rotundata* L.) minisetts by *Bacillus subtilis* isolated from culturable cowdung microflora. Polish Journal of Microbiology. 2007;**56**:1 03-1 1110

[109] Minorsky PV. On the inside. Plant Physiology. 2008;**146**:323-324

[110] Castro RO, Cornejo HAC, Rodriguez LM, Bucio JL. The role of microbial signals in plant growth and development. Plant Signaling & Behavior. 2009;**4**:701-712

[111] Zahir ZA, Muhammad A, Frankenberger WT. Plant growth promoting rhizobacteria: Applications and perspectives in agriculture. Advances in Agronomy. 2004;**81**:97-168

[112] Van Loon LC. Plant responses to plant growth-promoting rhizobacteria. European Journal of Plant Pathology. 2007;**119**:243-254

[113] Qin L, Jiang H, Tian J, Zhao J, Liao H. Rhizobia enhance acquisition of phosphorus from different sources by soybean plants. Plant and Soil. 2011;**349**: 25-36

[114] Ambrosini A, Beneduzi A, Stefanski T, Pinheiro FG, Vargas LK, Passaglia LMP. Screening of plant growth promoting rhizobacteria isolated from sunflower (*Helianthus annuus* L.). Plant and Soil. 2012;**356**:245-264

[115] Tranel PJ, Gealy DR, Kennedy AC. Inhibition of downy brome (*Bromus tectorum*) root growth by a phytotoxin from *Pseudomonas fluorescens* strain D7. Weed Technology. 1993;7:134139

[116] Gealy DR, Gurusiddah S, Ogg AG Jr, Kennedy AC. Metabolites from *Pseudomonas fluorescens* strain D7 inhibit downy brome (*Bromus tectorum*) seedling growth. Weed Technology. 1996;**10**:282-287

[117] Gamalero E, Lingua G, Tombolini R, Avidano L, Pivato B, Berta G. Colonization of tomato root seedling by *Pseudomonas fluorescens* 92rkG5:

spatio-temporal dynamics, localization, organization, viability, and culturability. Microbial Ecology. 2005;**50**:289-297

[118] Banowetz GM, Azevedo MD, Armstrong DJ, Halgrenand AB, Mills DI. Germination-arrest factor (GAF): Biological properties of a novel, naturally-occurring herbicide produced by selected isolates of rhizosphere bacteria. Biological Control. 2008;**46**: 380-390

[119] Kennedy AC, Elliott LF, Young FL, Douglas CL. Rhizobacteria suppressive to the weed downy brome (*Bromus tectorum* L.). Soil Science Society of America Journal. 1991;**55**:722-727

[120] Quail JW, Ismail N, Pedras MSC, Boyetchko SM. Pseudophomins A and B, a class of cyclic lipodepsipeptides isolated from a *Pseudomonas* species. Acta Crystallographica Section E: Crystallographic Communications. 2002;**58**:o268-o271

[121] Kennedy AC, Johnson BN, Stubbs TL. Host range of a deleterious rhizobacterium for biological control of downy brome. Weed Science. 2001;**49**: 792-797

[122] Kremer RJ, Begonia MFT, Stanley L, Lanham ET. Characterization of rhizobacteria associated with weed seedlings. Applied and Environmental Microbiology. 1990;**56**:1649-1655

[123] Kremer RJ. Growth suppression of annual weeds by deleterious rhizobacteria integrated with cover crops. In: Spencer NR, editor. Proceedings of the Xth International Symposium on Biological Control of Weeds; Bozeman, MT: USDA-ARS and Montana State University; 2000. pp. 931-940

[124] Adam O, Zdor R. Effect of cyanogenic rhizobacteria on the growth of velvetleaf (*Abutilon theophrasii*) and corn (*Zea mays*) in autoclaved soil and the influence of supplemented glycine. Soil Biology and Biochemistry. 2001;**33**: 801-809

[125] Weissmann R, Gerhardson B. Selective plant growth suppression by shoot application of soil bacteria. Plant and Soil. 2001;**234**:159-170

[126] Weissmann R, Uggla C, Gerhardson B. Field performance of a weed-suppressing *Serratia plymuthica* strain applied with conventional spraying equipment. Biological Control. 2003;**48**:725-742

[127] Li J, Kremer RJ. Growth response of weed and crop seedlings to deleterious rhizobacteria. Biological Control. 2006; **39**:58-65

[128] Zermane N, Souissi T, Kroschel J, Sikora R. Biocontrol of broom rape (*Orobanche crenata* Forsk. and *Orobanche foetida* Poir.) by *Pseudomonas fluorescens* isolate Bf7-9 from the faba bean rhizosphere. Biocontrol Science and Technology. 2007;**17**:487-497

[129] Patil VS. Isolation, characterization and identification of rhizospheric bacteria with the potential for biological control of *Sida acuta*. Journal of Environmental Research and Development. 2014;**8**:411-417

[130] Boyette CD, Hoagland RE. Bioherbicidal potential of *Xanthomonas campestris* for controlling *Conyza canadensis*. Biocontrol Science and Technology. 2015;**25**:229-237

[131] Schisler DA, Howard KM, Bothast RJ. Enhancement of disease caused by *Colletotrichum truncatum* in *Sesbania exaltata* by co inoculating with epiphytic bacteria. Biological Control. 1991;**1**:261-268

[132] Auld BA, McRae CF, Say MM. Possible control of *Xanthium spinosum* by a fungus. Agriculture, Ecosystems and Environment. 1988, 1988;**21**:219-223

[133] Gan P, Ikeda K, Irieda H, Narusaka M, O'Connell RJ, Narusaka Y. Comparative genomic and transcriptomic analyses reveal the hemibiotrophic stages shift of *Colletotrichum* fungi. The New Phytologist. 2013;**197**:1236-1249

[134] Grossmann K. Auxin: Current status of mechanism and mode of action. Pest Management Science. 2010; **66**:113-120

[135] Stewart-Wade SM, Boland GJ. Oil emulsions increase efficacy of *Phoma herbarum* to control dandelion but are phytotoxic. Biocontrol Science and Technology. 2005;**15**: 671-681

[136] Bailey KL, Falk S, Derby JA, Melzer M, Boland GJ. The effect of fertilizers on the efficacy of the bioherbicide, *Phoma macrostoma*, to control dandelions in turfgrass. Biological Control. 2013;**65**: 147-151

[137] Smith J, Wherley B, Reynolds C, White R, Senseman S, Falk S. Weed control spectrum and turfgrass tolerance to bioherbicide *Phoma macrostoma*. International Journal of Pest Management. 2015;**61**:91-98

[138] Graupner PR, Carr A, Clancy E, Gilbert J, Bailey KL, Derby JA. The macrocidins: Novel cyclic tetramic acids with herbicidal activity produced by *Phoma macrostoma*. Journal of Natural Products. 2003;**66**:1558-1561

[139] Quereshi S, Khan NA, Pandey AK. Anthraquinone pigment with herbicidal potential from *Phoma herbarum* FGCC#54. Chemistry of Natural Compounds. 2011;**47**:521-523

[140] Abu-Dieyeh MH, Watson AK. Efficacy of *Sclerotinia minor* for dandelion control: Effect of dandelion accession, age and grass competition. Weed Research. 2007;**47**:63-72

[141] Skipp RA, Bourdot GW, Hurrell GA, Chen LY, Wilson DJ, Saville DJ. *Verticillium dahliae* and other pathogenic fungi in *Cirsium arvense* from New Zealand pastures: Occurrence, pathogenicity and biological control potential. New Zealand Journal of Agricultural Research. 2013;**56**:1-21

[142] Briere SC, Watson AK, Hallett SG. Oxalic acid production and mycelial biomass yield of *Sclerotinia minor* for the formulation enhancement of a granular turf bioherbicide. Biocontrol Science and Technology. 2000;**10**:281-289

[143] Bailey KL. The bioherbicide approach to weed control using plant pathogens. In: Abrol DP, editor. Integrated Pest Management: Current Concepts and Ecological Perspective. San Diego, CA: Elsevier; 2014. pp. 245-266

[144] Kenny DS. DeVine the way it was developed: An industrialist's view. Weed Science. 1986;**34**:15-16

[145] Prasad R. Development of bioherbicides for integrated weed management in forestry. In: Brown H et al., editors. Proceedings of the 2nd International Weed Control Congress; Slagelse, Denmark: Department of Weed Control and Pesticide Ecology; 1996. pp. 1197-1203

[146] Jones RW, Hancock GJ. Soilborne fungi for biological control of weeds. In: Hoagland RE, editor. Microbes and Microbial Products as Microbial Herbicides. Washington, DC: American Chemical Society; 1990. pp. 276-286

[147] Khattak SU, Iqbal Z, Lutfullah G, Bacha N, Khan AA, Saeed M, et al. Phytotoxic and herbicidal activities of *Aspergillus* and *Penicillium* species isolated from rhizosphere and soil. Pakistan Journal of Weed Science Research. 2014;**20**:293-303

Chapter 16

Sustainable Development of Horticulture and Forestry through Bio-Inoculants

Easan Mohan and Kuppu Rajendran

Abstract

The role of microorganism is very critical in nutrient management of horticulture and plantation forestry. They are conductors of the nutrient management orchestra as they provide by inputs in terms of micro and macronutrients besides organic matter and can be called as bio-inoculants (biofertilizers). Biofertilizers play a vital role in fixing the atmospheric nitrogen and mobilization of phosphorous, sulfur, manganese, copper, and iron in the soil. Symbiotic (Rhizobium and Frankia) and nonsymbiotic microorganisms (*Azospirillum*) are known to improve the soil fertility by fixing the atmospheric nitrogen. Arbuscular mycorrhizae fungi (AM fungi) and phosphobacterium have ability to transfer insoluble phosphate into soluble form. Moving in this direction it becomes imperative to understand as forest microbiologist and pathologist, the roles played by microorganism in diverse plants-soil-microbe interaction to analyze their effectiveness in improving their efficiency. Biofertilizers are economy and environmentally safe, and there is a growing awakening among the tree growers and farmers. In agriculture, advantages of biofertilizer application are better known, but in tree crops, the utility of biofertilizers is still in an experimental stage. The review paper is collective evident for the compatibility of different biofertilizers and their augmentation effect on the production of quality seedling and nutrient management of tropical horticulture and plantation forestry.

Keywords: arbuscular mycorrhizae fungi (AM fungi), bio-inoculants (biofertilizers), horticulture, plantation forestry, sustainable nutrient management, Rhizobium, Frankia

1. Introduction

Plants are being an important component of socio-economic condition of human life and culture. Ever since the beginning, tree crops have furnished use with three of life's essentials, wood, food and oxygen. Besides these, they provide additional necessities to human being such as shelter, fuel wood, fodder for livestock, ethno medicine, architectural, agriculture implements, building construction tools, sound and wind barriers, soil improvement through litter production and nitrogen fixation in association with Rhizobium and Frankia. Many drugs which derived from plants generally have been replaced by more potent synthetic ones and trees remain a source for some drug ingredients for pharmaceutical industry. They play an important role in ecosystem services through carbon sequestration, improving air quality, climate amelioration, and conservation of water and supporting wildlife. They also reduce the atmospheric temperature and the impact of greenhouse gases by maintaining low levels of carbon dioxide.

Plant growth and productivity is generally regulated by the availability of soil nutrients. One of the major efforts to increase the plant productivity is through management of nutrients, which can be achieved by application of fertilizers. However, application of chemical fertilizers is not eco-friendly and economically viable in current scenario. Other alternative method is supplement of bio-inoculants (bio-fertilizers) for sustainable development of horticulture and forestry crops. Bio-inoculants are plant growth promoting beneficial microorganisms such as the species of *Azospirillum, Azotobacter, Bacillus, Ecto and Endo-mycorrhizal* fungi, *Frankia, Pseudomonas, Rhizobium*, *Trichoderma,* etc. Such microorganisms accelerate certain microbial process to augment the extent of availability of nutrients in the form, which can be assimilated by plants and also maintain the plant health by controlling diseases.

2. Rhizosphere and microbial interaction

Various types of microorganisms inhabit air, water and soil. They play an important role in restoring the physical, chemical and biological property of soil. Rhizosphere ecology and microbial interactions are responsible for key environmental processes, such as the bio-geo chemical cycling of nutrients, organic matter and maintenance of plant health and soil quality [1]. Among the microbial population, both beneficial and harmful bacteria as well as fungal species were found, but the microbial population was low when compared to rhizosphere soil [2].

Rhizosphere is the physical location in soil where plants and microorganisms interact. The interest in the rhizosphere microbiology derives from the ability of the soil microbiota to influence plant growth and vice versa. The presence of microorganisms in the rhizosphere will increase root exudation and it was found that 5–10% of the fixed carbon was exudates from the root under sterile condition, on the introduction of beneficial microorganisms, root exudation rate increases by 12–18%. The interaction between bacteria and fungi associated with plant roots may be beneficial, harmful or sometimes neutral for the plant, and effect of a particular bacterial species may very as a consequence of soil environmental conditions [3]. The beneficial microbes can be divided into two major types based on the living nature; free living (that live in soil) and symbiotic relationship with the plant root nodule of legume and actinomycete plants [4].

3. Bio-inoculants

Bio-inoculants are beneficial microorganisms for nutrient management, plant growth and are eco-friendly and natural inputs providing alternate source of plant nutrients, thus increasing farm income by providing extra yields and reducing input cost also. Bio-inoculants increase crop yield by 20–30%, replace chemical N and P by 25%, stimulate plant growth, enhance soil biodiversity, restore natural fertility and provide protection against drought and some soil borne plant pathogens. The role of bio-inoculants has already been proved extensively in enhancing the mineralization processes of organic matter and helping the release of nutrients, utility of soil organic matter contents and cations exchange capacity [5] and therefore, bio-inoculants are gaining importance in agriculture for the past few decades. However, the scientific exploitation of bio-inoculants in horticulture and forestry is scanty in developing countries like India.

4. Classification of bio-inoculants for tree crops

1. Nitrogen fixing symbiotic microorganisms (*Rhizobium* and *Frankia*)

2. Nitrogen fixing non-symbiotic microorganisms (*Azospirillum, Azotobacter* and blue-green algae)

3. Phosphate solubilizing microorganisms (*Arthrobacter, Pseudomonas, Bacillus, Aspergillus*)

4. Phosphate mobilizing microorganisms (Ecto and Endo—Mycorrhizal fungi)

5. Potash mobilizer (*Bacillus* sp., *Pseudomonas* sp.)

6. Sulfur uptake (*Pseudomonas, Klebsiella, Salmonella, Enterobacter, Serratia* and *Thiobacillus*)

7. Zinc solubilizer (*Bacillus subtilis, Thiobacillus thiooxidans* and *Saccharomyces* sp.)

8. Iron uptake (*Pseudomonas fluorescens*)

9. Plant growth promoters (*Pseudomonas* sp., *Bacillus* sp., *Serratia* sp.)

5. Plant growth promoting rhizobacteria (PGPR)

Plant growth promoting rhizobacteria are group of bacteria that actively colonize roots and increase plant growth and yield [6]. It enhances plant growth and productivity by synthesizing phytohormones, increasing the availability and facilitating the uptake of nutrients by decreasing heavy metal toxicity in the plants, antagonizing the plant pathogens [7]. The mechanisms by which PGPR promote growth are not fully understood [8], against phytopathogenic microorganisms by production of siderophores, the synthesis of antibiotics, enzymes and fungicidal compounds [9] and also solubilization of mineral phosphates and other nutrients [10].

5.1 Azospirillum

Azospirillum species are free-living N_2-fixing bacteria commonly found in soils and in association with roots of agriculture, horticulture and forestry species [11]. *Azospirillum* are known to act as plant growth promoting rhizobacteria (PGPR) and stimulate plant growth directly either by synthesizing phytohormones or by promoting improved N nutrition through biological nitrogen fixation (BNF). PGPR also produce several the growth promoting substances including IAA, GA3, Zeatin and ABA [12]. Presently there are seven species have been identified in this genus, *A. amazonense* [13], *A. brasilense, A. lipoferum,* [14], *A. doebereinerae* [15], *A. halopraeferens* [16], *A. irakense* [17] *and A. largimobile* [18].

Applications of plants with Azospirillum have promoted plant growth of agronomically important field crops by 10–30% in the field experiment [19, 20]. Nursery experiments proved that the inoculation of tree cops with *Azospirillum* could result in significant changes in various growth parameters, particularly shoot and root growth, biomass, nutrient uptake, tissue nitrogen content, leaf size of several shola tree species [21] and *Casuarina equisetifolia* [22, 23], *C. cunninghamiana* Mig. [24],

Moringa oleifera [25], *Acacia nilotica* [26], *Azadirachta indica* [37], *Delonix regia* [28], *Erythrina indica* [29], *Feronia elephantum* [30], *Jatropha curcas* [31]. Two years old *Casuarina equisetifolia* plants treated with bio-inoculants in field condition improve the growth of plants by 90% over uninoculated control [32]. *Azospirillum lipoferum* treated with *Jatropha curcas* under field conditions has increased the shoot length by 44.85% and primary and secondary root length by 39.3 and 37.5% respectively. Similarly, the root and shoot biomass also increased by 24.01 and 15.04% leaf area by 28.57% increase over control and the other *Azospirillum* species such as *A. brasilense*, *A. haloference* and *A. amazonense* [33]. The stimulatory effect exerted by *Azospirillum* has been attributed to several mechanisms including secretion of phytohormones (auxins and gibberellins), biological nitrogen fixation, and enhancement of mineral uptake of plants [8] due to the ability of synthesis of in vitro phyto-hormones such as IAA, gibberellins, cytokinin [34, 35] and produced by ethylene [36].

5.2 Effect of bio-inoculants and biochemical changes of tree crops

Plants inoculated with *A. brasilense* were always characterized by a higher chlorophyll concentration. Inoculation of crops caused a statistically significant increase of chlorophyll content in the case of oats in 1996 (15%) and wheat in 1997 (15%). Chlorophyll appeared to be a sensitive indicator of inoculation effect, which was also supported by Bashan et al. [37]. *A. lipoferum* inoculated *Jatropha curcas* seedlings has increase in level of chlorophyll a, b and carotene and such increase was maximum by 31.98, 14.5 and 18.9% and protein content (37.35%) amino acid (26.33), lipids (8.9) and carbohydrates (9.37) when compare to control plant under field conditions [31]. The total chlorophyll and soluble protein content was found to be higher in the *Moringa oleifera* seedlings inoculated with *A. brasilense* [25].

5.3 Pseudomonas

The genus *Pseudomonas* is one of the most diverse gram-negative non-spore forming, motile, rod shaped bacteria with an important metabolic versatility and pathogenicity [38]. Morphologically this genus is straight or slightly curved rods and produced yellowish green pigment in King's B. Medium. Plant growth promoting rhizobacteria consisting of primarily *Pseudomonas fluorescens* and *P. putida* were identified as important organisms with ability for plant growth promotion and effective disease management properties. The population density of fluorescent pseudomonas in the rhizosphere in usually reduced by AM fungi colonization [34, 39, 40]. Many strains of genus *Pseudomonas* possess the capability to promote plant growth [41], due to their 1-aminocyclopropane-I carboxylate deaminase activity, indole acetic acid (IAA) and siderophore production [42], PGPR can exert a beneficial effects on plant growth by suppressing soil borne pathogens [43], improving mineral nutrition [44] and phytohormone synthesis [45].

6. Arbuscular mycorrhizal (AM) fungi

The symbiotic association between fungus and root systems of higher plants is called mycorrhiza, which literally means root fungus. Ectomycorrhizae and entomycorrhizae or arbuscular mycorrhizae (AM) are playing important role in phosphorus and micronutrients uptake by tree species. The AM fungi association is endotrophic, and has previously been referred to as vesicular-arbuscular mycorrhiza (VAM), this name has been dropped since 1997 in favor of AM fungi, because all fungi are not produced vesicles [46]. Arbuscular mycorrhizal fungi belong to the division Zygomycetes and order Glomales. There are six genera of AM fungi have been identified and are *Glomus, Gigaspora, Aculospora, Scutellospora, Entrophosphora,* and *Sclerocystis. Acaulospora* and *Scutellospora* belong to Gigasporaceae; *Glomus* and *Sclerocystis* belong to Glomaceae [47]. Arbuscular mycorrhizal fungi (AMF), belonging to the phylum *Glomeromycota,* are obligate symbiotic fungi forming mutualistic associations with the roots of most of the tropical plants. Increased access to low-mobility soil mineral nutrients has been considered to be main beneficial effect of AMF on their host plants [48]. In addition, they have been shown to improve the uptake of Zn, Cu, S, Mg, Ca, K and other nutrients [49]. The AM fungal mycelia have been reported to stabilize soil through the formation of soil aggregated [50].

Arbuscular mycorrhizal (AM) fungi are the most widespread type and ecologically important root fungal that form symbiosis with 80% of land plant species which depend upon them for growth [51]. AM fungal symbiosis is characterized by fungal penetration of root cortical cells forming microscopic branched structures called arbuscules that increase that increase efficiency of plant-fungus metabolite exchange [48]. These microsymbionts occur widely under various environmental conditions with beneficial effects on soil structure improvement [52, 53] and have great importance due to their higher capacity to increase growth and yield through efficient nutrient uptake in infertile soils, water uptake and drought resistance in plants [54].

6.1 A combined effects AM fungi and *Pseudomonas* in tree species

The interaction between *Pseudomonas* and the arbuscular mycorrhizal fungus, *Glomus clarum* NT4 on spring wheat grown under gnotobiotic condition was investigated [55]. Although plant growth responses varied, positive response to Pseudomonad inoculants was obtained. Shoot biomass enhancement ranged from 16 to 48%, whereas enhancement ranged from 82 to 137% for roots. Typically, dual inoculation positively influenced the magnitude of response associated with any organism applied alone.

The highest mycorrhizal root colonization and number of AM fungal spores, and pseudomonas population were observed when *G. fasciculatum* and *P. monteilii* were coinoculated on to *Coleus forskohlii* plants [56] under organic field condition. Negative effects of *Glomus intraradices* on population of PGPR, *P. fluorescens* DF57 were shown by Ravnskov et al. [57] and suggested that competition for inorganic nutrients might explain the effect, since the mechanism did not require cell-to-cell contact. Marschner et al. [58, 59] suggested that similar negative effects of *Glomus intraradices* on *P. fluorescens* 2-79RL might be due to mycorrhizal induced decreases in root exudation, affecting the composition of the rhizosphere soil solution. *P. fluorescens* 92rk and P190r, and *G. mosseae* BEG12, inoculated alone, promoted tomato plant growth. Plant growth promotion by florescent pseudomonads has been ascribed to the suppression of phytopathogenic soil-borne microorganisms [43, 60]. Moreover, co-inoculation of three microorganisms showed synergistic effects compared with single inoculated plants and reports demonstrate additive effects on plants on plant growth of AMF and rhizobacteria [61, 62].

6.2 Effect of AM spores in rhizosphere of three species

The occurrence of AM spores depends upon the environmental conditions, plant species and soil type. There are two different types of AM spores such as *Acaulospora* and *Glomus* were observed in non-rhizosphere soil. Among the two different AM spore, Glomus was the dominant one. Spore density was very low 8 spore/100 g of soil [63, 64]. Analysis of root colonization was higher in mycorrhizal than non- mycorrhizal plants. Santhaguru et al. [65] reported that VAM infection was 100% in *Albizia amara, Peltophorum pterocarpum* and *Pongamia glabra, 80% in Derris scandens* 78% in *Erythrina variegata*, 18% in *Pterlobium* and16% in *Prosopis chilensis*. However, there is no VAM fungi infection in five plant species viz. *Albizia lebbeck, Bauhinia tomentosa, Cassia, Prosopis juliflora* and *Tamarindus indica* at Alagar Hills of Tamil Nadu, India. Similarly, AM Fungi colonized with several tree species semi-arid zone of South India, 1, 2 and 3 years old *Casuarina equisetifolia* [2], *Leucaena leucocephala* [66], *Feronia elephantum* with AM fungi (*Glomus fasciculatum*), *Samanea saman* [67]. Similarly, 16 different species of Arbuscular mycorrhizal fungi were isolated from rhizosphere of teak (*Tectona grandis*) among these *Glomus* and *Aculospora* found in dominant species and seedlings inoculated with combination of Arbuscular fungi had good quality seedlings and increased shoot height compared to with individual AM fungus in *Tectona grandis* [68].

6.3 Role of bio-inoculants on plant growth and metabolites

Leucaena leucocephala seedlings were inoculated with different types of vesicular-arbuscular mycorrhizal fungi found that the collar diameter increment of between 18 and 123% [66]. Similarly, *Pterocarpus indicus* inoculated with vesicular-arbuscular mycorrhizal fungi improve the shoot diameter [69], root collar diameter in sweet gum seedlings by 268% [70]. *Feronia elephantum* with AM fungi (*Glomus fasciculatum*) increase the plant growth especially root length and was recorded the root length increment was up to 84% [30]. Similarly shoot length was higher in *Samanea saman* [67] Mycorrhiza colonization also protect the roots from the soil pathogens [71]. AM fungi significantly increase the net photosynthesis by increasing total chlorophyll and carotenoid contents ultimately increasing carbohydrate accumulation. The chlorophyll content, fresh weight and leaf area are higher in mycorrhizal plants than in non-mycorrhizal plants but differences are significant only under draught stress conditions [72]. In mycorrhizal infected groundnut roots, high concentrations of ortho-hydroxy phenols were present. This type of phenols has been known to play an important role in plant disease resistance [73]. Inoculation of AM fungi is enhancing the plant quality by stimulating the synthesis of secondary metabolites which can be important for plant tolerance to abiotic and biotic stresses [74]. According to Morandi et al. [75] the Phenolic substances, such as phytotoxins are synthesized when the root is infected by a pathogen. They are non-specific toxic substances, which can be considered to play a role in disease resistance. Kapoor et al. [76] observed a significant increase in the density of glandular trichomes in the medicinal plant *Artemisia annua* following inoculation with the AM fungi *G. macrocarpum* and *G. fasciculatum* contributing to enhance artemisinin content in the plants.

The chlorophyll a, chlorophyll b, total chlorophyll and Carotenoid contents increased in mycorrhizal seedlings compared with non-mycorrhizal tree seedlings of *Cassia siamea*, *Delonix regia*, *Erythrina variegata*, *Samanea saman* and *Sterculia foetida* [77]. A significant enhancement in biochemical parameters like total chlorophyll content, soluble protein and NRase activity in *Pongamia pinnata* seedling 10.7, 48.5 and 43.6% increase over control with the combined inoculation of Rhizobium, Phosphobacteria and AM fungi [78]. Similarly, an increase in chlorophyll content and soluble protein was observed in *Ziziphus mauritiana* when inoculated with AM fungi [79] and *Dalbergia sissoo* inoculated with Rhizobium and mycorrhizae [80] and in Shola species inoculated with Azospirillum + Phosphobacteria and AM fungi [22]. Eucalyptus seedlings inoculated with mixed *Glomus mosseae, Trichoderma viride* and *Glomus fasciculatum* increases the phosphorous content of shoot and root over control. Then increased rate of P uptake and inflow in roots is regarded as the major contribution of AM infection [81]. The AM colonization increased initially up to 45 days but decreased thereafter [82].

6.4 Effect of AM fungi on growth and nutrient content

The fundamental importance of the mycorrhizal associations in restoration and to improve the revegetation is well recognized [83]. Arbuscular mycorrhiza colonized plants showed significant increment in height, biomass production and girth as compared to non mycorrhizal plats. Growth, biomass and P uptake were higher were higher on dual inoculation of *G. fasciculatum* and *G. macrocarpum* as compared to uninoculated tree species under both nursery and field condition. Tropical trees inoculated with AM fungi have shown increased nutrient uptake and growth, withstanding the transplant stock, hostile conditions like drought resistance and survival of *Acacia holosericea* [84]. *Casuarina equisetifolia* seedlings inoculated with AM (*Glomus fasciculatum*) increased shoot and root biomass [23, 24], *Eucalyptus tereticornis* [85] *Tectona grandis* [68] *Santalum album*, *Acacia auriculiformis*, *Grevillea robusta*, *Eucalyptus camaldulensis*, *Bombax ceiba* [86, 87] and *Albizia lebbeck* [88].

Inoculation with *Glomus mosseae* and *G. fasciculatum* along with other nitrogen fixing and phosphate solubilizing organism improved the quality and growth of neem seedlings, owing to greater absorption of nutrients, under nursery conditions in unsterilized soil [89], AM fungus (*G. fasciculatum*) and Rhizobium treated *Acacia nilotica* seedlings recorded an increase in shoot and root biomass [90]. Beneficial effects of AMF, such as growth promotion, increased root branching, lengths of lateral roots, specific root length and root diameter [91], protection against pathogens [92] and tolerance to abiotic stresses [93], could be due to positive interactions between mycorrhizae and associated microorganisms such as Pseudomonas, Arthrobacter and Burkholderia in a particular environment [94].

Combined inoculation of *Glomus fasciculatum* and Rhizobium on the growth of *Prosopis juliflora* seedlings showed better growth on shoot length and biomass. It was found that *G. fasciculatum*, Scutellospora sp., *G. leptotichum* and *G. mossease* were most efficient for *Dalbergia sissoo*, *Acacia auriculiformis*, *A. nilotica* and *Dalbergia latifolia*, respectively, and increase in plant biomass and height was to the extent of 34 and 24%, respectively, in *Dalbergia sissoo*, 126 and 50% in *A. auriculiformis*, 48 and 24% in *Dalbergia latifolia* and 100 and 112% in *Acacia nilotica* [95].

7. Trichoderma

The genus Trichoderma is the most common fungi found in all climatic condition. It can be isolated in all type of soil. It is also found in plant root, rotting wood, plant litter and seed. Fungi of the genus Trichoderma are important biocontrol agents (BCAs) of several soil borne phytopathogens. Trichoderma use different mechanisms for the control of phytopathogens which include mycoparasitism, competition for space and nutrients, secretion of antibiotics and fungal cell wall degrading enzymes. In addition, *Trichoderma* could have a stimulatory effect on plant growth 48 as a result of modification of soil conditions.

Shoot length and fresh weight were more in *Eucalyptus saligna* seedlings inoculated with *Trichoderma viride*. The greater height and fresh weight of *Acacia nilotica* inoculated with Trichoderma due to the Trichoderma species produce growth hormones which result in better growth of shoots. *Trichoderma* sp. co-inoculated with *Azotobacter* sp. and *Bacillus megaterium* showed a significant increase on the growth of Teak and Indian red wood under nursery condition [96]. The growth promoting substances are known to cause enhanced cell division and root development [97]. Similarly, many strains of Bacillus pseudomonas and Trichoderma have been implicated in improvement of overall growth of many crop plants [98].

8. Rhizobium

Rhizobium belongs to family Rhizobiaceae and the bacteria have the ability to reduce N_2 and thereby "fix" atmospheric nitrogen using the enzyme nitrogenase. It colonizes the roots of species legumes to form tumor like growths called root nodules, which act as biofactories of ammonia production (**Figure 1**). The process of biological nitrogen fixation was discovered the Dutch microbiologist Martinus Beijerinck. Rhizobia (e.g., Rhizobium, Mesorhizobium, Sinorhizobium) fix atmospheric nitrogen or dinitrogen, N_2 into inorganic nitrogen compounds such as ammonium, NH_4, Which is then incorporated into amino acids, which can be utilized by the plant. Plants cannot fix nitrogen on their own, but need it in one form or another to make amino acids and protein. Because legumes form nodules with rhizobia, they have high levels of nitrogen available to them. Rhizobium is a soil habitat bacterium, which is able to colonize the legume roots and fixes the atmospheric

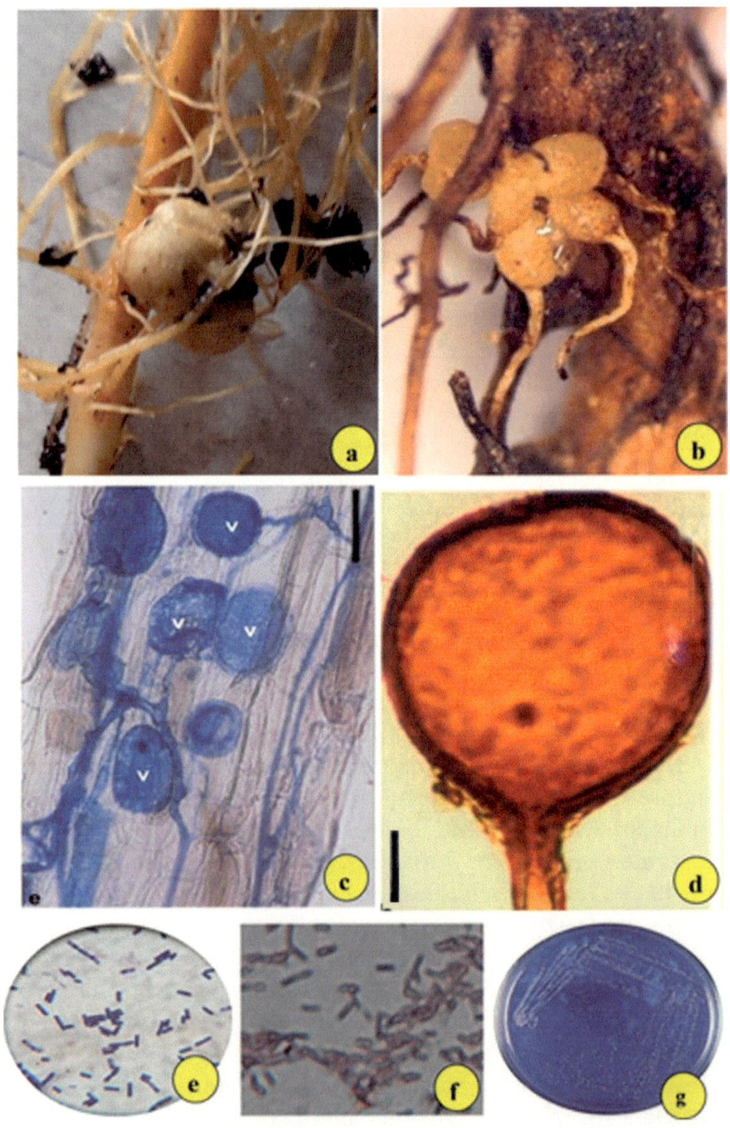

Figure 1.
(a) Rhizobium root nodule, (b) Frankia *root nodule, (c) AM fungi infection, (d) AM fungi spore, (e) phosphate solobilizing bacteria* Bacillus *sp., (f)* Paenibacillus polymyxa, *(g)* Azospirillum brasilense.

nitrogen. Rhizobium associated with nodulated legume trees have an outstanding potential for fixing atmospheric nitrogen (*Sesbania cannabina* and *Leucaena leucocephala*) can fix up to 75–584 kg N ha^{-1} yr^{-1} [99]. In recent year use of Rhizobium culture has been routinely recommended as an input in legume tree species cultivation. Rhizobium helps to boost up the tree growth by insoluble nutrients available for plant. Seedling treated with Rhizobium biofertilizer found to remarkable increase in growth and nodulation of *D. sissoo* [100], *A. nilotica* [26] *Albizzia* sp. [101].

8.1 Rhizobium with helper microbes

High nitrogen yield was estimated in the *Pongamia pinnata* seedling inoculated with Rhizobium + Phosphobacteria + VAM fungi [78]. Increased N content in the plant sample of various tree seedling, co-inoculated with different biofertilizers [102]. Similarly increase in biomass production due to VAM fungi inoculation with *Acacia* sp. [103] and in *Albizzia* sp. [104]. Rhizobium inoculation + PSB with 25% N significantly increase the average 53 nodule no./seedling was followed by only Rhizobium with 25% N (38 nodule/seedling) inoculation in *Acacia nilotica* shoot length increased from 58.50 to 78.75 cm, collar diameter from 5.05 to 6.15 mm and nodulation increased 0.071 to 0.342 g/seedling [105].

8.2 *Azospirillum* and *Rhizobium* interaction

Dual inoculation of *Azospirillum* and *Rhizobium* with legume plant has been found to increase plant-growth when compared with single inoculations. *Azospirillum* is considered a helper bacteria to *Rhizobium* by stimulating nodulation, nodule function, and possibly plant metabolism. Similarly, phytohormones produced by *Azospirillum* promoted epidermal-cell differentiation in root hairs that increased the number of potential sites for rhizobial infection and more nodule development [106]. Dual inoculation of AM fungi with *Rhizobium* improved nodulation, plant dry weight, N and P contents of *Leucaena leucocephala* in a P deficient soil compared to single inoculation with either organism [107] .

8.3 *Azotobacter* plant interaction

Azotobacter is a free living (non-symbiotic), aerobic, nitrogen fixing organism and these gram negative bacteria belongs to family Azotobacteriaceae. There are seven species of *Azotobacter* viz. *A. beijerinckii, A. chroococcum, A. vinelandii, A. paspali, A. agilis, A. insignis* and *A. macrocytogenes. A. chroococcum* appeared more in acidic soils and arable soils while *A. beijerinckii* in neutral and alkali soils. Apart from nitrogen, this organism is capable of producing antibacterial and antifungal compounds, hormones and siderophore [108]. Individual or combined inoculations stimulated the plant growth and significantly increased the concentrations of indole 3-acetic acid (IAA), P, Mg, N, and total soluble sugars in agri crop. Bioinoculants co-inoculation of nitrogen fixing organism *Azotobacter* and phosphate solubilizing microorganisms *Bacillus megaterium* showed a significant increase on the growth of teak and India red wood under nursery condition [96].

Azotobacter inoculated strawberry plants attained maximum height (24.92 cm) more number of leaves per plant (26.29), more leaf area (96.12 cm^2), number of runners per plant (18.70), heavier fruit (10.02gm), more fruit length (35.9 mm), and more fruit breadth (22.91 mm) as compared to all other treatment [109]. Similarly, combined application of manure + Azotobacter + wood ash + phosphorous solubilizing bacteria + oil cake improved significantly fruit diameter (3.11 cm), length (3.95 cm), volume (20.397 cm^3), weight (11.11 g), total sugars (7.95%), total soluble solids (9.01'B), acidity (0.857), TSS:acidity ratio (11:12) and yield (238.95 g/plant) [110].

Name of the species	Azospirillum/Azotobater			Frankia/Rhizobium			Phosphate solubilizing bacteria			AM fungi			Combination of bio-fertilizers			Reference
	CD	SL	BM	CD	SL	BM	CD	SL	BM	CD	SL	BM	CD	SL	BM	
Casuarina equisetifolia L.	14.2	55	19.6	17.6	23	23	6.8	406	11	18	19.5	23	63.5	62.2	115.0	[22]
Acacia nilotica L.	61.0	60	26	125	57	56	NA	NA	NA	96	60	48	236.3	131.2	156.8	[26]
Azadirachta indica (A) Juss	1.1	2.34	3.2	NA	NA	NA	NA	5.5	3.74	0.7	7.08	5.4	0.67	16.0	7.49	[27]
Moringa oleifera L.	75	22	276	NA	NA	NA	5.04	6	17	NA	NA	NA	5.6	11.7	176.5	[25]
Mangifera indica (L.) Delile	12.0/13.3	12.7/ 13.7	—	NA	NA	NA	NA	NA	NA	14.73	14.5	NA	11.8	12.9	NA	[117]
Delonix regia (Hook.) Raf.	5.8/4.2	5.2/ 4.3	17.8/5.8	NA	NA	NA	NA	NA	NA	3.5	2.42	0.9	13.0	7.9	21.2	[28]
Tectona grandis L.f.	--/69.8	--/ 0.65	--/28.2	NA	NA	NA	31.7	0.65	106.8	6.3	7.2	37.4	114.3	26.8	258.9	[116]
Samanea saman (Jacq.) Merr.	60.0	20.6	35.8	NA	NA	NA	54.4	6.68	19.5	78.9	16.9	30.9	108.6	46.9	74.9	[67]
Feronia elephantum L.	71.4	39.5	55.5	NA	NA	NA	48.8	6.68	15.9	82.9	20.7	41.7	122.8	47.0	92.4	[30]
Gmelina arborea (Roxb.)	11.6	13.9	38.8	NA	NA	NA	11.9	8.8	27.5	11.9	9.43	63.4	25.6	21.9	166.4	[118]

CD, collar diameter; SL, shoot length; BM, biomass; NA, not applicable.

Table 1.
Growth and biomass increases (percentage increased over control) of horticulture and forestry crops treated with bio-inoculants.

9. Frankia with actinorrhizal plants

Frankia is a genus of Actinomycetes, belongs to family Frankiaceae and an ability to fix the atmospheric nitrogen in symbiotic association with *Casuarina* species in tropical and temperate environmental condition. These microorganisms usually invade root hairs of Casuarina and developing within cortical cells in lobes of the resultant nodules. Frankia are able to convert the nitrogen gas in the atmosphere into amino acids, which are the building blocks of proteins. Frankia exchange nitrogen for carbohydrates from the plant. As the plant drop organic matter, or when the plants die, the nitrogen from their tissues is made available to other plants and organisms. This process of accumulating atmospheric nitrogen in plants and recycling it through organic matter is the major source of nitrogen in tropical ecosystems. Various agro-forestry practices such as alley cropping, improved fallow, and green manure/cover cropping exploit this natural fertility process by using nitrogen fixing plants.

Casuarina equisetifolia seedling inoculated with Frankia strains showed improved growth, biomass and tissue N content over control seedlings [24, 111, 112]. Nitrogenase activity of Frankia strains were significantly ($p < 0.05$) and negatively correlated with a tissue N content [111]. Similarly, under nursery experiments, the growth and biomass of *C. equisetifolia* rooted stem cuttings inoculated with Frankia showed three times higher growth and biomass than uninoculated control and improved growth in height (8.8 m), stem girth (9.6 cm) and tissue nitrogen content (3.3 mg/g) than uninoculated controls in field condition [112]. Frankia inoculated Casuarina seedlings planted in farm forestry improve the tree growth and biomass in the field condition [2, 112] and improve the nutrient cycling of actinorrhizal plants through high amount of litter production and decomposition [113]. Combined inoculation of Azospirillum, Phosphobacteria, AM fungi and *Frankia* produced excellent growth and biomass of *C. equisetifolia* seedlings due to co-inoculation with *Frankia* through improved nitrogen fixation [22, 114] (**Table 1**).

10. Methods of inoculation

10.1 Inoculation methods of *Azospirillum, Rhizobium* and phosphobacteria

Seed or nursery stage is best for application of bio-fertilizers. Suitable methods for forestry species is seed coating and inoculation in polythene bag. Two grams of carrier culture (10^{-8} cfu/g) can be applied in rhizosphere of seedlings in the polythene bags in the nursery.

10.2 Inoculation with seeds

Inoculation requirement varies from the size of the seeds. Normally 200 g of lignite/peat soil based culture (10^{8} cfu/g) is need for every 8–10 kg of seeds of the tree species. A slurry is formed by mixing the inoculant with cooled rice gruel (250 ml). The required quantity of seeds is added in the slurry and mixed thoroughly so that each seed is coated with the black colored inoculant. The treated seeds are then shade dried for 30 min and sown.

10.3 Inoculation in the nursery mother bed

Two hundred grams of lignite based carrier culture of *Rhizobium* or *Azospirillum* (10^{8} cfu/g) is required for 4 m × 1 m mother bed. It has to be spread uniformly and mixed thoroughly before sowing of seeds.

10.4 Seed treatment

Ten percent sugar or gum arabic solution or rice porridge is to be prepared to serve as a sticker for culture cells applied to seeds. This solution is to be sprinkled on required seeds and then the seeds spread on a polythene sheet and mixed uniformly. The peat based culture is sprinkled uniformly over the sticker-coated seeds and mixed simultaneously. After treatment the seeds are air dried in 1 h then the seed can be dipped in nursery mother bed.

10.5 Dipping seedlings

In case of transplanted seedlings, the seedlings from the nursery beds are uprooted and tipped in a suspension of biofertilizers before planting.

10.6 Inoculation in the nursery seedlings

Two gram of lignite based culture (10^8 cfu/g) is added to rhizosphere of the seedlings a week after transplanting. In the case of AM 5 g of vermiculate based culture can be used. The cultures may be mixed together and applied near the root zone. If necessary, the inoculant may be made bulk by mixing with the finely powered farm yard manure or sand for easy application.

10.7 Inoculation of out plantings

Ten grams of lignite based culture (10^8 cfu/g) is required per seedling which are to be planted in the field directly from the mother bed, in the form of naked root seedlings. Otherwise, 200 g of lignite based culture can be mixed with 10 l of water, and roots of seedlings can be dipped in it before planting.

11. Advantages of biofertilizers

- Biofertilizers have number of advantages than synthetic fertilizers. Bio fertilizers can facilitate not only supply of nutrients, but also produces vitamins and plant growth hormones. They prevent soil erosion by producing capsular polysaccharides and also control plant pathogens.

- Biofertilizers, will be isolated from the rhizosphere soil of host plant hence huge amount need not be spent for mother culture. It can be cultivated under normal laboratory condition using conventional media and fermentors within short span of time. Production method is very simple and production cost is cheaper than chemical fertilizers.

- Chemical fertilizers are required in huge quantity for land application. The physical optimum levels for getting the maximum grain yield for the medium duration rice hybrid $CORH_2$ was found to be 151:66:57 kg N, P_2O_5 and K_2O ha^{-1} [115]. But in case of biofertilizers, 1 g of carrier based inoculum of Azospirillum and phosphobacterium contain with a population load of 10^{-9} and 10^{-8} and approximately 12,500 infective propagule/10 g of soil [22]. Hence, very less quantity is sufficient and it may get multiplied into many fold as the optimum environmental conditions in the nursery and field. As the propagules multiply in the field they need not be applied repeatedly.

- It helps to improve the seed germination and induces the healthy seed emergence due to production of growth promoting hormones, gibberellins and cytokinin-like biologically active substances. Biofertilizers promote better root formation in trees for efficient absorption and assimilation of water and nutrients.

- Biofertilizers are involved in the litter decomposition and the breakdown of minerals into available form to plants. It directly facilitates the function of rhizoids in terms of absorption and translocation of minerals and water.

- Biofertilizers do not pollute the soil, whereas excess application of chemical fertilizers creates soil pollution. Biofertilizers are effective in promoting and maintaining the soil fertility which helps a better balance in the plantation forest ecosystem in terms of nutrient availability and cycling of nutrients.

- Due to the strong colonization of biocontrol microorganism and their secretory substances, the tree plants cultivated under this pattern will exhibit a strong resistance against an array of infectious disease caused by plant pathogens.

11.1 Limitations of bio-fertilizer utilization in forestry and horticulture

Apart from the advantages, biofertilizers have certain limitations. Lack of awareness on benefits of bio inoculants among the farmers and tree growers. Adequate availability and quality assurance of bioinoculants are being the limiting factors. Competition between native and introduced microbial population in the cultivated field also identified as a limiting factor. Hence, a preliminary analysis on the cultivable land about the native microflora, physico-chemical parameters is essential to overcome such limitations.

12. Conclusion

Bio-inoculants are renewable, cost effective, eco-friendly and economically viable population of beneficial microorganisms providing an alternate source of plant nutrients, thus increasing farm income by providing extra yields and reducing input cost. Bio-inoculants increase crop yield by 20–30%, replace synthetic fertilizers of N & P by 25%. Stimulate plant growth, activate soil biologically, restore natural fertility and provide protection against drought and some soil borne plant pathogens. Application of Bio-fertilizers in combined form in Horticulture and Forestry will play an important role in improving the soil fertility by supply of macro and micronutrients, organic carbon, accumulation of soil enzymes, suppression of plant pathogen by bioactive substances. This will have direct impact on socio-economy of tree growing farmers, maintain sustainability in natural soil ecosystem, wood and food crops availability in future. Therefore, the development of more efficient and sustainable agriculture strategies, guarantied food supply for an expanding world population and minimizing damage to the environment is one of the greatest challenges for humankind today. It is inferred that under appropriate management, the use of more efficient bioinoculants, co-inoculation with other bioinoculants lead to an increased growth and biomass of tree species in nutrient impoverished soil.

Author details

Easan Mohan and Kuppu Rajendran*
Centre for Research and P.G. Department of Botany, Thiagarajar College, Madurai, Tamil Nadu, India

*Address all correspondence to: kuppurajendran@rediffmail.com

References

[1] Barea JM, Aczon R, Azcon-Aguilar C. Mycorrhizal fungi and plant growth promoting rhizobacteria. In: Varma A, Abbott L, Werner D, Hampp R, editors. Plant Surface Microbiology. Heideberg, Germany: Springer-Verlag; 2004. pp. 351-371

[2] Rajendran K, Sugavanam V, Devaraj P. Assessment of beneficial microbial population in *Casuarina equisetifolia* plantation. Asian Journal of Microbiology, Biotechnology and Environmental Sciences. 1999;**1**(3-4):143-147

[3] Lynch JM. The Rhizosphere. Chichester, England: Wiley-Interscience; 1990

[4] Tilak KVPR, Ranganayaki N, Pal KK, De R, Saxena AK, Shekhar NC, et al. Diversity of plant growth and soil health supporting bacteria. Current Science. 2005;**89**:136-150

[5] Yadav SP. Effective micro-organisms, its efficacy in soil improvement and crop growth. In: Sixth International Conference on Kyusei, Nature Farming Pretoria, South Africa. 1999. pp. 28-31

[6] Wu SC, Cao ZH, LI Z, Cheung KC Wong MH. Effects of biofertilizer containing N-fixer, P and K solubilizers and AM fungi on maize growth: A greenhouse trial. Geoderma. 2005;**125**:155-166

[7] Burd GI, Dixon DG, Click BR. Plant growth promoting rhizobacteria that decrease heavy metal toxicity in plants. Canadian Journal of Microbiology. 2000;**33**:237-245

[8] Mrkovack N, Milic V. Use of *Azotobacter chrooccocum* as potential useful in agricultural allocation. Annales de Microbiologie. 2001;**51**:145-158

[9] Bharathi R, Vivekananthan R, Harish S, Ramanathan A, Samiyappan R. Rhizobacteria-based bio-formulations for the management of fruit rot infection in chillies. Crop Protection. 2004;**23**:835-843

[10] Cattelan A, Hartel P, Fuhrman JJ. Screening for plant growth promoting rhizobacteria to promote early soybean growth. Soil Science. 1999;**63**:1670-1680

[11] Bashan Y, de Bashan LE. Plant growth-promoting. In: Hillel D, editor. Encyclopedia of Soils in the Environment. Vol. 1, 2200. Oxford: Elsevier; 2005

[12] Perrig D, Boiero L, Masciarelli O, Penna C, Cassen F, Luna M. Plant growth promoting compounds produced by two agronomically important strains of *Azospirillum brasilense* and their implications for inoculants formulation. Applied Microbiology and Biotechnology. 2007

[13] Magalhaes FM, Baldani JI, Souto SM, Kuykendall JR, Dobereiner J. A acid-tolerant *Azospirillum* species. Journal of Anais da Academia Brasileira de Ciências. 1983;**55**:417-430

[14] Tarrand JJ, Krieg NR, Dobereiner J. A taxonomic study of the *Spirillum lipoferum* group with description of a new genus, *Azospirillum* gen nov and two species. *Azospirillum lipoferum* (Beijerink) comb nov and *Azospirillum brasilense* p nov. Canadian Journal of Microbiology. 1978;**24**:967-980

[15] Eckert B, Weber OB, Kirchhof G, Halbritter A, Stoffels M, Harmann A. *Azospirillum doebereinerae* sp. nov., a nitrogen-fixing bacterium associated with the C (4)-grass Miscanthus. International Journal of Systematic and Evolutionary Microbiology. 2001;**51**:17-26

[16] Reinhold B, Hurek T, Fendrik I, Pot B, Gillis M, Kersters K, et al.

Azospirillum halopraeferens sp.nov., a nitrogen-fixing organism associated with roots of Kallar grass (*Leptochloa fusca*) (L.) Kunth. International Journal of Systematic Bacteriology. 1987;**37**:43-51

[17] Khammas KM, Ageron E, Grimont PAD, Kaiser P. *Azospirillum irakense* sp. nov., a nitrogen fixing bacterium associated with rice roots and rhizosphere. Soil Research in Microbiology. 1989;**140**:679-693

[18] Skerman VBD, Sly LI, Williamson M. *Conglomeromonas largomobilis* gene. Nov. sp. nov., a sodium sensitive, mixed flagellated organism from fresh water. International Journal of Systematic Bacteriology. 1983;**33**:300-308

[19] Okon Y. *Azospirillum* as potential inoculants for agriculture. Trends in Biotechnology. 1985;**3**:223-228

[20] Sumner ME. Crop responses to *Azospirillum* inoculation. In: Stewart BA, editor. Advances in Soil Sciences. New York: Springer-Verlag; 1990. pp. 53-123

[21] Sekar I, Vanangamudi K, Suresh KK. Effect of biofertilizers on the seedling biomass, VAM colonization, enzyme activity and phosphorus uptake in the Shola tree species. My Forest. 1995;**31**(4):21-26

[22] Rajendran K, Sugavanam V, Devaraj P. Effect of biofertilizers on quality seedling production of *Casuarina equisetifolia*. Journal of Tropical Forest Science. 2003;**15**(1):82-96

[23] Saravanan TS, Rajendran K, Santhaguru K. Selection of suitable biofertilizers for production of quality seedlings of *Casuarina equisetifolia* (Forst.) using decomposed coir pith compost in root trainers. Asian Journal of Experimental Biological Sciences. 2012;**3**(4):752-761

[24] Rodriguez-Barrueco C, Cervantes E, Subba Rao NS, Rodriguez-Caceres E. Growth promoting effect of *Azospirillum brasilense* on *Casuarina cunninghamiana* Miq seedlings. Plant and Soil. 1991;**135**:121-124

[25] Kasthuri Rengamani S, Jothibasu M, Rajendran K. Effect of bioinoculants on quality seedlings production of drumstick (*Moringa oleifera* L.). Journal of Non-Timber Forest Products. 2006;**13**(1):41-46

[26] Rajendran K, Jeyashree R. Effect of biofertilizers on quality seedlings production of *Acacia nilotica*. Journal of Non-Timber Forest Products. 2007;**14**(1):5-11

[27] Rajendran K, Meenakshisundaram M. Microbial inoculants for quality seedlings production of Neem (*Azadirachta indica* A. Juss.). Journal of Non-Timber Forest Products. 2007;**14**(4):225-260

[28] Meenakshi Sundaram M, Santhaguru K, Rajendran K. Effects of bioinoculants on quality seedlings production of *Delonix regia* in tropical nursery conditions. Asian Journal of Biochemical and Pharmaceutical Research. 2011;**1**(1):98-107

[29] Rajendran K. Effect of bio inoculants on seedling growth, biochemical changes and nutrient uptake of *Erythrin aindica* L., in semi-arid region of South India. Journal of Biometrics & Biostatistics. 2012;**3**(2):134-140

[30] Mohan E, Rajendran K. Effect of plant growth-promoting microorganisms on quality seedling production of *Feronia elephantum* (Corr.) in semi-arid region of southern India. International Journal of Current Microbiology and Applied Sciences. 2014;**3**(7):103-116

[31] Kannan M, Rajendran K. A sustainable agro-biotechnology for

quality seedling production of *Jatropha curcas* L. in tropical nursery conditions. International Journal of Current Research and Academic Review. 2015;**3**(2):92-103

[32] Rajendran K, Deveraj P. Biomass and nutrient distribution and their return of *Casuarina equisetifolia* inoculated with biofertilizers in farm land. Biomass and Bioenergy. 2004;**26**:235-249

[33] Ravikumar S, Shanthy S, Kalaiarasi A, Sumaya S. Effect of halophilic phosphobacteria on *Avicennia officinalis* seedlings. Annals of Biological Research. 2010;**1**(4):254-260

[34] Patten C, Glick B. Bacterial biosynthesis of indole 3-acetic acid. Canadian Journal of Microbiology. 1996;**42**:207-220

[35] Rademacher W. Gibberellin formation in microorganisms. Plant Growth Regulation. 1994;**15**:303-314

[36] Strzeczyk E, Kamper M, Li C. Cytocinin-like-substances and ethylene production by *Azospirillum* in media with different carbon sources. Microbiological Research. 1994;**149**:55-60

[37] Bashan Y, Bustillos J, Leyva L, Hernandez JP, Bacilio M. Increase in auxiliary photoprotective photosynthetic pigments in wheat seedlings induced by *Azospirillum brasilense*. Biology and Fertility of Soils. 2006;**42**:279-285

[38] Palleroni NJ. Family I. Pseudomonadaceae. In: Krieg NR, Holt JG, editors. Bergey's Manual of Systematic Bacteriology. Vol. 1. Baltimore, US: Williams and Wilkins; 1984. pp. 141-199

[39] Waschkies C, Schropp A, Marschner H. Relations between grapevine replant disease and root colonization of grapevine (*Vitis* sp.) by *Fluorescent pseudomonads* and endomycorrhizal fungi. Plant and Soil. 1994;**162**:219-227

[40] Ames RN, Reid CPP, Ingham ER. Rhizosphere bacterial population responses to root colonization by a vesicular-arbuscular mycorrhizal fungus. The New Phytologist. 1984;**95**:555-563

[41] Glick BR, Karaturovic DM, Newell PC. A novel procedure for rapid isolation of plant growth promoting *Pseudomonads*. Canadian Journal of Microbiology. 1995;**41**:533-536

[42] Crowley DE, Wang YC, Reid CPP, Szansiszlo PJ. Mechanism of iron acquisition from siderophores by microorganisms and plants. Plant and Soil. 1991;**130**:179-198

[43] Weller DM. Biological control of soil-borne plant pathogens in the rhizosphere with bacteria. Annual Review of Phytopathology. 1988;**26**:379-407

[44] Kapulnik Y. Plant growth promotion by rhizsphere bacteria. In: Waisel Y, Eshel A, Kafkafi U, editors. Plant Roots. The Hidden Half. New York: Marcel Dekker; 1996. pp. 769-781

[45] Glick BR. The enhancement of plant growth by free-living bacteria. Canadian Journal of Microbiology. 1995;**41**:109-117

[46] Deacon JW. Modern Mycology. London, England: Blackwell Science; 1997

[47] Quilambo OA. The vesicular-arbuscular mycorrhizal symbiosis. African Journal of Biotechnology. 2003;**2**:539-546

[48] Smith SE, Read DJ. Mycorrhizal Symbiosis. London: Academic Press; 1997. p. 605

[49] Gildon A, Tinker PB. Interactions of vesicular arbuscular mycorrhizal

infections and heavy metals in plants. I. The effect of heavy metals on the development of vesicular-arbuscular mycorrhizas. The New Phytologist. 1983;**95**:247-261

[50] Andrade G, Linderman RG, Berthlenfalvay GJ. Bacterial associations with the mycorrhizosphere and hyphosphere of the arbuscular mycorrhizal fungus *Glomus mosseae*. Plant and Soil. 1998;**202**:79-87

[51] Wang B, Qiu YL. Phylogenetic distribution and evolution of mycorrhizas in land plants. Mycorrhiza. 2006;**16**:299-363

[52] Miller RM, Jastrow JD. Mycorrhizal fungi influence soil structure. In: Kapulnik Y, Douds DD Jr, editors. Arbuscular Mycorrhizas: Physiology and Function. Kluwer Academic Publication; 2000. pp. 3-18

[53] Stutz JC, Coperman R, Martin CA, Morton JB. Patterns of species composition and distribution of arbuscular mycorrhizal fungi in arid regions of southwestern North America and Namibia, Africa. Canadian Journal of Botany. 2000;**78**:237-245

[54] Nowak J. Effects of arbuscular mycorrhizal fungi and organic fertilization on growth, lowering, nutrient uptake, photosynthesis and transpiration of geranium (*Pelargonium hortorum* L.H. Bailey "Tango Orange"). Symbiosis. 2004;**37**:259-266

[55] Walley FL, Germida JJ. Response of spring wheat (*Triticum aestivum*) to interactions between *Pseudomonas* species and *Glomus clarum* NT4. Biology and Fertility of Soils. 1997;**24**:365-371

[56] Singh R, Gangwar SP, Singh D, Singh R, Pandey R, Kalra A. Medicinal plant *Coleus forskolii* Briq, disease and management (mini review). Medicinal Plants. 2011;**3**:1-7

[57] Ravnskov S, Nybroe O, Jakobsen I. Influence of an arbuscular mycorrhizal fungus on *Pseudomonas fluorescens* DF57 in rhizosphere and hyphosphere soil. The New Phytologist. 1999;**142**:113-122

[58] Marschner P, Crowley DE, Higashi RM. Root exudation and physiological status of a root-colonizing fluorescent pseudomonad in mycorrhizal and non-mycorrhizal pepper (*Capsicum annuum* L.). Plant and Soil. 1997;**189**(1):11-20

[59] Marschner H, Dell B. Nutrient uptake in mycorrhizal symbiosis. Plant and Soil. 1994;**159**:89-102

[60] Mazzola M. The potential use of natural and genetically engineered fluorescent *Pseudomonas* spp. as biological control agents. In: SubbaRao NS, Dommergues YR, editors. Microbial interactions in agriculture and forestry. Sci Phymouth. 1999;**1**:195-218

[61] Edwards SG, Young JPW, Fitter AH. Interactions between *Pseudomonas fluorescens* biocontrol agents and *Glomus mosseae*, an arbuscular mycorrhizal fungus, with in the rhizosphere. FEMS Microbiology Letters. 1998;**166**:297-303

[62] Galleguillors C, Aguirre C, Barea JM, Azcon R. Growth promoting effect of two *Sinorhizobium meliloti* strains (a wild type and its genetically modified derivative) on anon-legume plant species in specific interaction with two arbuscular mycorrhizal fungi. Plant Science. 2000;**159**:57-63

[63] Mohan V, Neelam V, Singh YP. Distribution of VAM fungi in nurseries and plantation of Neem tree (*Azadirachta indica*) in arid zone of Rajasthan. The Indian Forester. 1995;**121**:1069-1076

[64] Mohan E, Rajendran K. Assessment of microbial diversity

in non-rhizosphere soil of forest nurseries in Southern Tamil Nadu, India. International Journal of Current Microbiology and Applied Sciences. 2014;**3**(6):454-458

[65] Santhaguru K, Gladis Posimalar SB, Karunakaran R. Vesicular-arbuscular mycorrhizae in tree legumes and its rhizospheric soils in Alagar hills. The Indian Forester. 1995;**121**:817-822

[66] Huang RS, Smith WK, Post RS. Influence of vesicular-arbuscular mycorrhiza on growth, water relations and leaf orientation in *Leucaena leucocephala* (Lam.) DeWit. New Philologist. 1985;**99**:229-243

[67] Mohan E, Rajendran K. Effect of beneficial bioinoculants on the growth of monkey pod tree (*Samanea saman*) in nursery condition. Journal of Plant Development Sciences. 2012;**4**(3):379-382

[68] Verma R, Jamaluddin K. Association and activity of arbuscular mycorrhizae of teak (*Tectona grandis*) in Central India. The Indian Forester. 1995;**121**(6):533-539

[69] Castillo ET. Physiological responses of Narra (*Pterocarpus indicus* wild.) inoculated with VA Mycorrhiza and rhizobium in macolod soils [PhD thesis]. U.P. Los Banos; 1993

[70] Kormanik PP, Bryan WC, Schultz RC. Effects of three vesicular arbuscular mycorrhizal fungi on sweetgum seedlings from nine mother trees. Forest Science. 1981;**27**(2):327-335

[71] Perrin R. Interactions between mycorrhizae and diseases caused by soil borne fungi. Soil Use and Management. 1990;**6**:189-195

[72] Morte A, Lovisola C, Schubert A. Effect of drought stress on growth and water relations of the mycorrhizal association *Helianthemum almeriense*-Terfezia clavery. Mycorrhiza. 2000;**10**:115-119

[73] Krishna KR, Bagyaraj DJ. Phenolics of mycorrhizal and uninfected groundnut var. MGS.7. Current Research. 1986;**15**:51-52

[74] Gianinazzi S, Gollotte A, Binet MN, van Tuinen D, Redecker D, Wipf D. Agroecology: The key role of arbuscular mycorrhizas in ecosystem services. Mycorrhiza. 2010;**20**:519-530

[75] Morandi D, Bailey JA, Gianinazzi-Pearson V. Isoflavonoid accumulation in soybean roots infected with vesicular-arbuscular mycorrhizal fungi. Physiological Plant Pathology. 1984;**24**:357-364

[76] Kapoor R, Chaudhary V, Bhatnagar AK. Effects of arbuscular mycorrhiza and phosphorus application on artemisinin concentration in *Artemisia annua* L. Mycorrhiza. 2007;**17**:581-587

[77] Manoharan PT, Pandi M, Shanmugaiah V, Gomathinayagam S, Balasubramanian N. Effect of vesicular arbuscular mycorrhizal fungus on the physiological and biochemical changes of five different tree seedlings grown under nursery conditions. African Journal of Biotechnology. 2008;**19**:3431-3436

[78] Vanangamudi K, Venkatech AV, Mallika V, Ravichandran, Rai RSV. Impact of biofertilizers on morphological attributes in pungam (*Pongamia pinnata* (L.) Pierre) seedlings. Tropical Agriculture Research and Extension. 1998;**1**:7-11

[79] Mathur N, Vyas A. Physiological changes in *Ziziphus mauritiana* by different VAM fungi. The Indian Forester. 1996;**122**(6):501-505

[80] Niranjan R, Banwarilal S, Rao M. Studies on the effect of Rhizobium

and endomycorrhizal interaction in *Dalbergia sissoo*. In: Proceedings of a National Conference on Mycorrhiza. Hisar Agricultural University, Hisar, India; Feb. 14-16. 1990. pp. 205-207

[81] Kumar R, Jalali BL, Chand H. Influence of vesicular arbuscular mycorrhizal fungi on growth and nutrient uptake in chickpea. Journal of Mycology and Plant Pathology. 2002;**32**(1):11-15

[82] De Olieveiera VA, Schmidt VDB, Bellei M. Patterns of arbuscular and ectomycorrhizal colonization of *Eucalyptus dunnii* in Southen Brazil. Annsls of Diseases Sciences Forest. 1997;**54**:473-481

[83] Reynolds HL, Vogelsang KM, Hartley AE, Bever JD, Schultz PA. Variable responses of old-field perennials to arbuscular mycorrhizal fungi and phosphorus source. Oecologia. 2006;**147**:348-358

[84] Duponnois R, Plenchette C. A mycorrhiza helper bacterium enhances ectomycorrhizal and endomycorrhizal symbiosis of Australian *Acacia* species. Mycorrhiza. 2003;**13**:85-91

[85] Balasubramanian A, Ravichandran VK. Biofertilizer—An alternative source on nutrients for sustainable productivity of trees. Journal of Ecobiology. 1997;**9**(3):203-206

[86] Nagaveni HC, Ananthapadmanabha HS, Singh A, ChodankarDN.Prospectusforafforestation in mine dumps using VAM fungi. My Forest. 1997;**33**(2):465-468

[87] Nagaveni HC, Ananthapadmanabha HS, Vijayalakshmi G, Somashekar PV. Comparative effect of inorganic fertilizer and VAM fungi on growth of teak plants. My Forest. 1998;**34**(1):697-700

[88] Pavan KP. Influence of bioinoculants on the growth of *Albizia lebbeck* in nursery condition. Research Journal of Agricultural Sciences. 2011;**2**:265-268

[89] Muthukumar T, Udaiyan K, Rajeshkannan V. Response of Neem (*Azadirachta indica* A. Juss) to indigenous arbuscular mycorrhizal fungi, phosphate-solubilizing and asymbiotic nitrogen-fixing bacteria under tropical nursery conditions. Biology and Fertility of Soils. 2001;**34**:417-426

[90] Priya Rani M, Aggarwal A, Mehrotra RS. Growth responses in *Acacia nilotica* inoculated with VAM fungus (*Glomus fasciculatum*), *Rhizobium* sp and *Trichoderma harzianum*. Journal of Mycopathological Research. 1998;**36**(1):13-16

[91] Koide RT. The physiology of the mycorrhizal plant. In: Tommerup IC, editor. Advances in Plant Pathology. New York: Academic Press; 1993. pp. 33-54

[92] Sharma AK, Johri BN, Gianinazzi S. Vesicular arbuscular mycorrhizae in relation to plant diseases. World Journal of Microbiology and Biotechnology. 1992;**8**:559-563

[93] Gohre V, Paszkowski U. Contribution of the arbuscular mycorrhizal symbiosis to heavy metal phytoremediation. Planta. 2006;**223**:1115-1122

[94] Mansfeld-Giese K, Larsen J, Dker LB. Bacterial populations associated with mycelium of the arbuscular mycorrhizal fungus *Glomus intraradices*. FEMS Microbiology Ecology. 2002;**41**:133-140

[95] Uniyal K, Thapar HS. Growth responses in *Prosopis juliflora* inoculated with VAM fungi and *Rhizobium* in sodic soil. Van Vigyan. 1995;**33**(344):182-184

[96] Aditya B, Ghosh A, Chattopadhyay D. Co-inoculation effects of Nitogen fixing and phosphate solublising microorganisms on teak (*Tectona grandis*) and Indian redwood (*Chukrasia tubularis*). Journal of Biological Sciences. 2009;**1**:23-32

[97] Arshad M, Frankenberger W T. Microbial production of plant growth regulators. In: Blaine F Jr, editor. Soil Microbial Ecology – Applications in Agricultural and Environmental Management Meeting. New York: Marcel Dekker; 1993

[98] Eneback SA, Wel G, Kloepper JW. Effects of plant growth promoting rhizobacterial on lobly pine seedlings. Forest Science. 1998;**44**:139-144

[99] Nutman PS. IBP field experiments on nitrogen fixation by nodulated legumes. In: Nutman PS, editor. Symbiotic Nitrogen Fixation in Plants. London: Cambridge Univ. Press; 1976. pp. 211-237

[100] Kumar A, Dash D, Jhariya MK. Impact of rhizobium on growth, biomass accumulation and nodulation in *Dalbergia sissoo*. The BioScan. 2013;**8**(2):553-560

[101] Chauhan YS, Pokhriyal TC. Nodulation and nitrogen fixation as influenced by *Rhizobium* and mycorrhizal inoculation in *Albizia lebbek* (L.) Benth seedlings. Annals of Forestry. 2002;**10**:243-251

[102] Sekar I. Response of certain Shola tree species to the inoculation of biofertilizers and application of growth stimulants under nursery conditions [M.Sc.(For.) thesis]. Coimbatore, Tamil Nadu, India: Tamil Nadu Agricultural University; 1992

[103] Reena J, Bagyaraj DJ. Response of *Acacia nilotica* and *Calliandra calothyrsus* to different VAM. Arid Soil Research and Rehabilitation. 1990;**4**:261-268

[104] Purohit I, Prasad P, Nautiyal AR. Influence of rhizobial inoculation on nodulation and seedling growth in three nitrogen fixing tree species. Indian Journal of Forestry. 1995;**18**:337-340

[105] Rakesh Kumar D, Dash SB, Gupta R, Soni, Anup KU, Singh Inoculation effects of rhizobium and phosphorous solubulizing bacteria on growth and nodulation of *Acacia nilotica*. International Journal of Current Microbiology and Applied Sciences. 2017;**6**:2444-2453

[106] Andreeva IN, Red'Kina TV, Ismailov SF. The involvement of indole acetic acid in the stimulation of rhizobium-legume symbiosis by *Azospirillum brasilense*. Russian Journal of Plant Physiology. 1993;**40**:901-906

[107] Manjunath A, Bagyaraj DJ, Gopala gowala HS. Dual inoculation with VA mycorrhiza and *Rhizobium* is beneficial to *Leucaena*. Plant and Soil. 1984;**78**:445-448

[108] Sharma AK. A Handbook of Organic Farming. Jodhpur: Agrobios India; 2002. p. 90

[109] Rana RK, Chandel JS. Effect of biofertilizer and nitrogen on growth, yield and fruit quality of strawberry. Progressive Horticulture. 2003;**35**(1):25-30

[110] Dadoshpor AJ, Mohammad. Impact of integrated organic nutrient handling on fruit yield quality of strawberry. Journal of Ornamental and Horticultural Plants. 2012;**2**(4):251-256

[111] Karthikeyan A. *Frankia* strains for improving growth, biomass and nitrogen fixation in *Casuarina equisetifolia* seedlings. Journal of Tropical Forest Science. 2016;**28**(3):235-242

[112] Saravanan TS, Rajendran K, Uma M, Chezhian P. Effect of bio

inoculants on quality seedling production and nutrient uptake of *Casuarina equisetifolia* Forst. grown in decomposed coir pith. In: Velu RK, editor. Microbiological Research in Agroecosystem Management. India: Springer; 2013. pp. 141-154

[113] Karthikeyan A, Chandrasekaran K, Geetha M, Kalaiselvi R. Growth response of *Casuarina equisetifolia* Forst. rooted stem cuttings to Frankia in nursery and field conditions. Journal of Biosciences. 2013;**38**:741-747

[114] Uma M, Saravanan TS, Rajendran K. A sustainable bio-resourse technology for quality seedling production of *Casuarina equisetifolia* in tropical nursery condition. The Indian Forester. 2012;**138**(12):1091-1099

[115] Krishnakumar S, Nagarajan R, Natarajan SK, Jawahar D, Pandian BJ. NPK fertilizers for hybrid rice (*Oryza sativa* L.) productivity in Alfisols of southern districts of Tamil Nadu. Asian Journal of Plant Sciences. 2005;**4**:574-576

[116] Kundu S, Datta P, Mishra J, Rashmi K, Ghosh B. Influence of biofertilizer and inorganic fertilizer in pruned mango orchard cv. Amrapali. Journal of Crop and Weed. 2011;**7**:100-103

[117] Paroha S, Chandra KK, Yadav R. Integrated effect of biofertilizers (AM, azatobactor and PSB) fertilizers on growth and nutrient acquisition by *Tectona grandis*. Journal of Tropical Forestry. 2009;**25**(1-2):52-61

[118] Maharana R, Dobriyal MJ, Behera LK, Sukhadiya M. Enhancement of seedling vigour through biofertilizers application in gamhar (*Gmelinaarborea* Roxb.). International Journal of Chemical Studies. 2018;**6**(5):54-60

Chapter 17

Nano-Fertilizers for Sustainable Crop Production under Changing Climate: A Global Perspective

Muhammad Aamir Iqbal

Abstract

Since green revolution, chemical fertilizers are deemed an indispensable input of modern crop production systems, but these have associated environmental and ecological consequences. Loss of nutrients from agricultural fields in the form of leaching and gaseous emissions has been the leading cause of environmental pollution and climate change. Ensuring the sustainability of crop production necessitates exploring other sources of nutrients and modifying prevalent nutrient sources. Nanotechnology, which utilizes nanomaterials of less than 100 nm size, may offer an unprecedented opportunity to develop concentrated sources of plant nutrients having higher-absorption rate, utilization efficacy, and minimum losses. Nanofertilizers are being prepared by encapsulating plant nutrients into nanomaterials, employing thin coating of nanomaterials on plant nutrients, and delivering in the form of nano-sized emulsions. Nanopores and stomatal openings in plant leaves facilitate nanomaterial uptake and their penetration deep inside leaves leading to higher nutrient use efficiency (NUE). Nanofertilizers have higher transport and delivery of nutrients through plasmodesmata, which are nanosized (50–60 nm) channels between cells. The higher NUE and significantly lesser nutrient losses of nanofertilizers lead to higher productivity (6–17%) and nutritional quality of field crops. However, production and availability, their sufficient effective legislation, and associated risk management are the prime limiting factors in their general adoption as plant nutrient sources.

Keywords: controlled release fertilizers, eutrophication, nanogels, encapsulated nutrients, slow released fertilizers

1. Introduction

Intensive farming practices introduced and evolved since the inception of green revolution have been deemed unsustainable as the utilization efficacy of applied chemicals including mineral fertilizers has remained below 30% [1]. Fertilizers have taken axial role with respect to boosting crops yield and nutritional quality especially after the development of fertilizer responsive crop varieties. Among mineral nutrients, nitrogen is the first and foremost nutrient required for crop plants as it is the constituent of chlorophyll and many proteins and enzymes and thus plays a significant role during the vegetative growth of crops. Nitrogen is absorbed by the plants in the form of nitrate (NO^{-}_{3}) and ammonium (NH^{+}_{4}) [2]. Nitrogen is lost through the processes of nitrate leaching, de-nitrification and ammonia volatilization. Loss of mineral nutrients through leaching and runoff to surface and ground water along with abundant volatilization constitute growing concerns owing to economic losses and environmental pollution. Conventional application techniques are resulting in seriously overdosing of chemical fertilizers which has become evident through the phenomenon of eutrophication (algal growth on the surface of water bodies due to nutrients enriched water, which hampers oxygen supply to fish) in many European and North American countries. Moreover, nitrogen volatilization results in the release of nitrous oxides and thus being the greenhouse gases, contribute to the global warming. It is really unfortunate that modern profit-oriented farming systems encompass nitrogenous fertilizers use efficiency of only 45–50%, while the corresponding figure for phosphorous fertilizers has been reported to be only 10–25% [3].

It is also pertinent to mention that ammonium ions react with alkaline rain water which leads to the formation of ammonia gas that escapes into the atmosphere and thus becoming a source of environmental pollution. Whenever, there is excess of nitrogen, more and more nitrates and ammonium ions get accumulated in the leaves of crops especially leafy vegetables and become detrimental to human health. In addition, nitrate rich diets have been reported to be associated with numerous human diseases such as bladder and gastric cancer as well as methemoglobinemia [4]. It is being stressed to deliver the required quantities of active agents only where they are direly needed. Environmentalists and consumers call for reducing the use of synthetic fertilizers to decrease pollution and residue effect on form produces along with conserving agro-ecosystems.

Nanotechnology is a promising field of research which has the potential to offer sustainable remedies to pressing challenges confronted to modern intensive agriculture. Nanotechnology employs nanomaterials which typically have the size of 1–100 nm and this small size imparts unique characteristics and benefits to nanomaterials. In addition to numerous other benefits, large surface area offers opportunity for better and effective interaction of nanoparticles to target sites. Nanofertilizers hold potential to fulfill plant nutrition requirements along with imparting sustainability to crop production systems and that too without compromising the crops yield [5].

This chapter entails and attempts to fulfill the need to periodically compile and review the present state and advances on nanofertilizers and to spur interest for conducting further in-depth research. The ultimate goal is to synthesize and assess the role of nanofertilizers in boosting nutrients uptake and nutrients use efficiency, reducing losses through leaching and gaseous emissions along with reducing the risk of nutrient toxicity for ensuring food security achieved through higher productivity and economic turn outs by practicing the sustainable farming practices. This chapter briefly sheds light on the critical role of nanotechnology pertaining to modern farming practices, its potential in developing smart fertilizers, nanofertilizers and their different types of formulations, biological mechanism of nanofertilizers in plants, numerous advantages offered by nanofertilizers and field evidences of superior performances of nanoparticles in imparting critical characteristics to crop plants leading to higher productivity. Lastly, few limitations pertaining to the development and use of nanoparticles as plant nutrient source have also been described.

2. Critical applications of nanomaterials in agriculture

Nanotechnology encompasses controlling matter at 1–100 nm dimensions for utilization in taking images, measurements and preparing models for making virtual predictions along with manipulation of matter at nanoscale. Like all other fields, the solid impact of nanomaterials is also being felt in agriculture sector. Previously, nanoencapsulation entailing encapsulation of active agents by microspheres of starch on a matrix having nanopores proved its resilience in accurately delivering the active agents to target sites [6]. These nanocapsules or micro-beads become attached to heir of bees in the similar fashion to pollens and keep parasites at bay owing to slow release of active agents gradually and slowly. Thus, nanoencapsulation resulted in minimum use of active agents and offered the maximum protection to bees against parasites. On the similar fashion, nanogels were developed which assist in controlled release of pheromones from insects to offer them protection against diversified pests. Nanoencapsulation has also yielded encouraging results for improving the fertilizer use efficacy with significant reduction of active ingredients use [7].

In order to detect pathogen and to prolong the shelf life of packaged foods, nanosensors and nanobiosensors have given encouraging results. However, development of nanomaterials using nanotechnology is an evolving field of research and future is destined to witness extensive and multidimensional benefits in food production and preservation. In future, it will be impossible to ensure food and nutritional security without developing nanomaterials based technologies for food production and agriculture.

3. Nanotechnology's strategic potential in developing fertilizers of future

Modern intensive farming systems utilize organic and mineral manures in order to supply essential plant nutrients, but this approach has resulted in serious deterioration of ecosystems and environment [8]. Loss of nitrogen as nitrous oxide and nitrates leaching has resulted in eutrophication and manifesting the impacts of global warming and climate change. Phosphate fertilizers have even lesser nutrient use efficacy (NUE) that has been reported to be below 20% [9]. Nanofertilizers have the potential to enhance NUE owing to higher nutrients uptake caused by smaller surface area of nanomaterials which increases nutrient-surface interaction. Along with boosting crops yield on sustainable basis, nanofertilizers hold potential to put a halt to environmental pollution caused by fertilizers. Slow release fertilizers (chemical compounds having slight solubility in water or other solvents and get broken down gradually and slowly by soil microbial population) coated with nanoparticles significantly reduced nitrate leaching and de-nitrification [10]. Moreover, controlled releasing fertilizers (have higher solubility in contrast to slow release fertilizers but are coated with materials which significantly reduce the exposure of active ingredient with the solvent resulting in controlled liberation of nutrients through diffusion) coated with nanomaterials for reducing surface area my provide excellent of source of supplying plant nutrients in times to come.

4. Nanoscale fertilizers and their formulations

Different fertilizers inputs have been reported to be resized into smaller fractions through mechanical means or by employing specific chemical methods, which may increase nutrients uptake and reduce losses as well as nutrient toxicity. Nano-sized particles have been prepared from urea, ammonia, peat and other synthetic fertilizers as well as plant wastes. A formulation process involving urea deposition on calcium cyanamide resulted in nano-sized N fertilizer [11]. In another formulation, grinded urea was mixed with different biofertilizers to prepare an effective nanofertilizer to supply nutrients slowly and gradually for a longer period of time [12]. In similar way, ammonium humate, peat and other synthetic materials were mixed to prepare nanosized fertilizers. Mechanical cum biochemical approach is being employed to prepare such nanofertizers where materials are grinded to nanosized particles through mechanical means and then biochemical techniques are put in action to prepare effective nanoscale formulations. In addition, nano-emulsions are also being prepared by adding nanosized colloids to emulsions [13]. In short, fertilizers encapsulation with nanoparticles offers wide perspective for developing plant nutrient sources with greater absorption and nutrient use efficiency. The encapsulation of nutrients with nanomaterials can be performed in three distinct ways;

1. Plant nutrients can be encapsulated within the nanomaterials of varying nature and chemical composition.

2. Nutrient particles may be coated with a thin layer of nanomaterials such as polymer film.

3. Nutrients may also be delivered in the form of emulsions and particles having dimension in the range of nanoparticles.

5. Biological mechanisms of nanofertilizers action

Nanofertilizers have been advocated owing to higher NUE as plants cell walls have small pore sizes (up to 20 nm) which result in higher nutrient uptake [14]. Plant roots which act as the gateways for nutrients, have been reported to be significantly porous to nanomaterials compared to conventional manuring materials. The uptake of nanofertilizers can be improved by utilizing root exudates and molecular transporters through the ionic channels and creation of new micro-pores [15]. Nano-pores and stomatal openings in leaves have also been reported to felicitate nanomaterials uptake and their penetration deep inside leaves. It was concluded that in broad/faba bean (*Vicia faba*), nano-sized particles (43 nm) were instrumental in penetrating deep to leaf interior in large number compared to larger particles of more than 1.0 micrometer size [16]. Similarly, the leaf stomatal radii of Arabian coffee (*C. arabica*) was below 2.5 nm, while that of sour cherry (*P. cerasus*) were also below 100 nm [17] and thus effectiveness of nanofertilizers in enhancing nutrient uptake was suggested.

Nanofertilizers have also been supported to have higher NUE owing to higher transport and delivery of nutrients through plasmodesmata which are nanosized (50–60 nm) channels for transportation of ions between cells [18]. Carbon nanotubes transported fluorescent dyes to tobacco cells through enhanced penetration of cell membranes and effectively played the role of molecular transporters [19]. The nanoparticles of silica were also instrumental in transporting and delivering different cargoes to target sites in different plants [20].

6. Nanofertilizers advantages over conventional mineral fertilizers

Mineral nutrients if applied to crops in the form of nanofertilizers hold potential to offer numerous benefits for making the crop production more sustainable and eco-friendly [21]. Some of salient advantages are;

1. Nanofertilizers feed the crop plants gradually in a controlled manner in contradiction to rapid and spontaneous release of nutrients from chemical fertilizers.

2. Nanofertilizers are more efficacious in terms of nutrients absorption and utilization owing to considerably lesser losses in the form of leaching and volatilization.

3. Nanoparticles record significantly higher uptake owing to free passage from nano sized pores and by molecular transporters as well as root exudates. Nanoparticles also utilize various ion channels which lead to higher nutrient uptake by crop plants. Within the plant, nanoparticles may pass through plasmodesmata that results in effective delivery on nutrient to sink sites.

4. Due to considerably small losses of nanofertilizers, these can be applied in smaller amounts in comparison to synthetic fertilizers which are being applied in greater quantities keeping in view their major chunk that gets lost owing to leaching and emission.

5. Nanofertilizers offer the biggest benefit in terms of small losses which lead to lower risk of environmental pollution.

6. Comparatively higher solubility and diffusion impart superiority to nanofertilizers over conventional synthetic fertilizers.

7. Smart nanofertilizers such as polymer coated fertilizers avoid premature contact with soil and water owing to thin coating encapsulation of nanoparticles such as leading to negligible loss of nutrients. On the other hand, these become available as soon as plants are in position to internalize the released nutrients.

7. Field evidences of nanofertilizers use for sustainable crops production

The research findings of a field investigation proved in line with the postulated hypothesis where nano nitrogen fertilizers proved instrumental in boosting the productivity of rice. It was inferred that nano nitrogen fertilizer hold potential to be used in place of mineral urea and it can also reduce environmental pollution caused by leaching, de-nitrification and volatilization of chemical fertilizers [22]. Similarly, exogenously applied nutrients as nanomaterials increased the vegetative growth of cereals including barley [23] (man), while in contrast, nanofertilizers applied in conjunction with reduced doses of mineral fertilizers were found to be instrumental in boosting yield attributes and grain yield of cereals [24]. Nanofertilizer of zinc applied as ZnO was found to be instrumental in boosting peanut yield due to robust plant growth, increased chlorophyll content of leaves and significantly better root growth [25]. The growth and yield boosting impact of different nanomaterials is depicted in **Table 1**.

Nanofertilizers	Crops	Yield increment (%)
Nanofertilizer + urea	Rice	10.2
Nanofertilizer + urea	Rice	8.5
Nanofertilizer + urea	Wheat	6.5
Nanofertilizer + urea	Wheat	7.3
Nano-encapsulated phosphorous	Maize	10.9
Nano-encapsulated phosphorous	Soybean	16.7
Nano-encapsulated phosphorous	Wheat	28.8
Nano-encapsulated phosphorous	Vegetables	12.0–19.7
Nano chitosan-NPK fertilizers	Wheat	14.6
Nano chitosan	Tomato	20.0
Nano chitosan	Cucumber	9.3
Nano chitosan	Capsicum	11.5
Nano chitosan	Beet-root	8.4
Nano chitosan	Pea	20
Nanopowder of cotton seed and ammonium fertilizer	Sweet potato	16
Aqueous solution on nanoiron	Cereals	8–17
Nanoparticles of ZnO	Cucumber	6.3
Nanoparticles of ZnO	Peanut	4.8
Nanoparticles of ZnO	Cabbage	9.1
Nanoparticles of ZnO	Cauliflower	8.3
Nanoparticles of ZnO	Chickpea	14.9
Rare earth oxides nanoparticles	Vegetables	7–45
Nanosilver + allicin	Cereals	4–8.5
Iron oxide nanoparticles + calcium carbonate nanoparticles + peat	Cereals	14.8–23.1
Sulfur nanoparticles + silicon dioxide nanoparticles + synthetic fertilizer	Cereals	3.4–45%

Table 1.
Impact of nanofertilizers on productivity of different crops under varying pedo-climatic conditions [32–40].

In agreement to these findings, it was also reported that nanofertilizers of zinc improved the seed production of vegetables [26]. Similarly, nano carbon incorporated fertilizers effectively reduced the days to germination and promoted root development of rice seedling. It was inferred that nano-composites have the potential to promote vital processes such as germination, radicle and plumule growth and development [27]. Another aspect of nanofertilizers was explored regarding crop cycle as nanoparticles which were loaded with NPK, reduced the crop cycle of wheat up to 40 days, while grain yield was also increased in comparison to mineral fertilizers applied at recommended rates [28]. Slow release fertilizer coated with nanoparticles boosted the productivity of wheat-maize cropping system [29]. In addition to soil applied nanofertilizers, foliar application of chitosan was reported to be instrumental in boosting tomato yield by 20%, while it remained non-significant as far as carrot yield was concerned [30]. However, growth promoting effect of foliar applied chitosan was

Nanofertilizers	Crops	Imparted characteristics
Nanoparticles of ZnO	Chickpea	Increased germination, better root development, higher indoleacetic acid synthesis.
Nano silicon dioxide	Maize	Drought resistance, increment in lateral root roots number along with and shoot length.
Nano silicon dioxide	Maize	Increased leaf chlorophyll.
Nano silicon dioxide	Tomato	Taller plants and increased tuber diameter.
Colloidal silica + NPK fertilizers	Tomato	Increased resistance to pathogens.
Nano-TiO_2	Spinach	Improved vigor indices and 28% increased chlorophyll.
Polyethylene + indium oxide	Vegetables	Increased sunlight absorption
Polypropylene + indium–tin oxide	Vegetables	Increased sunlight utilization
Gold nanoparticles + sulfur	Grapes	Antioxidants and other human health benefits.
Kaolin + SiO_2	Vegetables	Improved water retention.
Bentonite + N-fixing bacteria inoculation	Legumes	Improved soil fertility and resistance to insect-pest.
Nanocarbon + rare earth metals + N fertilizers	Cereals	Improved nitrogen use efficiency
Stevia extract + nanoparticles of Se + organo-Ca + rare-earth elements + chitosan	Vegetables	Enhanced root networking and root diameter.
Nano-iron slag powder	Maize	Reduced incidence of insect-pest
Nano-iron + organic manures	Cotton	Controlled release of nutrients acts as an effective insecticide and improves soil fertility status.

Table 2.
Impact of nanofertilizers on different crops under varying pedo-climatic conditions [34–46].

also recorded for horticultural crops such as cucumber, beet-root etc. The significantly higher selenium uptake by many crops including green tea was observed when it was applied as nanosized particles [31]. There are various other impacts that can be imparted by nanomaterials in different crops and some of these have been described in **Table 2**.

8. Limitations of nano fertilizers

Despite offering numerous benefits pertaining to sustainable crop production, nanofertilizers have some limitations regarding research gaps, absence of rigorous monitoring and lack of legislation which are currently hampering the rapid development and adoption of nanoparticles as a source of plant nutrients [47]. A few of the limitations and drawbacks associated to nanofertilizers use for sustainable crop production are enlisted below.

1. Nano fertilizers related legislation and associated risk management continue to remain the prime limitation in advocating and promoting nano fertilizers for sustainable crop production.

2. Another limiting factor is the production and availability of nano fertilizers in required quantities and this is the foremost limitation in wider scale adoption of nano fertilizers as a source of plant nutrients.

3. The higher cost of nano fertilizers constitutes another hurdle in the way of promulgating them for crop production under varying pedo-climatic conditions across the globe.

4. Another major limitation pertaining to nanofertilizers is the lack of recognized formulation and standardization which may lead to contrasting effects of the same nanomaterials under various pedoclimatic conditions.

5. There are many products being claimed to be nano but in fact are submicron and micron in size. This dilemma is feared to remain persistent until and unless uniform size of nanoparticles (1–100 nm) gets implemented.

9. Conclusions

Nanofertilizers applied alone and in conjunction with organic materials have the potential to reduce environmental pollution owing to significant less losses and higher absorption rate. In addition, nanomaterials were recorded to improve germination rate, plant height, root development and number of roots, leaf chlorophyll and fruits antioxidant contents. Moreover, controlled and slow released fertilizers having coating of nanoparticles, boost nutrient use efficiency and absorption of photosynthetically active radiation along with considerably lower wastage of nutrients. The future of nanofertilizers for sustainable crop production and time period needed for their general adaptation as a source of plant nutrients depend on varied factors such as effective legislation, production of novel nanofertilizers products as per requirement and associated risk management. There is a dire need for standardization of nanomaterials formulations and subsequently conducting rigorous field and greenhouse studies for performance evaluation. For sustainable crop production, smart nanofertilizers having the potential to release nutrients as per plants requirement in temporal and spatial dimensions must be formulated. Lastly, researchers and regulators need to shoulder the responsibility by providing further insights in order to take full advantage of the nanofertilizers for sustainable crop production under changing climate with the risk of causing environmental pollution.

Author details

Muhammad Aamir Iqbal
Faculty of Agriculture, Department of Agronomy, University of the Poonch Rawalakot (AJK), Pakistan

*Address all correspondence to: muhammadaamir@upr.edu.pk

References

[1] Albanese A, Tang PS, Chan CW. The effect of nanoparticle size, shape and surface chemistry on biological systems. Annual Review of Biomedical Engineering. 2012;**14**:1-16

[2] Dubchak S, Ogar A, Mietelski JW, Turnau K. Influence of silver and titanium nanoparticles on arbuscular mycorrhiza colonization and accumulation of radiocaesium in the *Helianthus annuus*. Spanish Journal of Agricultural Research. 2010;**8**:S103-S108

[3] Haverkamp RG, Marshall AT. The mechanism of metal nanoparticle formation in plants: Limits on accumulation. Journal of Nanoparticle Research. 2009;**11**:1453-1463

[4] Hong F, Zhou J, Liu C, Yang F, Wu C, Zheng L, et al. Effect of nano-TiO2 on photochemical reaction of chloroplasts of spinach. Biological Trace Element Research. 2005;**105**:269-279

[5] Janmohammadi M, Amanzadeh T, Sabaghnia N, Dashti S. Impact of foliar application of nano micronutrient fertilizers and titanium dioxide nanoparticles on the growth and yield components of barley under supplemental irrigation. Acta agriculturae Slovenica. 2016;**107**:265-276

[6] Larue C, Laurette J, Herlin-Biome N, Khodja H, Fayard B, Flank AM, et al. Accumulation, translocation and impact of TiO2 nanoparticles in wheat: Influence of diameter and crystal phase. Science of the Total Environment. 2012;**431**:197-208

[7] Larue C, Veronesi G, Flank AM, Surble S, Herlin-Boime N, Carriere M. Comparative uptake and impact of TiO2 nanoparticles in wheat and rapeseed. Journal of Toxicology and Environmental Health. Part A. 2012;**75**:722-734

[8] Lin S, Reppert J, Hu Q, Hunson JS, Reid ML, Ratnikova T, et al. Uptake, translocation and transmission of carbon nanomaterials in rice plants. Small. 2009;**5**:1128-1132

[9] Lopez-Moreno ML, de la Rosa G, Hernandez-Viezcas JA, Castillo-Michel H, Botez CE, Peralta Videa JR, et al. Evidence of the differential biotransformation and genotoxicity of ZnO and CeO2 nanoparticles on soybean (Glycine max) plants. Environmental Science and Technology. 2010;**44**:7315-7320

[10] Ma X, Geisler-Lee J, Deng Y, Kolmakov A. Interactions between engineered nanoparticles and plants: Phytotoxicity, uptake and accumulation. Science of the Total Environment. 2010;**408**:3053-3061

[11] Wan A, Gao Q, Li H. Effects of molecular weight and degree of acetylation on the release of nitric oxide from chitosan–nitric oxide adducts. Journal of Applied Polymer Science. 2010;**117**:2183-2188

[12] Wang J, Koo Y, Alexander A, Yang Y, Westerhof S, Zhang Q, et al. Phytostimulation of poplars and Arabidopsis exposed to silver nanoparticles and Ag+ at sublethal concentrations. Environmental Science and Technology. 2013;**47**:5442-5449

[13] Taiz L, Zeiger E. Plant Physiology. 5th ed. Sunderland, MA, USA: Sinauer Associates Inc; 2010. pp. 67-86

[14] Fleischer M, O'Neill R. Ehwald. The pore size of non-graminaceous plant cell wall is rapidly decreased by borate ester cross-linking of the pectic polysaccharide rhamnogalacturonan II. Plant Physiology. 2014;**121**:829-838

[15] Rico CM, Majumdar S, Duarte-Gardea M, Peralta-Videa JR,

Gardea-Torresdey JL. Interaction of nanoparticles with edible plants and their possible implications in the food chain. Journal of Agricultural and Food Chemistry. 2011;**59**:3485-3498

[16] Eichert T, Goldbach HE. Equivalent pore radii of hydrophilic foliar uptake routes in stomatous and astomatous leaf surfaces – Further evidence for a stomatal pathway. Physiologia Plantarum. 2008;**132**:491-502

[17] Eichert T, Kurtz A, Steiner U, Goldbach HE. Size exclusion limits and lateral heterogeneity of the stomatal foliar uptake pathway for aqueous solutes and water-suspended nanoparticles. Physiologia Plantarum. 2008;**134**:151-160

[18] Zambryski P. Cell-to-cell transport of proteins and fluorescent tracers via plasmodesmata during plant development. The Journal of Cell Biology. 2004;**162**:165-168

[19] Liu J, Zhang YD, Zhang ZM. The application research of nano-biotechnology to promote increasing of vegetable production. Hubei Agricultural Sciences. 2009;**48**:123-127

[20] Torney F, Trewyn BG, Lin SY, Wang K. Mesoporous silica nanoparticles deliver DNA and chemicals into plants. Nature Nanotechnology. 2007;**2**:295-300

[21] Subramanian KS, Manikandan A, Thirunavukkarasu M, Sharmila RC. Nano-fertilizers for balanced crop nutrition. In: Rai M, Ribeiro C, Mattoso L, Duran N, editors. Nanotechnologies in Food and Agriculture. Switzerland: Springer International Publishing; 2015. pp. 69-80

[22] Milani N, McLaughlin MJ, Stacey SP, Kirby JK, Hettiarachchi GM, Beak DG, et al. Dissolution kinetics of macronutrient fertilizers coated with manufactured zinc oxide nanoparticles. Journal of Agricultural and Food Chemistry. 2012;**60**:3991-3998

[23] Rubio-Covarrubias OA, Brown PH, Weinbaum SA, Johnson RS, Cabrera RI. 2009. Evaluating foliar nitrogen compounds as indicators of nitrogen status in Prunus persica trees. Scientia Horticulturae. 2009;**120**:27-33

[24] Benzon HRL, MRU R, Ultra V Jr, Lee SC. Nano-fertilizer affects the growth, development and chemical properties of rice. International Journal of Agronomy and Agricultural Research. 2015, 2015;**7**:105-117

[25] Prasad TNVKV, Sudhakar P, Sreenivasulu Y, Latha P, Munaswamya V, RajaReddy K, et al. Effect of nanoscalezinc oxide particles on the germination, growth and yield of peanut. Journal of Plant Nutrition. 2012;**35**:905-927

[26] Singh NB, Amist N, Yadav K, Singh D, Pandey JK, Singh SC. Zinc oxidenanoparticles as fertilizer for the germination, growth and metabolism ofvegetable crops. Journal of Nanoengineering and Nanomanufacturing. 2013;**3**:353-364

[27] Liu Y, Tong Z, Prud'homme RK. Stabilized polymeric nanoparticles for controlled and efficient release of bifenthrin. Pest Management Science. 2008;**64**:808-812

[28] Abdel-Aziz HMM, Mohammed NAH, Aya MO. Nano chitosan-NPK fertilizer enhances the growth and productivity of wheat plants grown in sandy soil. Spanish Journal of Agricultural Research. 2016;**14**:e0902. DOI: 10.5424/sjar/2016141-8205

[29] Xiao Q, Zhang FD, Wang YJ. Effects of slow/controlled release fertilizers felted and coated by nano-materials on nitrogen recovery and loss of crops. Plant Nutrition and Fertilizer Science. 2008;**14**:778-784

[30] Walker R, Morris S, Brown P, Gracie A. Evaluation of potential for chitosan to enhance plant defence. Australian Government, RIRDC Report No. 2004. 4. 49

[31] Yoo HS, Kim TG, Park TG. Surface-functionalized electrospun nanofibers for tissue engineering and drug delivery. Advanced Drug Delivery Reviews. 2009;**61**:1033-1042

[32] Yang H, Liu C, Yang D, Zhang H, Xi Z. Comparative study of cytotoxicity, oxidative stress and genotoxicity induced by four typical nanomaterials: The role of particle size, shape and composition. Journal of Applied Toxicology. 2009;**29**:69-78

[33] Zebarth BJ, Drury CF, Tremblay N, Cambouris AN. Opportunities for improved fertilizer nitrogen management in production of arable crops in eastern Canada: A review. Canadian Journal of Soil Science. 2012;**89**:113-132

[34] Zhang Z, He X, Zhang H, Ma Y, Zhang P, Ding Y, et al. Uptake and distribution of ceria nanoparticles in cucumber plants. Metallomics. 2011;**3**:816-822

[35] Zhao L, Sun Y, Hernandez-Viezcas JA, Servin AD, Hong J, Niu G, et al. Influence of CeO2 and ZnO nanoparticles on cucumber physiological markers and bioaccumulation of Ce and Zn: A life cycle study. Journal of Agricultural and Food Chemistry. 2013;**61**:11945-11951

[36] Zheng L, Hong F, Lu S, Liu C. Effect of nano-TiO2 on strength of naturally aged seeds and growth of spinach. Biological Trace Element Research. 2005;**104**:83-91

[37] Mandeh M, Omidi M, Rahaie M. In vitro influences of TiO2 nanoparticles on barley (*Hordeum vulgare* L.) tissue culture. Biological Trace Element Research. 2012;**150**:376-380

[38] Manikandan A, Subramanian KS. Evaluation of zeolite based nitrogen nano-fertilizers on maize growth, yield and quality on inceptisols and alfisols. International Journal of Plant & Soil Science. 2016;**9**:1-9

[39] Mazumdar H, Ahmed GU. Phytotoxicity effect of silver nanoparticles on *Oryza sativa*. International Journal of ChemTech Research. 2011;**3**:1494-1500

[40] Rathnayaka RMNN, Mahendran S, Iqbal YB, Rifnas LM. Influence of urea and nano-nitrogen fertilizers on the growth and yield of rice (*Oryza sativa* L.). International Journal of Research Publications. 2018;**5**:7

[41] Sohrab D, Ali T, Gholamhossein D, Javier A. Reza Khorasani. 2016. Effects of foliar applications of zinc and boron nano-fertilizers onpomegranate (Punica granatum cv. Ardestani) fruit yield and quality. Scientia Horticulturae. 2016;**210**:57-64. DOI: 10.1016/j.scienta.2016.07.003

[42] Servin AD, Morales MI, Castillo-Michel H, Hernandez-Viezcas JA, Munoz B, Zhao L, et al. Synchrotron verification of TiO2 accumulation in cucumber fruit: A possible pathway of TiO2 nanoparticle transfer from soil into the food chain. Environmental Science and Technology. 2013;**47**:11592-11598

[43] Ma Y, Kuang L, He X, Bai W, Ding Y, Zhang Z, et al. Effects of rare earth oxide nanoparticles on root elongation of plants. Chemosphere. 2010;**78**:273-279

[44] Mohammadi R, Maali-Amiri R, Abbasi A. Effect of TiO2 nanoparticles on chickpea response to cold stress. Biological Trace Element Research. 2013;**152**:403-410

[45] Parks SE, Irving DE, Milham PJ. A critical evaluation of on-farm rapid tests for measuring nitrate in leafy vegetables. Scientia Horticulturae. 2012;**134**:1-6

[46] Reed RB, Ladner DA, Higgins CP, Westerhoff P, Ranville JF. Solubility of nano-zinc oxide in environmentally and biologically important matrices. Environmental Toxicology and Chemistry. 2012;**3**:93-99

[47] Remedios C, Rosario F, Bastos V. Environmental nanoparticles interactions with plants: Morphological, physiological and genotoxic aspects. Journal of Botany. 2012;**8**:1-8. DOI: 10.1155/2012/751686

Chapter 18

Urban Horticulture and Its Modernization by Using LED Lightning in Indoors Vegetable Production

Žarko M. Ilin and Dubravka M. Savić

Abstract

Urban horticulture also includes the production of vegetables, mostly leafy vegetables, in high tech protected areas with or without daylight. Vegetable crop growing is a scientific discipline that studies biology and technology in growing vegetable crops in either the open-field and greenhouse environment. The objective is to gain high-yield agricultural crops, good quality edible parts that are safe for human consumption and a minimal environmental pollution. Vegetables are annual, biannual or perennial herbaceous plants that rarely develop a woody stem during its vegetative period, mostly in the lower section of the stem. The vegetable edible parts are rich in water and are used either fresh and raw or processed. Once picked, the edible parts may be stored for a short period of time (several weeks, up to 9 months at the most). The vegetable edible parts are: roots and tubers, stems and stalks, sprouts, bulbs, leaves (cruciferous or headed vegetables), leaf stems, immature flower heads, fruits (mature or immature), and seed (mature or immature). Vegetables could be grown in urban areas, in protected areas with or without daylight. LED lightning represent one of the most important modernizations and implementation of vegetable production in urban areas.

Keywords: horticulture, vegetable production, modernization, LED lightning

1. Introduction

According to a new UN DESA report (United Nations, Department of Economic and Social Affairs "World Population Prospects: The 2015 Revision," 2015), the current world population of 7.3 billion is expected to reach 8.5 billion by 2030, then 9.7 billion in 2050, and 11.2 billion in 2100.

Due to mentioned data of UN DESA report, it is necessary to consider new solutions in food production, generally. Plant production is a part of food production and it requires innovative types of plant growing, especially in modern glasshouses or other types of protected areas without daylight. In such protected areas full equipment which would provide optimal climate conditions for successful plant growing is necessary. It means developing and applying the newest crops production technologies in modern greenhouses. On that way, it would be possible to get healthy, high quality, and safe food, which is connected with high protection of

IntechOpen

environment. It is very important for smaller and urban areas. Plant production in urban areas is a great contribution in food production for growing population, generally. One of such kind of plant production is vertical crop production in modern greenhouses or closed systems, both equipped with all necessary installation for providing optimal climate control. In such environment pests, control is efficient, too.

Vegetables have an important role in human consumption as fresh products, food, and pharmaceutical industry. They are mostly low calorie, low fat, and low protein foods, but are a significant source of some of the most needed vitamins, minerals, and microelements. Average recommended daily intake for vegetables is about 400 g [1]. Interesting dietary guidelines vegetable consumption USDA posted on their website (USDA Dietary Guidelines for Americans). In the mentioned post there is a detailed description of daily servings of vegetables for people living in different parts of world.

Vegetables reaction on the global climate changes is very sensitive and vegetable production is becoming more difficult by time. At the same time, with the increase in population on the planet, demand for vegetables is increasing. Therefore, the modernization of horticultural production has been catching the attention of scientists for years. One way to modernize horticultural production generally is to use LED lighting in a protected space, with and without daylight in urban areas.

Urban farming comprises production of various crops in urban areas, in objects with artificial light and without daylight, completely controlled environment in order to provide successful process of photosynthesis and crop productivity. It is one of the solutions for food production for a growing population on the planet. It is solution for the food production in the near future, and nowadays.

Such objects are provided with all necessary installations for optimal crops growth, high productivity, clean, and safe fresh food.

High-tech urban farming can be conducted in various spaces, like special rooms, chambers, buildings intended for such purpose, removable grow-trainers. It is sustainable production without pesticide usage.

Some of very important equipment is light emitting diodes (LED) for horticulture, which are made to provide needed light recipe for every phenophase of crop growth. It is important for photosynthesis and crop productivity. The other important installation, especially in closed systems without daylight, is possibility to provide needed CO_2 implementation, which is also very important for process of photosynthesis and crop productivity. Thus, it is possible to get, for example, high quality various vegetables and safe for human consumption.

According to Sprecht *et al.* [2] greenhouses could be placed on the roofs of the buildings in the cities, for example, what belongs to urban horticulture.

2. Supplemental lighting

Supplemental lighting is necessary in seedling production, as well as in growing vegetable crops that have a longer vegetative period and high demands for light (tomatoes, peppers, and cucumber), and for vegetables with smaller habitus growing in climate chambers in various urban facilities. In greenhouses, it is usually needed during long winter months and periods of overcast. The supplemental lighting prolongs a day, compensates for a natural light limiting effect in winter, and enhances the amount of the available light. Supplemental lighting support photosynthesis in plants and empower plant growth, so they become more resistive on various diseases. Supplemental lighting should not be confused with photoperiodic lighting, which is applied to create long days, thus controlling the plant growth and development processes.

Fundamental issue is lamps and diodes production technologies because they have to be adapted for horticultural crop production. Some of the lighting technologies include incandescent bulbs, halogen incandescent bulbs, fluorescent lamps, compact fluorescent lamps, high-intensity discharge lamps, and light-emitting diodes.

For now, in the vegetable crop production practice, the High-Intensity Discharge (HID) lamps and LED lamps are in use. The HID lamps emit high heat (up to 50%), so they must be placed at about 2 m or more above the crops. The LED lamps are placed at about 40 cm or more above the crops (depends on the species) or in between the crop's rows in the greenhouses, as they do not produce heat that could damage plants.

Nowadays, there are very intensive studies in the ways the plants use the incoming photosynthetic active radiation (PAR) which is based on a principle of an exponential increase of absorbed photosynthetic active radiation with the increase in leaf area index [3].

3. Importance of LED lightning

Importance of LED lightning in urban horticulture and generally, horticultural production in protected areas could be seen on many official and representative movies of companies which have been cooperating in official scientific projects with eminent universities around the world.

Many scientific examinations in the recent years showed that LED lightning save energy and improve and empower plat growth and modern urban vegetable growing—in protected areas with and without daylight.

Light-emitting diodes (LED) represent a promising technology for the greenhouse industry that has technical advantages over traditional lighting sources, as well as a significantly positive impact on the plant photosynthesis process and therefore on the crops yield. They are only recently being tested for horticultural applications, both in greenhouses and in special chambers with a total control of climatic and other conditions necessary for the crop's growth and development. For the time being, they are mostly used in growing leafy vegetables and herbs. LEDs are solid-state light-emitting devices that emit broad-band (white) spectrum light that is necessary for both the vegetative and the reproductive crops phases. Depending on the vegetable varieties and their edible parts (vegetative part, fruit, and immature flower heads), LEDs could be designed to emit light for each phenophase of the crops, so as to adjust it to particular crops and production goals. One of the most important features of LEDs for horticultural application is that the generation of light in LEDs does not produce heat in the beam of light, and LEDs are cool to the touch. So, LEDs can be placed at 40–50 cm above the rows of crops or in between the rows in the greenhouse, and there will be no damage to the plants coming from excess heat.

Besides the crucial role of light and its special qualities for the process of photosynthesis, a combination of light or photosynthetically active radiation (PAR) and supplemental carbon dioxide is also very important in the greenhouse. It is vital that both light and carbon dioxide are provided in sufficient amounts within the greenhouse, or otherwise, a lack of either may pose a limiting factor for the photosynthetic process and consequently for the crop's productivity. Therefore, when supplemental lighting is applied for purposes of increasing the crops productivity, it is necessary at the same time to maintain a suitable carbon dioxide concentration in the greenhouse.

In this way, it is possible to grow vegetables in the greenhouses without natural daylight but with the application of the suitable LED lighting (depending on the crops variety, its edible parts, growing requirements and other) and other controlled climatic conditions for optimal plant growth. LED lighting lasts approximately 18 hours and 6 hours plants are in the dark.

4. Light quality in supplemental lighting in vegetable growing in completely controlled environment without daylight

The suitable light quality in the greenhouses actually refers to the wavelengths (colors) that are efficient in inducing photosynthesis in plants and other growing processes. The light wavelengths are expressed in nanometers (nm). The visible spectrum wavelengths range from about 390 to 760 nm, which is only a small portion of the sunlight (radiation) electromagnetic spectrum. The visible light consists of: violet (380–430 nm), blue (430–500 nm), green (500–570 nm), yellow (570–590 nm), orange (590–630 nm), and red light (630–760 nm). The visible light range mostly corresponds to the Photosynthetically Active Radiation (PAR) from about 400 to 700 nm. The stated wavelengths have the right amount of energy for the biochemical processes, while their ratio in the available light is of crucial importance for determining the quality of light. About half of the sunlight energy participates in the photosynthetic processes. The rest of the energy comes from the sunlight short wavelength spectrum (UV—ultraviolet radiation) and sunlight long wavelength spectrum (IR—infrared radiation).

Blue section of the spectrum, also known as cool light, induces these wavelengths that encourage vegetative and leaf growth through strong root growth and intense photosynthesis.

Red section of the spectrum induces stem growth, tuber and bulb formation, flowering, and fruit production, and chlorophyll production.

Far-red light may cause plants to stretch (elongate) and may trigger flowering in some long-day plants. The plants are exposed more to the far-red than to the red light, which may become a problem with the greenhouse vegetable crop production due to possible shading (for whatever reason) or due to the reduced plants vegetative space.

Green and yellow sections of the spectrum that reach the plants are reflected, thus giving them their green color. Most of the absorbed sunlight wavelength belongs to the blue and red range of the spectrum. However, the recent studies have shown that plants do also absorb some green and yellow light, using it in the process of photosynthesis [4]. Generally, a light source that provides light in the entire visible range will better meet the needs of the plant.

For the time being, in the greenhouse vegetable crops growing practice, the high-pressure sodium (HPS) lamps are used, but also the LED lamps are gaining (**Figure 1**) an increasing significance in the plastic and glass greenhouses and in special chambers vegetable production. Also, in The Netherlands, the latest studies at the Wageningen and Maastricht universities research centers have their guidelines for greenhouse lighting with little or no natural daylight for special feature vegetable crops growing—increased vitamin C content, reduced nitrates content, increased sugar content, and higher yield.

With red, white, and far red light, it is possible to prepare ideal light recipe for particular vegetable species and improve process of photosynthesis and production of assimilates which empower plants. The most important is how plants response on various recipes. So, plants become more resistant toward unfavorable conditions for its growth and toward diseases. In case of adding combination of red, blue, and far red light to combination of red, white, and far red light, it is possible to reach more

Figure 1.
LED lightning in between rows of cucumber plants with daylight in modern glasshouse.

than 20 various recipes of plants lightning. Then, in combination with CO_2, temperature, various substrates, and humidity, it is possible to obtain ideal light recipes for optimal plant growth. Such kind of experiments could be expensive, but with good plan and expertise (know-how) costs could be lower.

According to Goldammer [5] besides optimal light recipe for the particular crops, it is necessary to understand process of plant growth in order to apply it in the practice, and to do optimizing all the other parameters like climate, irrigation, nutrition, software, sensors, seeds, substrates. Actually, all parameters in indoor plant environment are gathering via sensors and special software in computer where it is possible to control and manage them. The right interaction between all mentioned parameters and growing crops give the best results in indoor completely controlled environment in vegetable crop production. Vegetable crops are kept out of bugs and pests, taste optimized, could be produced all year around in natural way, with less waste in fresh food production, generally. On that way, food is clean, healthy, and nutritious, and production is efficient.

The most suitable for urban horticulture are usage of NFT systems, combined NFT system, and rockwool cubes, and LEDs above the crops, or LEDs could be used between plants rows grown on rockwool substrates. Which type of crops growing and type of LEDs which would be applied depend on morphology of plants species [4].

Generally, nowadays, trend in horticulture is vegetable production under the LED lightning because of numerous advantages in comparison with other types of supplemental lighting.

5. Supplemental carbon dioxide in vegetable production in completely controlled environment and in the greenhouses with LED lightning installation

Carbon dioxide (CO_2) gas is the essential component for the process of photosynthesis, and the plants uptake it through their stomata on the leaves.

Photosynthesis is a chemical process occurring in plants in which the light energy is used to convert carbon dioxide into water and sugars (carbohydrates) and oxygen (O_2) gas. The sugars in plants, obtained in the process of photosynthesis, are then used for the plant development and growth through the process of respiration.

In the objects with LED lightening and completely controlled environment, without daylight, enrichment of the air with CO_2 is necessary because of process of photosynthesis and crop productivity. Due to lack of sun light, in such objects is very important to define right amount of CO_2.

In the air (outside the greenhouse), there is about 400 ppm of carbon dioxide. The CO_2 concentration is increased when coal, natural gas, oil, and kerosene are burnt. Inside the greenhouse, the amount of CO_2 may be significantly depleted as plants use it intensively in the process of photosynthesis (**Figure 2**; [4]), which may lead to a decreased crops productivity or yield. For that reason, "CO_2 fertilization" or "CO_2 enrichment" is a standard practice in modern greenhouses.

Since there is about 500 times more oxygen in the air than carbon dioxide [4], it makes sense to increase the CO_2 concentrations in the greenhouse (particularly in highly equipped glasshouses). It has a positive effect on the oxygen-carbon dioxide ratio. The photosynthesis is higher by 30–50% at CO_2 concentrations of about 1000 ppm, regardless of the amount of light.

The increased concentrations of carbon dioxide are good as long as they do not limit the process of photosynthesis. Photosynthesis depends on light, temperature, air humidity, and carbon dioxide contents in the greenhouse. There is often a question of what is the optimal concentration, but it is hard to give a correct answer to it as the process of photosynthesis does not depend solely on CO_2. Also, a point should be made that climatic factors affect the stomatal opening mechanism (through which the plants uptake CO_2). Generally, a small increase in the plant photosynthesis process may be achieved at 1000–1200 ppm, but then, there is also an increased possibility of damage to the crops. One experiment done on eggplant crops showed that the first damage to the plants occurred at a constant CO_2 level of 800 ppm [4]. Quite often, the intensity of the photosynthesis may be higher at lower doses of carbon dioxide and higher intensity of light, and the other way around.

Supplementing the greenhouse air with carbon dioxide may not be necessary at all as long as the processes of the crops development and growth are quite satisfactory for the vegetable grower. Also, in a case of intensive greenhouse ventilation, the carbon dioxide concentration may drop below a level that is necessary for the normal photosynthesis process, so increasing the CO_2 concentration may not be an economical measure (unless the greenhouse ventilation rate is lowered).

If the crops quality and production are below the satisfactory level, however, carbon dioxide supplementing should be the next measure. Generally, the production period from late autumn to early spring increases the potential need for CO_2 supplementing the greenhouse air, which actually corresponds to a lower ventilation rate due to low outdoor temperatures.

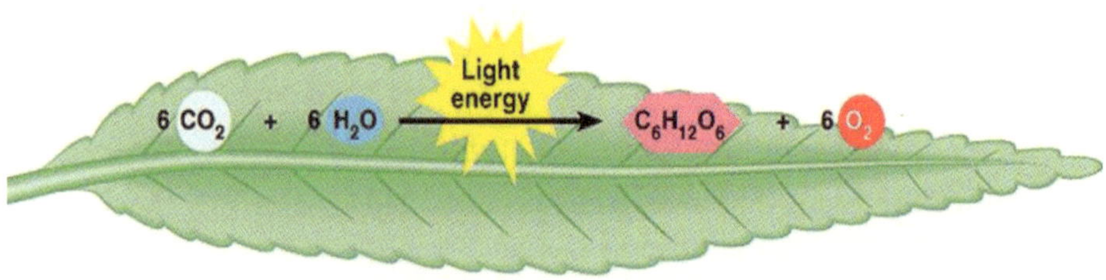

Figure 2.
Photosynthetic process equation.

Normal ventilation provides an amount of carbon dioxide that is similar to its levels in the outdoor air (350–400 ppm). But then, frequent ventilation in the greenhouse is not desirable, so that CO_2 supplementing has long been a common practice in vegetable crops growing. The necessary greenhouse carbon dioxide concentration is determined upon the type of the crops grown in the greenhouse, the greenhouse total volume and ventilation, lighting, temperature, air humidity, and stomatal opening [6].

Since carbon dioxide is one of the products of burning (e.g., fuel for greenhouse heating system), this segment of the heating process can be used for supplementing the greenhouse air. There are various ways of extracting carbon dioxide from other products of burning (fuel), so that the CO_2 from the boiler room can be dosed and at certain times directed and distributed into the greenhouse.

Also, pure carbon dioxide can be used, which is delivered to growers in special tanks, in liquid form and then can be converted into gas and distributed in the greenhouse. This way of supplementing the CO_2 has become increasingly popular as it eliminates any potential damage to the crops, allows control of other greenhouse climatic conditions that regulate the process of photosynthesis and crops productivity, provides easy control of the carbon dioxide levels, and is more flexible for supplementing the CO_2 when necessary.

One disadvantage of the liquid CO_2 is that it is usually more expensive than that obtained from burning fuel, e.g., natural gas.

Also, it would be advisable to install a proper system that registers the CO_2 concentration and then distributes it in the greenhouse. Such a system, like in other greenhouse installation operations, has corresponding sensors that are linked to a special computer software that registers, monitors, and controls all the greenhouse environment parameters. In this way, it is possible to detect a cause of each change and correct it in a short period of time.

The distribution of CO_2 depends mainly on the air movement within the greenhouse, as CO_2 does not travel very far through diffusion. One of the pure CO_2 distribution ways is by a central pump that pushes it into a system of flexible perforated plastic pipes (made of polyethylene or other plastic material). The pipes for CO_2 distribution are placed below the substrate special gutters with plants (if crops are grown in such gutters) or in the lower sections of the crops (if the plants are not grown in gutters). Then, through the pipe perforation, the carbon dioxide is distributed in the air around the plants. Very important is to obtain conditions that keep the leaf stomata open in order to uptake carbon dioxide [4].

In greenhouses, LED lightning could be placed on the top of the crops or in between rows of the particular crop. LED light does not have high emission of heat and cannot damage plants if they are placed in between rows in the crop. Even, that type of LED lightning is possible to be moved up and down, what depends on the crop development. Combination of LED lightning and sun light in greenhouses is an excellent combination for saving electrical energy and to empower crops growth.

With the progress of the scientific research about LED lightning (**Figure 3**) for horticulture and its interaction with plants in order to achieve better quality of edible parts and improve energy efficiency [7] in crop production, it could be high yield, uniform color, firmness, nitrate control in edible parts of plants. Intensity of LEDS which is enough for various recipes is approximately 600 $\mu mol\ m^{-2}\ s^{-1}$ [8, 9].

LED lightning [10] is an efficient source of light in horticulture needed for photosynthesis and plant productivity. Advantages of LED lightning in horticulture than the other types of lightning are relatively low energy consumption, lower radiation heat, long lifetime, flexibility in positioning above or inside a crop, the ability to control the light spectrum and produce high light levels. Important characteristic of LEDs is possibility to control and make various light recipes which participate

Figure 3.
LED lightning in protected area without day light.

in optimizing photosynthesis, photomorphogenesis, and nutrient contents. LEDs make easier monitoring of nitrogen absorption by leafy plants and avoid harmful nitrate concentration in leafy plants and its influence on human health.

Urban farming is carried out mainly hydroponically (NFT system or rockwool) with or without daylight with complete controlled environment [10], which bring some benefits like controlled usage of water, nutrients, pesticides multiple crops per year, high quality of edible parts, less labor, and easier harvesting.

Urban farming provides remarkable reduction in electricity cost for transplants production by using thermally insulated walls, multi-shelves, advanced lighting and air conditioning systems, etc.

Urban farming enables vertical production of propagation plant material and regular crop production in fully controlled environment. It means that the area for urban farms can be various. It can be placed in supermarkets, or other places where people gather and want to refresh with fresh vegetables or fruits, for example. Urban farms can be smaller or larger areas (e.g., of several square meters), and with vertical cultivation, the crop yield is achieved as in larger areas (e.g., in modern greenhouses or completely controlled objects without sun light).

6. Conclusion

Further development of LED lightning in horticulture, especially in vegetable production, would bring many advantages in producing clean, safe, health food for humans, but for animals, too.

Due to significant climate change and growing population on the planet, there is a need for new solutions regarding safe and high-quality food production. One of the solutions is improvement and usage of LED lightning in controlled environment without sun light or greenhouses with LED light as a supplemental light in the particular crop production.

In order to achieve global food security goals, it is possible to implement alternative farming methods that could increase horticultural outputs and reduce negative climate impacts on the food production. It means that urban areas could be used for high quality food production and using LED lightning.

Acknowledgements

Dubravka Savić would like to express her special gratitude to her colleagues Prof. Dr. Milivoj Belić (Faculty of Agriculture, University of Novi Sad) and Prof. Dr. Žarko M. Ilin (Faculty of Agriculture, University of Novi Sad) for the participation in the project titled "Development of new technologies for modern and sustainable production of vegetables" (TR31036) funded by the Ministry of Education and Science, Republic of Serbia. Part of the project has been dedicated to investigation of LED lightning in vegetable production indoors and in the greenhouses.

Author details

Žarko M. Ilin[1] and Dubravka M. Savić[2]*

1 Department for Field and Vegetable Crops, Faculty of Agriculture, University of Novi Sad, Novi Sad, Serbia

2 Department for Farming and Vegetable Crop Science, Faculty of Agriculture, University of Belgrade, Belgrade, Serbia

*Address all correspondence to: dubravkas@sezampro.rs; dubravkas@agrif.bg.ac.rs

References

[1] Popović M. Povrtarstvo. Beograd: Nolit; 1989

[2] Sprecht K, Hartmann S, Ulf B, Magdalena S, Werner A, Thomaier S, et al. Urban agriculture of the future: An overview of sustainability aspects of food production in and on buildings. Agriculture and Human Values. 2014;**31**(1):33-51

[3] Booij R, Kreuzer ADH, Smit AL, Van der Werf A. Effect of nitrogen availability on dry matter production, nitrogen uptake and light interception of Brssels sprouts and leeks. Netherlands Journal of Agricultural Science. 1996;**44**:3-19

[4] Heuvelink E, Kierkels T. Plant Physiology in Greenhouses. Horti-Text B.V; 2017

[5] Goldammer T. Greenhouse Management: A Guide to Operations and Technology. USA: Apex Publishers; 2017

[6] Benton JJ Jr. Hydroponics. A Practical Guide for the Soilless Grower. Florida. USA: CRC Press; 2005

[7] Nicole CCS, Charalambous F, Martinakos S, van de Voort S, Li Z, Verhoog M, et al. Lettuce growth and quality optimization in a plant factory. Acta Horticulturae. 2016;**1134**:231-238

[8] Ntagkas N, Min Q, Woltering EJ, Labrie C, Nicole CCS, Marcelis LFM. Illuminating tomato fruit enhances fruit vitamin C content. Acta Horticulturae. 2016;**1134**:351-356

[9] Xu Y. Seven dimensions of light in regulating plant growth. Acta Horticulturae. 2016;**1134**:445-452

[10] Kozai T, Ohyama K. Commercialized closed systems with artificial lighting for plant production. In: Moe R, editor. Proceedings. Vth International Symposium on Artificial Lighting. Acta Horticulturae. 2006;**711**:61-70

Chapter 19

Remote Sensing: Useful Approach for Crop Nitrogen Management and Sustainable Agriculture

Salima Yousfi, José Fernando Marin Peira,
Gregorio Rincón De La Horra and Pedro V. Mauri Ablanque

Abstract

Soil fertility is among the most important criteria that affect crop yield and quality. Nitrogen stress due to the low soil fertility and the lack of nitrogen availability is a major factor limiting the crop productivity in arid and semiarid environments, where fertilization is not optimized in terms of timing and quantity. Managing nitrogen fertilization is one of the most important criteria in the precision agriculture, which helps to improve crop production, environment conditions, and farmer's economy. It is very important to apply N fertilizers with efficient methods allowing to the nutrient use efficiency and avoiding nitrogen losses and environment contamination. Nowadays, remote sensing methods using spectral and thermal approaches have been proposed as potential indicators to rapid identification of crop nitrogen status by providing information about vegetation canopy properties across large areas. The use of remote sensing methods to schedule nitrogen fertilization can help farmers to practice a more sustainable agriculture, minimizing risks of losing the harvest by providing an adequate rate of nitrogen when the crops' needs and at a specific location.

Keywords: nitrogen fertilization, remote sensing, smart N management, farmer's decisions, precision agriculture, sustainable agriculture

1. Introduction

Soil nitrogen amount is among the most important criteria that affect crop yields and quality. Plant growth and development need nitrogen in greater quantity, since it is involved in various physiological processes. Nitrogen is a part of many components of plant cells, including amino acids, nucleic acids [1], proteins, and chlorophyll in plant leaves. Likewise, nitrogen availability produces rapid and early crops' growth, increases protein content of crops, facilitates the uptake and utilization of other nutrients as potassium and phosphorous, improves fruit quality, and controls overall growth of plant [2, 3]. However, nitrogen deficiency rapidly inhibits the growth of plants and alters many metabolic processes. The lack of nitrogen decreases photosynthesis [4], causes appearances of chlorosis [5], reduces chloroplast size [6], and provokes a high decrease in crop quality and yields.

Consequently, analyzing nitrogen amount in soil and crops and the application of N fertilizer in the event of a deficit are essential to improve crop production.

 IntechOpen

A key factor in the efficiency of nitrogen application is to adjust N input to N crop demand [7]. The addition of N fertilizer when crops' needs, with the right dose, may increase yields and reduce the farmer's input costs. In this way, precision agriculture permits the distribution of the correct quantity of agricultural inputs (fertilization and water irrigation) in real time and at a specific location. Mulla and Schepers [8] reported that precision agriculture aims to improve site-specific agricultural decision-making through collection and analysis of data, formulation of site-specific management recommendations, and implementation of management practices to correct the factors that limit crop growth, productivity, and quality. Moreover, Gebbers and Adamchuk [9] described the precision agriculture as a key to optimize the use of available resources to increase the profitability and sustainability of agricultural operations, to reduce negative environmental impact, and to improve the quality of the work environment and the social aspects of farming.

Therefore, monitoring nitrogen fertilization with a high coverture of crop, high spatial variability, and right timing of applications are very important to improve crop's production and to help farmers to take decisions. Nowadays, crop water and nitrogen managements use many indices acquired by remote sensing techniques. Mulla and Miao [10] informed that proximal sensing of crops is currently the primary tool used to detect nutrient deficiencies for variable rate application of fertilizer. This is based on researches that showed nitrogen deficiencies could be detected using spectral reflectance in the green, red, red edge, and near-infrared portions of the spectrum.

In this chapter, we explain the usefulness of the use of remote sensing techniques in precision agriculture in managing nitrogen fertilization. Remote sensing methods are rapid and nondestructive ways of permitting multiple optical measurement indicators of plant greenness and crop nitrogen status. Fox and Walthall and Hunt et al. [11, 12] showed that the greenness of plants is strongly related to leaf chlorophyll content and to N status, and so it has been used as an indicator of N availability. Moreover, remote sensing techniques can be generally defined as doing the right management practices at the right location, in the right rate, and at the right time [10]. This would reduce surplus N in the crop production system without reducing crop yield, which would in turn reduce N losses to surface and groundwaters [13].

2. Remote sensing systems and smart nitrogen fertilization

Remote sensing methods using spectral and thermal approaches have been proposed as potential indicators to allow rapid identification of crop nitrogen status by providing information about vegetation canopy properties. Guérif et al. [14] reported that reflective sensors represent a new approach showing great potential to provide quick and easy, nondestructive estimates of plant nitrogen status. Remote sensing observations in the visible and near-infrared spectral may provide information of leaf chlorophyll content, and such information permit the early detection of plant nutrient deficiency. Guérif et al. [14] have demonstrated that the canopy chlorophyll content is more strongly related to the canopy nitrogen content. This provides the necessary link between remote sensing observations and the canopy state variables used as indicators of nitrogen status. Moreover, nitrogen stress decreases canopy reflectance in near-infrared [15, 16] and increases canopy reflectance over all visible wavelengths, because of a shortage of chlorophyll and other light-absorbing pigments [15]. Therefore, vegetation indexes combining information from visible and near-infrared regions may maximize sensitivity to N stress and are used as tools in nitrogen fertilization.

2.1 Ground remote sensing to detect nitrogen deficiency

The use of low-cost ground remote sensing methods to schedule nitrogen fertilization may contribute to more sustainable agriculture [17]. Ground remote sensing instruments are very useful for small-scale operational field monitoring of biotic and abiotic stress agents. This technology has better temporal, spectral, and spatial resolutions than satellite remote sensing [18]. In this category, the effective and easy-to-measure Trimble GreenSeeker is the most used; it is equipped with an active sensor and emits its own light to measure canopy reflectance corresponding to the red and near-infrared. GreenSeeker measures the normalized difference vegetation index (NDVI), which is formulated using the following equation: (NIR − R)/(NIR + R), where R is the reflectance in the red band and NIR is the reflectance in the near-infrared band. In addition, Hunt et al. [12] informed that NDVI value varies with absorption of red light by plant chlorophyll and the reflection of NIR radiation by water-filled leaf cells.

NDVI is one of the most well-known vegetation indices used in precision agriculture in managing crops' fertilization. NDVI values indicate N uptake, plant health, and yield prediction [19] and correlate positively with intercepted photosynthetically active radiation and also correlate well with N content [20]. In this regard, NDVI readings are used to assess the effect of nitrogen fertilization [21, 22], since its values depend on two factors, nitrogen content and total biomass [23].

In the last decades, several companies offer equipment with N-sensor for proximal sensing of crop nitrogen and nutrient deficiencies as Trimble's GreenSeeker, Ag Leader's OptRx, and Yara's N-sensor. Farmers can use these services to analyze the level of nitrogen in their crops and make decisions before providing nitrogen fertilization. All these companies and many others help the farmer to analyze the levels of nitrogen deficiency in crops and to calculate the exact amount of N fertilizer for each crop.

2.2 Nitrogen management by airborne and satellite imagery

One of the most active applications in the nitrogen fertilization managing is the use of aerial remote sensing services. Multispectral, hyperspectral, and thermal aerial imagery obtained by unmanned aerial vehicle (UAV) flights is a useful tool to detect nitrogen N crop needs. According to Ref. [24], there are various categories of imaging systems derived from remote sensing and used in fertilization application. Among them we cited the RGB/CIR cameras, which combine infrared (CIR), red, green, and blue light imagery (visible or RGB) and enable to estimate green biomass [24] and N status (NDVI type of information). The multispectral cameras, which can acquire a limited number of spectral bands at once in the VIS-NIR regions, are widely used for evaluating green biomass, nutrient status, pigment degradation, and photosynthetic efficiency [24]. The infrared cameras or thermal imaging cameras have a potential use in predicting nutrients stress in crops.

Nowadays, various companies provide farmers aerial remote sensing services through multispectral and hyperspectral or thermal aerial imagery, which is used for the diagnostics of crop nutrient deficiency in different crops (wheat, rice, cotton, horticultures, and other crops). Several models of nitrogen applications are developed by the use of aerial platform imagery, permitting to improve farm nitrogen management. Nitrogen algorithm models, developed by the information obtained through N-sensor, can help farmers to calculate the correct dose of N supply needs by crops and location and thus increase crops' production and decrease environmental contamination due to excessive N fertilization.

Additionally, multispectral and hyperspectral satellite imagery also has a major role in managing crop growth. Data of satellite imagery (sometimes with free access) are frequently used in fertilization management and soil analyses on large spatial and temporal scale. Söderström et al. [25] explained that the advantage of using satellite data for N management within fields compared with handheld or vehicle-mounted sensors is that the data collected cover huge areas and can be used on a multitude of scales, from watersheds and landscapes to fields. New low-cost or publicly available satellite systems such as Sentinel-2 with high temporal resolution and with additional wavebands targeted for assessment of crop properties open up exciting possibilities for improved N management and nutrient use efficiency for more efficient food chains.

The CropSAT, Sentinel, and Fertisat are some satellites that offer farmers the possibility to improve the efficiency of nitrogen fertilization, using variable rate application (VRA) technologies. In addition, nitrogen maps created by satellite imagery can be used for site-specific adjustment of N fertilizer in the fields and are often adapted to specific requirements of crops, since each species has a different phenology and thus a different quantity and critical time for nitrogen application. Nevertheless, sometimes the climatic conditions are the main problems using satellite imagery to crop managements. In addition, weather conditions also influence the absorption of nitrogen by plants; hence, information derived from remote sensing imagery consider all these elements when defining precise doses of nutrient fertilizer.

3. Smart nitrogen fertilization for sustainable agriculture

In the recent years, remote sensing techniques are being considered as a key factor in the sustainability of agriculture. Information managed through NDVI and thermal and multispectral imageries are the most used in the precision agriculture for vegetation monitoring, since it permits to correct in real time problems found in the fields as the lack of nutrients or overfertilization; thus, avoiding losses of production and environment damages.

3.1 Timing of fertilization and plant needs

Both deficit and excess of N fertilizer have negative effects on plant growth. Guérif et al. [14] informed that too much nitrogen is not good either, as nitrogen toxicity can occur in overfertilized plants, leading to stunted growth and a poor-quality plant. In addition, Rubio et al. [26] showed that an excessive amount of N in ammonium form may adversely affect plant growth, causing a rapid development of the crop, with rapid stem elongation that makes plants too soft and blocks the absorption of Ca^{2+}. Moreover, Dynarski [27] added that overfertilized crops permit to take up more nitrogen than they need, which disrupts the balance of nutrients in plant tissue, and the result is that crops will be deficient in other necessary nutrients, such as sulfur and zinc, reducing in this way crop quality. Other authors added that inadequate quantity of N fertilizer could decrease fruit production [28, 29], increase susceptibility to insect pests [11, 30] and pathogens [28, 30, 31], and reduce nutritional quality of harvested products [28, 32].

However, the efficiency of N fertilization does not only depend on the contribution of the appropriate amount but also on providing this amount to crops at the right time, since each species has critical points in the cycle of growth where nitrogen input is primordial. For example, wheat cultivars need N supply in the spring and early summer, while corn absorbs most nitrogen fertilizer in midsummer, and other crops need N fertilizer just after crop emergence or at seedling. In this regard, nitrogen fertilizer should be applied when the crop needs and with adequate quantity.

In this regard, many researches showed that leaf spectral reflectance properties are closely related to growth environments (water and nutrients' availability), but it is also strongly associated to crop needs at different growth stages. In recent years, remote sensing methods are frequently used in managing N fertilizer in various farms; these techniques can help farmers to calculate the exact dose of N fertilizer necessary for crops and most importantly apply it at the right period. Many researchers have developed several models and algorithms using remote sensing information (through NDVI, drones, and satellite imagery) to determine N rates for different species and locations.

3.2 Remote sensing and soil fertility

The excess of N fertilizer causes negative effect on soil, since it affects composition and fertility of soils. The long-term use of fertilizers has become a significant source of soil and water pollution [33, 34]. The application of nitrogen above the appropriate levels may cause nitrate accumulation in lower parts of the root expansion, and consequently there is a risk for soil nitrogen leaching [35, 36]. In addition, [37] affirmed that soil acidity is developed in response to nitrogen fertilizer addition when addition of N exceeds the assimilation or N storage by biotic components or soil organic matter, respectively. The same authors added that excessive N fertilizer input could affect soil chemical and biological health, as well as the soil organic matter.

Hence, it is important first to know the current soil nutrient levels before any supplement of N fertilizer. In this way, soil parameter reflectance can provide information on the compositions and properties of soil. The study of spectral reflectance of soils has the ability to provide nondestructive and rapid prediction of soil physical, chemical, and biological properties [38]; soil texture, structure, and moisture [39]; and soil mineralogy and organic matter [40]. Therefore, potential information acquired through remote sensing technologies can help avoid soil degradation due to overfertilization. Moreover, mapping and analysis of soil fertility using remote sensing imagery or N-sensors before N supply can diminish soil compaction and increase N efficiency and absorption.

3.3 Environment protection

Another important effect of remote sensing techniques in nitrogen management is the protection of the environment. Excessive and long period of nitrogen fertilization accumulates contaminants in the soil and provoked environment damages. Guérif et al. [14] informed that overfertilizing can be a source of unnecessary extra costs as well as an environmental hazard in the case of nutrient runoff. Moreover, Saggar et al. and Vistoso et al. [41, 42] indicated that N fertilizer application at levels exceeding plant requirements leads to significant environmental consequences in many parts of the world due to N losses, such as nitrate NO_3 leaching, NH_3 volatilization, and nitrous oxide (N_2O) emission. Deterioration of water quality is also one of the most serious global environmental problems derived from the excessive crop nitrogen fertilization. Groundwater or surface water is being polluted mainly by nitrates when crop overfertilization occurs. Riley et al. [43] showed that the transport of N from agricultural soils to surface waters has been linked to eutrophication of freshwater and estuaries. High fertilization rates lead to N losses with negative impacts not only on atmospheric greenhouse gas (GHG) concentrations but also on water quality [44].

The excess of nitrogen fertilizer can be leached downward into groundwater, be mixed with surface waters, or be released into the atmosphere as gases, causing a high rate of environmental pollution. In addition, matching N application and crop requirements decrease deleterious environmental effects of excessive fertilization, either by nitrate pollution of water [45] or by gaseous emissions [46].

Consequently, all these negative consequences for the environment, associated with excessive nitrogen fertilizer, need new technological approaches to improve nutrient management. The use of remote sensing data to control dose and timing of nitrogen fertilizer can protect environment and permit best management of crops to more sustainable agriculture.

3.4 Impact of the intelligent fertilization in farmer's economy

In addition, excessive application of fertilizers also affects the farmer economy negatively. Efficiency of nitrogen fertilization can help farmers to improve control of incomes and reduce costs, avoiding unnecessary supply of N fertilizer. The application of the right dose of fertilizer (and sometimes no fertilization) helps farmers for best crop management, since the application of N does not always increase performance. The estimation of N plant and soil status prior to fertilization is important, particularly when fertilizer rates are above the optimal farmer's economic level and crop needs; in this case, farmers can reduce the unnecessary N fertilization and maintain yield at a lower cost. At many times, the minimum fertilization can optimize the yield and income of farms and permit the sustainability of agriculture. At present, the number of farmers who accept the use of new technologies in their crop management has increased. Farmers have realized that the better use of fertilizer through remote sensing information can greatly improve their income, protect their crops, and develop the rural environment.

4. Farmer's decisions to sustainable agriculture

In the past, farmers were not customary to the applications of new technologies in their farms. The farmers used classical methods to manage their crops and frequently applied irrigation and fertilized without having information on plant needs and soil composition. Traditional crop management leads to harvest loss, particularly when the different types of stresses are detected very late. In addition, the excessive use of fertilizers by farmers provokes often soil degradation and environmental pollution. Rosea et al. [47] indicated that as a response to the environmentally and socially destructive practices of postwar mechanization and intensification, the concept of sustainable agriculture has become prominent in research, policy, and practice. Sustainable agriculture aims to balance the economic, environmental, and social aspects of farming, creating a resilient farming system in the long term.

However, in recent years, remote sensing techniques to the sustainable agriculture are applied successfully by numerous farmers and in different category of crops including cereals, viticulture, horticulture, and grassland. Farmers using remote sensing information in their crop management can increase the efficiency of resource use and reduce the uncertainty of decisions required in the field.

At present, smart devices and intelligent systems interact flexibly with the precision agriculture. Remote sensing platforms that provide data storage and interpretation permit the intelligent analysis of crop status and accurate farmer's decisions. Cambra Baseca et al. [48] informed that systems for precision agriculture can be based on satellite navigation systems or terrestrial systems for geographic information and sensors located in the plot. These systems collect information to be used to make decisions with greater precision and to optimize crop yields.

Smart strategies, used by farmers to the sustainable nitrogen management, can help farmers to take the right decisions to reduce nutrient loss in the environment, maximize uptake of N by crops, reduce fertilizer costs, and protect environmental conditions. These strategies can be as simple as applying the right fertilizers in the right period and with the exact quantity at a precise location. Fertilization decision system is commonly designed for soil nutrient evaluation, management, and crop fertilization by integrating modern information technology, with soil quality evaluation and crop fertilization theory [49]. Consequently, precision agriculture using new technologies has powerful to improved farmer's decisions (appropriate use of fertilizer). Therefore, decisions taken by farmers afterward remote sensing diagnostics have significant effect in the sustainability of agriculture, the high crop productivity and quality, the prevention of environmental degradation, the farmer's economy (costs and income), and the rural improvement.

5. Conclusions

Nitrogen fertilization with the correct dose and timing is essential for successful crop production. Sustainable nutrient management strategies have been highly successful in various farms through the world. Monitoring fertilization has become a valuable tool in farm crop management, helping farmers to improve crop productions and to increase incomes.

Smart fertilization based on remote sensing techniques has shown various advantages, such as improvement of crop productivity and quality and agroecosystem protection. In the latest years, the costs of remote sensing application in nitrogen management have decreased and have become more common in agricultural sectors.

However, the economy of small farmers or farmers from developing countries is limited and doesnot permit the use of remote sensing services. For these reasons, the use of new technologies in monitoring nitrogen fertilization should be reinforced by the assistance and grants of the state in many regions of the world. Technical and economic support to improve the level of knowledge of the new technologies between farmers permits the sustainability of agriculture, the improvement of the global production, and environment protection.

Author details

Salima Yousfi[1*], José Fernando Marin Peira[2], Gregorio Rincón De La Horra[2] and Pedro V. Mauri Ablanque[1]

1 Agro-Environmental Research Department, Madrid Institute for Rural, Agrarian and Food Research and Development (IMIDRA), Madrid, Spain

2 Area Verde MG Projects S.L., Madrid, Spain

*Address all correspondence to: salima.yousfi@madrid.org

References

[1] Bernard SM, Habash DZ. The importance of cytosolic glutamine synthetase in nitrogen assimilation and recycling. The New Phytologist. 2009;**182**:608-620

[2] Bloom AJ. The increasing importance of distinguishing among plant nitrogen sources. Plant Biology. 2005;**25**:10-16

[3] Hemerly A. Genetic controls of biomass increase in sugarcane by association with beneficial nitrogen-fixing bacteria. In: Plant and Animal Genome XXIV Conference; Plant and Animal Genome. January 2016

[4] Brix H. Effects of thinning and nitrogen fertilization on growth of Douglas-fir: relative contribution of foliage quantity and efficiency. Canadian Journal of Forest Research. 1983;**13**:167-175

[5] Fernandes MS, Rossiello R. Mineral nitrogen in plant physiology and plant nutrition. Plant Science. 1995;**14**:111-148

[6] Laza RC, Bergman B, Vergara BS. Cultivar differences in growth and chloroplast ultrastructure in rice as affected by nitrogen. Journal of Experimental Botany. 1993;**44**:1643-1648

[7] Arregui LM, Quemada M. Strategies to improve nitrogen-use efficiency in winter cereal crops under rainfed Mediterranean conditions. Agronomy Journal. 2008;**100**:277-284

[8] Mulla DJ, Schepers JS. In: Pierce FJ, Sadler EJ, editors. Key processes and properties for site-specific soil and crop management, The State of Site Specific Management for Agriculture. Madison, WI: ASA/CSSA/SSSA; 1997. pp. 1-18

[9] Gebbers R, Adamchuk VI. Precision agriculture and food security. Science. 2010;**327**:828-831

[10] Mulla D, Miao Y. Precision farming. In: Land Resources Monitoring, Modeling, and Mapping with Remote Sensing. Taylor and Francis Group, LLCS; 2016. pp. 161-178

[11] Fox RH, Walthall CL. Crop monitoring technologies to assess nitrogen status. In: Schepers JS, Raun WR, editors. Nitrogen in Agricultural Systems, Agronomy Monograph 49. Madison, USA: ASA, CSSA, SSSA; 2008. pp. 647-674

[12] Hunt ER, Doraiswamy PC, McMurtrey JE, Daughtry CST, Perry EM, Akhmedov B. A visible band index for remote sensing leaf chlorophyll content at the canopy scale. International Journal of Applied Earth Observation and Geoinformation. 2013;**21**:103-112

[13] Scharf PC, Schmidt JP, Kitchen NR, Sudduth KA, Hong SY, Lory JA, et al. Remote sensing for nitrogen management. Journal of Soil and Water Conservation. 2002;**57**:518-524

[14] Guérif M, Houlès V, Baret F. Remote sensing and detection of nitrogen status in crops. Application to precise nitrogen fertilization. In: 4th International Symposium on Intelligent Information Technology in Agriculture, Beijing. October 2007. pp. 26-29

[15] Beatty MK, Johannsen CJ, Ross K. In situ detection of leaf chlorophyll content and leaf nitrogen content in *Zea mays* L. with remote sensing. Agronomy Abstract. 2000:281

[16] McMurtrey JE III, Chappelle EW, Kim MS, Meisinger JJ, Corp LA. Distinguishing nitrogen fertilization levels in field corn (*Zea mays* L.) with actively induced fluorescence and passive reflectance measurements. Remote Sensing of Environment. 1994;**47**:36-44

[17] Araus JL, Kefauver SC. Breeding to adapt agriculture to climate change: Affordable phenotyping solutions. Current Opinion in Plant Biology. 2018;**45**:237-247

[18] Jackson TJ. Soil water modeling and remote sensing. IEEE Transactions on Geoscience and Remote Sensing. 1986;**24**:37-46

[19] Teboh JM, Tubaña BS, Udeigwe TK, Emendack YY, Lofton J. Applicability of ground-based remote sensors for crop N management in Sub Saharan Africa. Journal of Agricultural Science. 2012;**4**:175-188

[20] Chen D, Brutsaert W. Satellite-sensed distribution and spatial patterns of vegetation parameters over tall grass prairie. Journal of the Atmospheric Sciences. 1998;**55**:1225-1238

[21] Stone ML, Solie JB, Whifney RW, Raun WR, Lees HL. Sensor for detection of nitrogen in winter wheat. In: SAE Symposium on Off-Highway Equipment, Indianapolis, IN; SAE, Warrendale, PA; August. 1996

[22] Shanahan JF, Holland K, Schepers JS, Francis DD, Schlemmer MR, Caldwell R. Use of crop reflectance sensors to assess corn leaf chlorophyll content. In: Digital Imaging and Spectral Techniques: Applications to Precision Agriculture and Crop Physiology. Madison, WI: ASA; 2003. pp. 135-150. ASA Spec. Publ. 66

[23] Cabrera-Bosquet L, Molero G, Stellacci A, Bort J, Nogués S, Araus JL. NDVI as a potential tool for predicting biomass, plant nitrogen content and growth in wheat genotypes subjected to different water and nitrogen conditions. Cereal Research Communications. 2011;**39**:147-159

[24] Araus JL, Cairns JE. Field high-throughput phenotyping: The new crop breeding. Frontier Trends in Plant Science. 2014;**19**:52-61

[25] Söderström M, Piikki K, Stenberg M, Stadig H, Martinsson JM. Producing nitrogen (N) uptake maps in winter wheat by combining proximal crop measurements with Sentinel-2 and DMC satellite images in a decision support system for farmers. Acta Agriculturae Scandinavica Section B: Soil and Plant Science. 2017;**67**:637-650

[26] Rubio OA, Grünwald NJ, Cadena MA. Influence of nitrogen on late blight infection in potato cultivation in Toluca, Mexico. American TERRA. 2005;**23**:487-493

[27] Dynarski K. Preventing over-fertilization for better crop quality and yield. 19 December 2018. Available from: https://blog.teralytic.com/preventing-over-fertilization/

[28] González-Raya E, Benavides A, Ramírez H, Robledo V, Ratikanta M, Reyes A, et al. Growth of tomato and fruit quality with different concentrations of nitrate. American TERRA. 2005;**23**:105-111

[29] He Y, Terabayashi S, Asaka T, Namiki T. Effect of restricted supply of nitrate on fruit growth and nutrient concentration in the petiole sap of tomato cultured hydroponically. Journal of Plant Nutrition. 1999;**22**:799-811

[30] Jauset M, Sarasúa MJ, Avilla J, Albajes R. Effect of nitrogen fertilization level applied to tomato on the greenhouse whitefly. Crop Protection. 2000;**19**:255-261

[31] Duffy K, Défago G. Macro and microelement fertilizers influence the severity of Fusarium crown and root rot of tomato in a soilless production system. HortScience. 1999;**34**:287-291

[32] Maynard N, Barker AV, Minotti PL, Peck NH. Nitrate accumulation in

vegetables. Advances in Agronomy. 1976;**28**:71-118

[33] Hanson CR. Nitrate concentrations in Canterbury groundwater: A review of existing data. In: Environment Canterbury. Report. 2002

[34] Almasri MN, Kaluarachchi JJ. Assessment and management of long-term nitrate pollution of ground water in agriculture-dominated watersheds. Journal of Hydrology. 2004;**295**:225-245

[35] Ferguson RB, Shapiro CA, Hergert GW, Kranz WL, Klocke NL, Krull DH. Nitrogen and irrigation management practices to minimize nitrate leaching from irrigated corn. Journal of Production Agriculture. 1991;**4**:186-192

[36] Sogbedji JM, Van Es HM, Yang CL, Geohring LD, Magdoff F. Nitrate leaching and nitrogen budget as affected by maize nitrogen rate and soil type. Journal of Environmental Quality. 2000;**29**:1813-1820

[37] Khan MN, Mobin M, Abbas ZK. Fertilizers and their contaminants in soils, surface and groundwater. In: Encyclopedia of the Anthropocene. 2017. pp. 225-240

[38] Ben-Dor E, Inbar Y, Chen Y. The reflectance spectra of the organic matter in the visible near-infrared and short wave infrared region (400-2500 nm) during a combine decomposition process. Remote Sensing of Environment. 1997;**61**:1-15

[39] Stoner ER, Baumgardner MF. Characteristics variation in reflectance of surface soils. Soil Science Society of America Journal. 1981;**45**:1161-1165

[40] Irons JR, Weismiller RA, Peterson GW. Soil reflectance. In: Asrar G, editor. Theory and Application of Optical Remote Sensing. Wiley Series on Remote Sensing. New York: Wiley; 1989. pp. 66-106

[41] Saggar S, Jha N, Deslippe J, Bolan NS, Luo J, Giltrap DL, et al. Denitrification and N_2O:N_2 production in temperate grasslands: processes, measurements, modelling and mitigating negative impacts. Science of the Total Environment. 2013;**465**:173-195

[42] Vistoso E, Alfaro M, Saggar S, Salazar F. Effect of nitrogen inhibitors on nitrous oxide emissions and pasture growth after an autumn application in volcanic soil. Chilean Journal of Agricultural Research. 2012;**72**:133-139

[43] Riley WJ, Ortiz-Monasterio I, Matson PA. Nitrogen leaching and soil nitrate, nitrite, and ammonium levels under irrigated wheat in Northern Mexico. Nutrient Cycling in Agroecosystems. 2001;**61**:223-236

[44] Haygarth PM, Bardgett RD, Condron LM. Phosphorus and nitrogen cycles and their management. In: Gregory PJ, Nortcliff S, editors. Soil Conditions and Plant Growth. West Sussex, UK: Wiley-Blackwell; 2013. pp. 132-159

[45] Quemada M, Gabriel JL, Zarco-Tejada P. Airborne hyperspectral images and ground-level optical sensors as assessment tools for maize nitrogen fertilization. Remote Sensing. 2014;**6**:2940-2962

[46] Snyder CS, Bruulsema TW, Jensen TL, Fixen PE. Review of greenhouse gas emissions from crop production systems and fertilizer management effects. Agriculture, Ecosystems and Environment. 2009;**133**:247-266

[47] Rose DC, Sutherland WJ, Barnes AP, Borthwick F, Ffoulkes C, Hall C, et al. Integrated farm management for sustainable agriculture: Lessons for

knowledge exchange and policy. Land Use Policy. 2019;**81**:834-842

[48] Cambra Baseca C, Sendra S, Jaime Lloret J, Tomas J. Smart decision system for digital farming. Agronomy. 2019;**9**:216. DOI: 10.3390/agronomy9050216

[49] Priyanka P, Chandak P, Agrawal AJ. Smart farming system using data mining. International Journal of Applied Engineering Research. 2017;**12**(11):2788-2791. ISSN 0973-4562. Research India Publications. http://www.ripublication.com

Chapter 20

Risk Assessment of Sunflower Production Using In-Field Rainwater Harvesting on Semi-Arid Ecotope in South Africa

Jestinos Mzezewa and Leon Daniel van Rensburg

Abstract

Risk assessment of sunflower production was carried out using an empirical model. The crop yield prediction for semi-arid areas (CYP-SA) was used to simulate sunflower yield using 26 years (1984–2010) climatic data. Scenarios of crop yield simulation included production techniques associated with in-field rainwater harvesting (IRWH), and conventional tillage (CT). IRWH is a no-till (NT) crop production practice that promotes runoff from a crusted runoff strip into basins where water infiltrates beyond evaporation. The study focused on the effect of initial soil water content at planting *viz.* empty profile (water content near the lower limit of plant available water (LL)); half profile (water content between LL and the drained upper limit (DUL)); full profile (water content near DUL) and planting dates (November, December and January). Yield difference at 80% probability was 74% higher under IRWH compared to CT with empty initial soil water content at planting. Results indicated that IRWH is more sustainable compared to the CT.

Keywords: ecotope, crop yield simulation, empirical model, production risk, rainwater harvesting

1. Introduction

Water scarcity is a major constraint in semi-arid areas, leading to a natural focus on in-field rainwater conservation [1]. However, field experiments to assess water harvesting techniques are very expensive and laborious [2]. As a result several models of water harvesting have been developed in order to quantify risk for different production techniques. Models can be used as research tools to conduct research faster and cost-effectively [3]. In addition, a valuable property of models is their ability to utilize long-term climate data to provide long-term yield simulations, which can serve to quantify risk [4–6]. The modeling in this study simulate the in-field rainwater harvesting (IRWH) production technique (**Figure 1**) that was implemented during the field experiments [7]. This modeling approach has been described previously [2–4, 8]. By constructing cumulative probability functions (CPFs), risk associated with various production systems can be quantified [2, 4, 5, 8–11].

Crop Yield Predictor for Semi-Arid Areas (CYP-SA) model was developed to simulate crop yield on a semi-arid ecotope in South Africa [3, 4]. An ecotope is defined as a

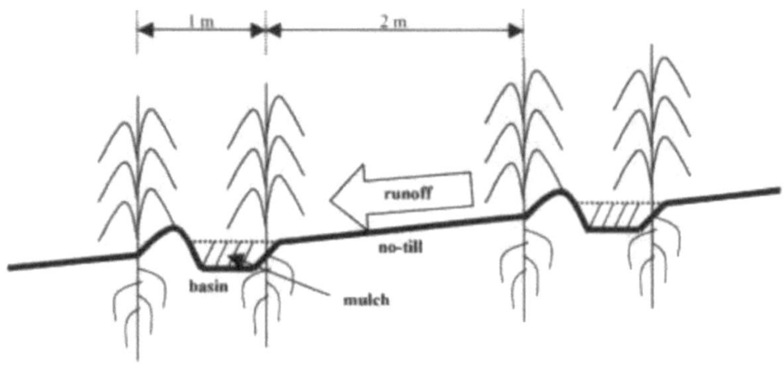

Figure 1.
A diagrammatic representation of the in-field rainwater harvesting (IRWH) production technique.

homogenous piece of land with a unique combination of climate, topographic and soil characteristics [4]. The CYP-SA model has potential to be applied for assessing risk for crop production in other ecotopes and therefore was chosen for this study to assist in decision making. The IRWH was introduced on this ecotope during the 2007/08 and 2008/09 growing seasons. The objective of this study was to simulate long-term (26 years) sunflower yield to quantify risk for two production techniques (IRWH versus CT) on a semi-arid ecotope in the Limpopo Province of South Africa: University of Venda, Thohoyandou (22° 58′ S; 30° 26′ E, 596 m) whose 23-year average rainfall is 781 mm with coefficient of variation (CV) of 315% [12]. The daily temperatures at the University of Venda vary from 25 to 40°C in summer and between approximately 12 and 26°C in winter. Soil at the University of Venda was classified as a Hutton form [13], equivalent to a Rhodic Ferralsol [14]. The soil is deep (>1500 mm) with clay concentration of 60% [7].

1.1 Model description

CYPA-SA runs on daily time-step [8, 15]. The inputs required by the model are crop modified upper limit (CMUL) of plant available water (PAW); drained upper limit (DUL) of PAW; lower limit (LL) of PAW; rainfall (P); evaporative demand (ET_0) and soil water content at planting (θ_p) (**Figure 2**). More symbols are explained in **Table 1**. It was assumed that runoff (R) would be zero if the precipitation was less than 8 mm. Runoff from rainfall events of more than 8 mm can be calculated using Eqs. (1) and (2).

CT :

$$P < 8 : Pe = P$$

$$P > 8 : P = P - [\{0.473 \times P) - 2.168\} \times 0.4] \quad R^2 = 0.60 \tag{1}$$

(after [4])

where P is the rainfall for a day (mm) and P_e is the effective rainfall for a particular day (mm)

IRWH :

$$P < 8 : P_e = P$$

$$P > 8 : P_e = P + [(0.474 \times) - 0.8791] \quad R^2 = 0.64 \tag{2}$$

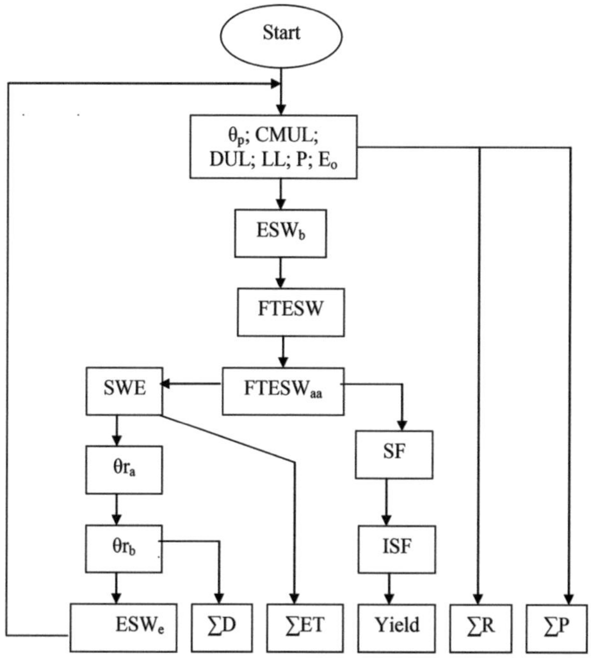

Figure 2.
Flow diagram of the CYP-SA model.

Symbol	Explanation
ESW_b	Extractable soil water at the beginning of a day
FTESW	Fraction of total extractable soil water
$FTESW_{aa}$	Adapted fraction of total extractable soil water
SWE	Soil water extraction
θr_a	Water content of rootzone, not adapted to cater for values above CMUL
θr_b	Adapted water content of rootzone, to cater for values not to exceed CMUL
ESW_e	Extractable soil water at the end of a day
SF	Stress factor
ISF	Integrated stress factor and the stress weighting factor (λ)
D	Deep drainage
ET	Evapotranspiration
R	Runoff

Table 1.
CYP-SA model flow diagram symbols.

1.2 Calibration and verification of the CYP-SA model

The calibration process can provide important insight into both local conditions and model performance [16]. The original model algorithms rules were evaluated by combining available data for ecotopes in the Free State Province [4, 8]. Modifications of the model were necessary to adapt to the soil and climatic conditions of this ecotope. Model calibration was achieved by inputting soil and climate data as detailed in Section 2.1. The original model runoff Eqs. (1) and (2) were replaced with Eqs. (3) and (4) which were developed for this ecotope [17].

CT :

$$R = 0.421 \times P\text{–}3.008 \quad R^2 = 0.769 \tag{3}$$

IRWH :

$$R = 0.59 \times P\text{–}2.203 \quad R^2 = 0.947 \tag{4}$$

The predicted yield was compared with the measured sunflower yield for that season. The model parameters were adjusted stepwise until the predicted yield matched with the measured yield. This was repeated for all replications for that season. The averaged correction factors were then used for model verification. Separate calibrations were done for CT and IRWH treatments. Model verification test were designed to evaluate the model performance. After calibration of the model, it was verified using another set of field data from 2008/08 growing season. The verified model was then used for simulation of long-term sunflower yield.

1.3 Statistical analyses for model verification

Model reliability tests were performed following the procedures proposed by Wilmott [18] who recommended use of the index of agreement (D-index), root mean square error systematic ($RMSE_s$), root mean square error unsystematic ($RMSU_u$) and root mean square error (RMSE) for model evaluation. The mean absolute error (MAE) and RMSE are among the best overall measures of model performance.

1.4 Model application

Meteorological data for both Thohoyandou and University of Venda were used for long-term (26 years) sunflower yield simulation. Rainfall and class A-pan evaporation data have been recorded for Thohoyandou meteorological station in the period 1984–2004. Calculated E_o values. For University of Venda meteorological records were used. The data was obtained from Agricultural Research Council-Institute for Soil, Climate and Water- Pretoria. Long-term evaluation of production techniques was achieved by comparing cumulative probability functions (CPFs) of yield using the CYP-SA model and long-term climate data.

The following θ_p were used in the simulations: 0% (soil water content of the profile near empty, but enough in the top soil for germination of seeds, defined as the difference between DUL and LL), 50% (half full) and 100% (full). Total DUL and LL of the soil profile are 448 and 250 mm, respectively. The amount of plant-extractable soil water at planting in the effective rooting zone is 0 mm for empty profile, 99 mm for half full profile and 198 for full profile. The simulations were run for 26 seasons from 1984 to 2010 with different production techniques (CT and IRWH) and three planting dates: 1 November (early planting), 1 December (intermediate) and 1 January (late planting).

All statistical tests on cumulative probability functions were carried out using the Kolmogorov–Smirnov test for two samples ($P < 0.05$). This test is about the agreement between two empirical cumulative distributions [11]. The null hypothesis is that the two groups are the same, and the test statistic D for two data sets x (the maximum distance between the two distributions) is defined as:

$$D = \max|S_{1(x)} - S_{2(x)}|, \tag{5}$$

where $S_{1(x)}$, $S_{2(x)}$ are the cumulative distributions, $S(x)$, for the two samples.

2. Results and discussion

2.1 Evaluation of the model performance

Results of model verification test using the procedure of [18] are presented in **Table 2**. The prediction performance was reasonable. The D-indices for both CT and IRWH were high (>0.80), indicating good model performance. Furthermore, crop yield was correctly predicted ($R^2 > 0.8$) for IRWH and reasonably predicted for CT ($R^2 = 0.68$; **Table 2**), confirming a positive association and good agreement between measured and simulated yield. On the whole crop yield was underestimated by less than 25% in the CT whilst the model overestimated crop yield by less than 15% in the IRWH treatment (**Table 2**). The models for both CT and IRWH showed low $RMSE_u$/RMSE values (<0.5), indicating that a high level of bias was associated with the models. The bias was also indicated by the large $RMSE_s$ relative to RMSE. Poor prediction of runoff could account for less satisfactory statistical indices in the model [19].

2.2 Effects of initial soil water at planting

Yield variation over 26 years predicted under CT and IRWH management practices were compared using cumulative probability curves (**Figure 3**). The curves were constructed by averaging across all scenario factors. The curves were compared with for each scenario factor using the Kolmogorov–Smirnov test ($P \leq 0.05$). Sunflower yield was significantly higher in full initial water than in the other two water contents in the CT. Similarly, empty profile water gave the lowest yield compared to both half and full water contents in the IRWH (**Table 3**). At 80% probability the yield in the empty, half and full profile was 355, 691 and 1088 kg ha^{-1} for the CT, respectively. In the IRWH the yield was 1357, 1984 and 2601 kg ha^{-1} in the empty, half and full profile, respectively (**Figure 3**). In general, the full profile gave the greatest yield, followed by half profile and then the empty profile, illustrating the importance of adequate profile water content at planting in semi-arid environment. This is akin to the observation made in the Free State Province South Africa [2, 4, 8, 11]. They reported that sufficient soil water content at planting is important for a good harvest. In addition, water stored between DUL and LL before planting contributes mainly to transpiration [8]. Recently Garcia-Lopez et al. [20] reported that sunflower yield ranged between 1333 and 2622 kg ha^{-1} corresponding to irrigation water that ranged from 146 to 326 mm, respectively. The lowest yield was obtained under deficit irrigation volume, emphasizing the importance of adequate soil water content on sunflower yield. More recently, sunflower yield reduction was observed under four levels of irrigation with the lowest level of water availability accounting for the lowest yield loss [21]. They reported sunflower seed yield of 2371, 2173, 2018 and 1764 kg ha^{-1} corresponding to 100 potential evapotranspiration (PET), 80 PET, 60 PET and 40% PET, respectively. Similar results were obtained when the seed yield of different sunflower inbred lines were compared under limited and full irrigation [22]. There were no effects of planting dates on yield in

Treatment	MAE	RMSE	$RMSE_s$	$RMSE_u$	D-index	R^2	Measured mean yield	Predicted mean yield
CT	405	415	405	91	0.993	0.68	1685	1280
IRWH	370	403	384	121	0.994	0.96	1844	2062

Table 2.
Statistical analysis of CYP-SA model performance predicting yields produced with conventional tillage (CT) and in-field rainwater harvesting (IRWH) (kg ha^{-1}).

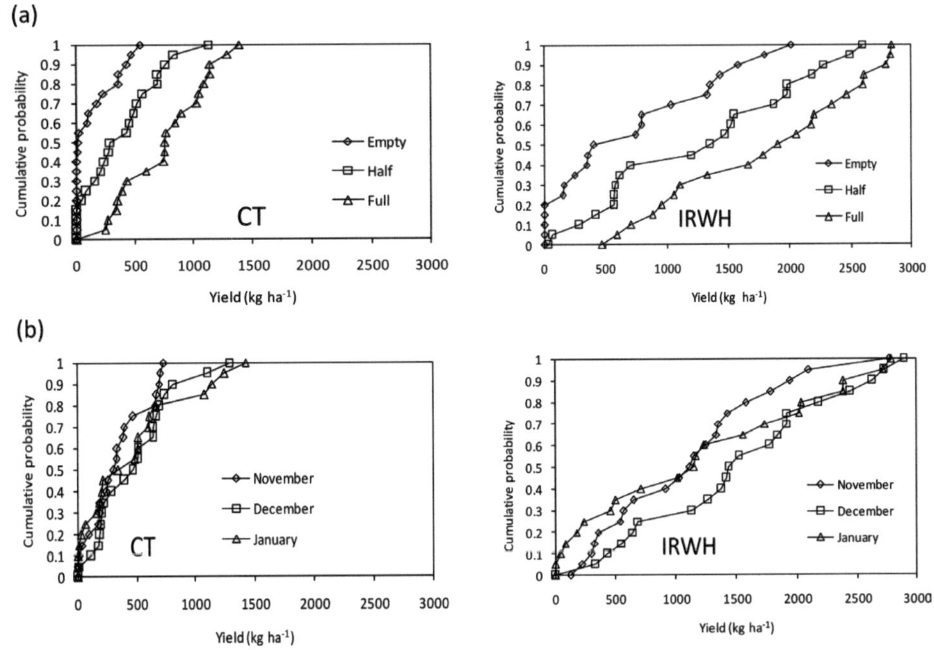

Figure 3.
Cumulative probability of simulated long-term (1984–2010) sunflower for conventional (CT) and in-field rainwater harvesting (IRWH): (a) different profiles of initial soil water content (averaged over three planting dates) and (b) different planting dates.

Initial soil water	Tillage treatments		Planting date	Tillage treatments	
	CT	IRWH		CT	IRWH
Empty	a	a	November	a	a
Half	a	ab	December	a	a
Full	b	b	January	a	a

The same letter within columns indicates not significant difference at $P \leq 0.05$.

Table 3.
The Kolmogorov–Smirnov (KS) test comparing tillage treatments (CT and IRWH) at different initial soil water levels and planting dates.

either production technique although differences were noted between techniques (**Table 3** and **Figure 3**). This may be attributed to high variability of rainfall on the ecotope. In another study [12] reported that the coefficient of variation (CV) for November, December and January were 1.27, 1.56 and 1.35, Using the CYP-SA model [8] reported that late planting (January) was significantly better ($P \leq 0.01$) than December planting in the semi-arid Free State Province. He attributed this to the fact that when the sunflower crop is planted in November its flowering period may coincide with favorable rainfall conditions in December. **Figure 3** also shows that at 80% probability farmers are likely to harvest about 700 kg ha^{-1} of sunflower in CT regardless of the planting date. However, in the Mediterranean region it is reported that earlier planting date resulted in higher seed yield in all the 4 years of the study [23]. The graphs also show that farmers planting sunflower using IRWH in December, are likely to expect a 50% chance of getting higher yields (1432 kg ha^{-1}) than planting either in November (1118 kg ha^{-1}) or January (1148 kg ha^{-1}). The results show that IRWH has a yield advantage in sunflower productivity, as reported

Initial soil water	Statistics		Planting date	Statistics	
	D-statistic	Probability level		D-statistic	Probability level
Empty	0.48	0.011	November	0.62	0.000
Half	0.57	0.001	December	0.67	0.000
Full	0.62	0.000	January	0.43	0.029

Table 4.
The Kolmogorov–Smirnov (KS) test for cumulated sunflower yield with CT and IRWH production techniques produced at different initial soil water levels and planting dates.

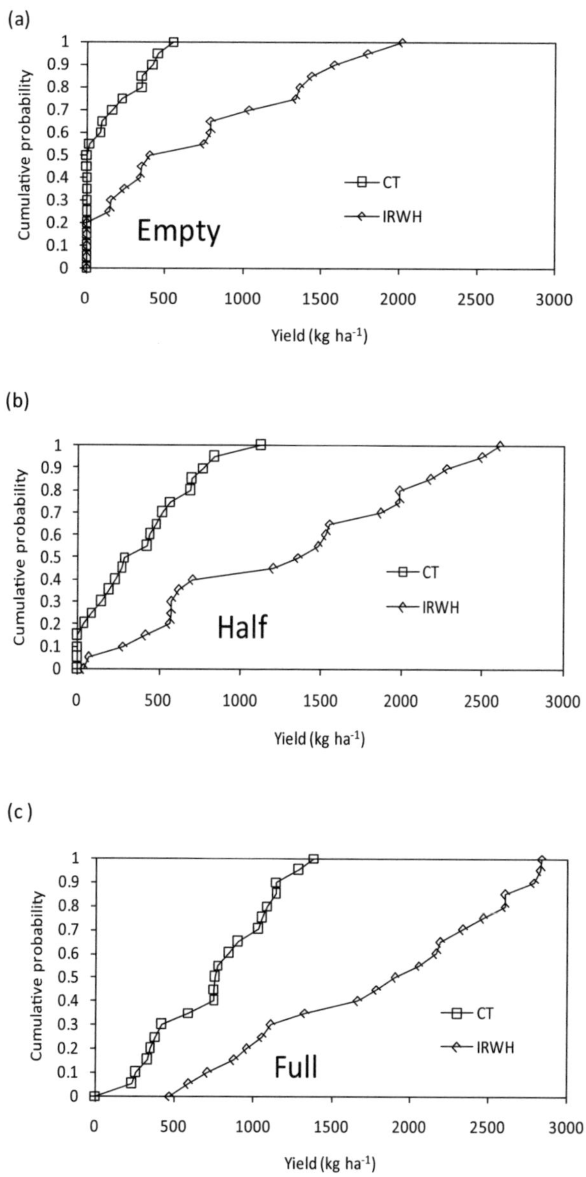

Figure 4.
Cumulative probabilities of simulated long-term (1984–2010) sunflower yield with conventional (CT) and in-field rainwater harvesting (IRWH) for three water contents (averaged over three planting dates); (a) empty, (b) half and (c) full.

earlier [2–4]. It was clear that the most important factor is initial soil water content when IRWH was compared with CT, a result as also obtained by Walker et al. [2].

Further comparison of cumulative probability curves for each scenario across different techniques was carried using the Kolmogorov–Smirnov test. There was significant yield difference between CT and IRWH production at all levels of initial profile water content and planting dates, confirming the advantages of IRWH over CT (**Table 4** and **Figures 4** and 5). At 80% probability the yield difference between CT and IRWH was 74, 65 and 58% in empty, half and full profile, respectively (**Figure 4**). This finding confirms the finding of [2] who reported that the lower the initial water content at planting, the greater the yield difference between the IRWH and CT. Similar results were reported by [24]. From **Figure 5** it could be deduced that the yield difference between CT and IRWH was the same (68%) for the months of December and January, whilst the difference was 58% for the months of November, further confirming that planting dates may play a less important role than profile water content on this ecotope. The results strongly suggest that a farmer

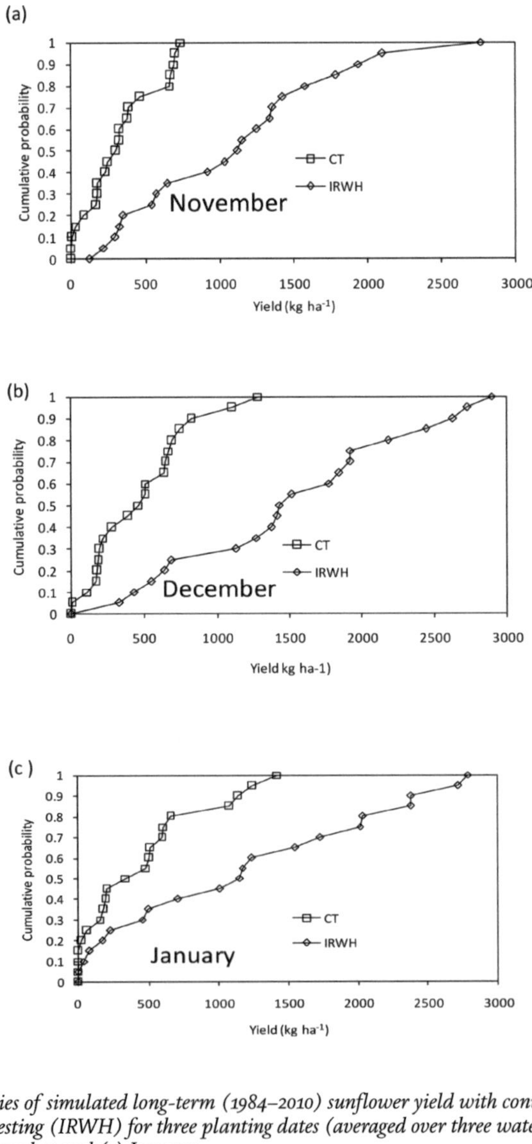

Figure 5.
Cumulative probabilities of simulated long-term (1984–2010) sunflower yield with conventional (CT) and infield rainwater harvesting (IRWH) for three planting dates (averaged over three water contents); (a) November, (b) December and (c) January.

who adopts IRWH and plants on empty profile water content is likely to get higher yields than who uses CT production technique. The findings could be important in semi-arid environments where water for agriculture is a constraint. Earlier [2] reported that for all scenarios with empty initial soil water, the curves for IRWH were significantly different from those for CT while the difference was not significant under half and full initial soil water when they simulated maize yield in the Free State Province. In his study [8] concluded that simulated sunflower production risk was significantly less under IRWH compared to CT when CYP-SA was run between three-fourth full and full profile water content.

3. Conclusions

In this study the CYP-SA model was applied to simulate long-term sunflower yields to quantify the risk of crop production at the ecotope. The conventional tillage (CT) was compared with the in-field rain water harvesting (IRWH) production technique. Results from this study indicated that farmers who choose to adopt the IRWH technique and plant when profile water content is empty can get higher yields compared to those who choose the CT technique. The IRWH technique consistently gave higher sunflower yield than the CT regardless of the planting date although sowing date was not significant within each crop production technique. Therefore, it could be concluded that the IRWH is a sustainable crop production technique compared to the CT.

Acknowledgements

We acknowledge the kind gesture of Dr. J.J. Botha of the Agricultural Research Council (ARC) of South Africa for providing CYP-SA model and for his contribution towards this manuscript. We would also like to thank the ARC for making available the long-term climatic data for the study area.

Author details

Jestinos Mzezewa[1*] and Leon Daniel van Rensburg[2]

1 Department of Soil Science, University of Venda, Thohoyandou, South Africa

2 Department of Crop, Soil and Climate Sciences, University of the Free State, Bloemfontein, South Africa

*Address all correspondence to: jestinos.mzezewa@univen.ac.za

References

[1] Jensen JR, Bernhard RH, Hasen S, McDonagh J, Boberg JP, Nielsen NE, et al. Productivity in maize based cropping systems under various soil-water-nutrient management strategies in a semi-arid, Alfisol environment in East Africa. Agricultural Water Management. 2003;**59**:217-237

[2] Walker S, Tsubo M, Hensley M. Quantifying risk for water harvesting under semi-arid conditions. Part II. Crop yield simulation. Agricultural Water Management. 2005;**76**:94-107

[3] Botha JJ, van Rensburg LD, Anderson JJ, Hensley M, Machelli MS, van Staden PP, et al. Water conservation techniques on small plots in semi-arid areas to enhance rainfall use efficiency, food security, and sustainable food production. Pretoria, South Africa: Water Research Commission; 2003. WRC Report No 1176/1/03

[4] Hensley M, Anderson JJ, van Staden PP, Du Toit A. Optimizing rainfall use efficiency for developing farmers with limited access to irrigation water. Water Research Commission, Pretoria, South Africa; 2000. WRC Report No 878/1/00

[5] Popova Z, Kercheva M. CERES model application for increasing preparedness to climate variability in agricultural planning-risk analyses. Physics and Chemistry of the Earth. 2005;**30**:117-124

[6] Singels A, Annandale JG, de Jager JM, Schulze RE, Inman-Mamber NG, Durand W, et al. Modelling crop growth and crop water relations in South Africa: Past achievements and lessons for the future. South African Journal of Plant and Soil. 2010;**27**(1):49-65

[7] Mzezewa J, Gwata ET, van Rensburg LD. Yield and seasonal water productivity of sunflower as affected by tillage and cropping systems under dryland conditions in the Limpopo Province of South Africa. Agricultural Water Management. 2011;**98**:1641-1648

[8] Botha JJ. Evaluation of maize and sunflower production in a semi-arid area using in-field rainwater harvesting [PhD thesis]. Bloemfontein, South Africa: University of the Free State; 2006

[9] Diaz-Ambrona CGH, O'Leary GJ, Sadras VO, O'Connell MG, Connor DJ. Environmental risk analysis of farming systems in a semi-arid environment: Effect of rotations and management practices on deep drainage. Field Crops Research. 2005;**94**:257-271

[10] Jagtap SS, Abamu FJ, Kling JG. Long-term assessment of nitrogen and variety technologies on attainable maize yields in Nigeria using CERES-maize. Agricultural Systems. 1999;**60**:77-86

[11] Tsubo M, Walker S. An assessment of productivity of maize grown under water harvesting system in a semi-arid region with special reference to ENSO. Journal of Arid Environments. 2007;**71**:299-311

[12] Mzezewa J, Misi T, van Rensburg LD. Characterisation of rainfall for sustainable crop production at a semi-arid ecotope in the Limopopo Province (SA). Water SA. 2010;**36**(1):19-26

[13] Soil Classification Working Group. Soil Classification—A Taxonomic System for South Africa. Pretoria, South Africa: Department of Agricultural Development; 1991

[14] World Reference Base for Soil Resources. A framework for International Classification, Correlation and Communication. World Soil Resources; Report No. 103; IUSS/ISRIC/FAO; 2006

[15] Rasmussen VP, Hanks RJ. Spring wheat yield model for limited moisture conditions. Agronomy Journal. 1978;**70**:940-944

[16] Muthukrishnan S, Harbor J, Lim KL, Engel BA. Calibration of a simple rainfall-runoff model for long-term hydrological impact evaluation. URISA Journal. 2006;**18**(2):35-42

[17] Mzezewa J, van Rensburg LD. Effects of tillage on runoff from a bare clayey soil on semi-arid ecotope in the Limpopo Province of South Africa. Water SA. 2011;**37**(2):1-8

[18] Wilmott CJ. On the validation of models. Physical Geography. 1981;**2**:184-194

[19] Anderson JJ. Rainfall-runoff relationships and yield response of maize and dry beans on the Glen/Bonhein ecotope using conventional tillage and in-field rainwater harvesting [PhD thesis]. Bloemfontein, South Africa: University of the Free State; 2007

[20] Garcia-Lopez J, Lorite IJ, Garcia-Luiz R, Ordonez R, Dominguez J. Yield response of sunflower to irrigation and fertilization under semi-arid conditions. Agricultural Water Management. 2016;**176**:151-162

[21] Ebrahim E, Seyyed SM, Bybordi A, Damalas CA. Seed yield and oil quality of sunflower, safflower and sesame under different levels of irrigation water availability. Agricultural Water Management. 2019;**218**:149-157

[22] Farshad N, Soodabeh J, Mehdi G, Ali E. Studying the physiological and yield responses of sunflower inbred lines to full and limited irrigation. Journal of Integrative Agriculture. 2018;**17**(7):1605-1611

[23] Barros JFC, De Carvalho M, Basch G. Response of sunflower (*Helianthus annuus* L.) to sowing date and plant density under Mediterranean conditions. European Journal of Agronomy. 2004;**21**:347-356

[24] Anderson JJ, Botha JJ, van Rensburg LD. The use of crop modeling to compare different tillage techniques. In: Beukes D, de Villiers M, Mkhize S, Sally H, van Rensburg LD, editors. Proceedings of the Symposium and Workshop on Water Conservation Technologies for Sustainable Dryland Agriculture in Sub-Saharan Africa (WCT); Bloemfontein, South Africa; 2003. pp. 162-168